Progress in Inflammation Research

For further volumes:
http://www.springer.com/series/4983

Isabelle Couillin • Virginie Pétrilli •
Fabio Martinon
Editors

The Inflammasomes

Springer

Editors
Dr. Isabelle Couillin
Molecular Immunology and Embryology
UMR6218, University of Orleans and
CNRS
3B rue de la Férollerie
45071 Orleans
France
couillin@cnrs-orleans.fr

Dr. Virginie Pétrilli
Department Immunity, Microenvironment
and Virus
Centre de Recherche en Cancérologie de
Lyon
28 rue Laennec
69008 Lyon
France
petrilli@lyon.fnclcc.fr

Dr. Fabio Martinon
Harvard School of Public Health
Department of Immunology & Infectious Diseases
Huntington Avenue 651
02115-6017 Boston
Massachusetts
USA
Fabio.Martinon@unil.ch

Series Editor
Prof. Dr. Michael J. Parnham
Director of Science
MediMlijeko d.o.o.
Pozarinje 7
HR-10000 Zagreb
Croatia

ISBN 978-3-0348-0147-8 e-ISBN 978-3-0348-0148-5
DOI 10.1007/978-3-0348-0148-5

Library of Congress Control Number: 2011932722

Cover design: deblik, Berlin

Printed on acid-free paper

Springer Basel AG is part of Springer Science+Business Media (www.springer.com)

Preface

Innate immunity has been viewed as the first line of defense discriminating "self" (host proteins) from "non-self" (microorganisms). However, emerging literature suggests that innate immunity actually serves as a sophisticated system for sensing signals of "danger" such as pathogenic microbes or host-derived signals of cellular stress, while remaining unresponsive to non-dangerous motifs, such as normal host molecules. The innate immune system engages an array of pattern-recognition receptors (PRRs) to detect invariant microbial motifs. PRRs are expressed by cells at the frontline of defense against infection, including macrophages, monocytes, dendritic cells, neutrophils, and epithelial cells, as well as cells of the adaptive immune system. The discovery of Toll-like receptors (TLRs) provided a class of membrane receptors that sense extracellular microbes and trigger anti-pathogen signaling cascades. More recently, intracellular microbial sensors have been identified, including NOD-like receptors (NLRs). NLRs detect/sense conserved structures of the microorganisms, the pathogen-associated molecular patterns (PAMPs). PRRs activate intracellular signals that collaborate for efficient activation of host defense. One such specific collaboration is the interaction between TLRs and intracellular NLRs that recognize PAMPs, as well as host-derived danger signals danger associated molecular patterns (DAMPs). Importantly, NLR members are involved in the assembly of molecular platforms, the inflammasomes, activated upon cellular infection, or stress that trigger the maturation of pro-inflammatory cytokines such as interleukin-1β (IL-1β) and IL-18 to engage innate immune defenses.

Inflammasomes are not only involved in cytokine maturation but also in a highly inflammatory form of cell death called pyroptosis. Moreover, the inflammatory response to infection appears to be coupled to cell death as an important mediator of host defence. Occurring in the context of infection, pyroptosis is morphologically, mechanistically, and physiologically distinct from other forms of cell death. The physiological significance of this unique form of cell death is not clear since pyroptosis can favor pathogen elimination on one hand, while contributing to the pathophysiology of disease on the other.

Different inflammasomes are activated by various bacterial and viral activators through different receptors/sensors in particular NLRP1, NLRP3, and NLRC4, as well as the PYHIN (pyrin and HIN200 domain-containing) family member AIM2 have all been demonstrated to form functional inflammasomes capable of activating caspase-1.

Recent genetic studies using mice deficient in inflammasome components demonstrate the involvement of the inflammasome in the outcome of infection with the fungus *Candida albicans*, the bacteria *Staphylococcus aureus*, *Salmonella typhimurium* and *Legionella pneumophila*, the helminth *Schistosoma mansoni* as well as the malarial parasite *Plasmodium berghei*. Live fungi, schistosomal egg antigen or malarial hemozoin have the ability to activate the inflammasome and induce secretion of mature IL-1β.

The most fully characterized inflammasome is the Nlrp3 inflammasome, which is activated by various endogenous and exogenous danger signals such as environmental irritants, signals of tissue damage, and pathogens. Danger signals are endogenous host molecules that are *not in place*, such signals include the presence of ATP or uric acid in the extracellular space, the breakdown or release of soluble extracellular matrix components or the presence of DNA in the cytosol of the cell. All these danger signals have been proven to activate the release of mature IL-1β through caspase-1 activation.The NLRP3 inflammasome is activated by crystals and particles of varied size and structure. These inflammasome-activating particles have a causative role in human diseases such as gout, asbestosis, silicosis, and Alzheimer's, or are useful as vaccine adjuvants, like in the case of alum. The molecular aspects involved in activation of the NLRP3 inflammasome by particles are being rapidly elucidated and emphasize the importance of phagocytosis, K^+ efflux, and generation of reactive oxygen species in this process. Knowledge of the involvement of inflammasome activation by danger signals in different diseases is increasing every year and has emerged as responsible for the sterile inflammatory response.

The broad spectrum of activators is reflected at the physiological level in its implication in normal and dysregulated immune responses, including various autoinflammatory diseases, cancer, skin and lung inflammation, and in the defence against pathogens. The skin is constantly subjected to microbial, chemical, and physical insults. Not surprisingly, it is endowed with the capacity to detect these events and respond accordingly, alerting the immune system when needed, often through the inflammasome. Cancer progression is associated with chronic inflammation and the dampening of antitumor immune responses. Although pro-inflammatory cytokines such as IL-1β have been proposed to be involved in the initial development of cancer, there is emerging evidence that these pro-inflammatory molecules can also act as potent adjuvants of T-cell-mediated immune responses. Strikingly, the activation of the NLRP3 inflammasome was recently shown to be instrumental in the initiation of an immune anti-cancer response that was required for the success of chemotherapy.

Non regulated inflammation in lung often leads to chronic inflammatory diseases such as asthma, chronic obstructive pulmonary disease, and pulmonary fibrosis. Acute cigarette smoke exposure induces inflammasome-dependent maturation of IL-1β and ASC-associated protein complex is necessary to lung emphysema.

Bleomycin- or crystalline silica particles- induced pulmonary inflammation and fibrosis are dependent on NLRP3 inflammasome. NLRP3-containing inflammasome appears to be an essential mediator for allergic lung inflammation. Pulmonary exposure to nanomaterials may lead to pulmonary inflammation and fibrosis.

The dysregulated secretion of IL-1 that occurs upon the loss of one or more roadblocks is indeed the cause of a number of severe chronic human diseases, characterized by massive inflammation. These disorders, collectively called "Auto-inflammatory diseases," differ in pathogenesis and clinical manifestations but share a dramatic therapeutic response to IL-1 blocking. The identification of the gene responsible for Familial cold auto-inflammatory syndrome (FCAS), Muckle–Wells syndrome (MWS) and Chronic Infantile Neurological Cutaneous and Articular Syndrome (CINCA) shed light on the possible link between this class of monogenic inflammatory disorders and the pivotal role pathogenic role of IL-1β. Missense mutations of the *NALP3/CIAS1* gene provide a gain of function to the NLRP3 protein resulting in increased secretion of active IL-1β.

Inflammatory reactions must be well synchronized and controlled, as the dysregulation of inflammatory processes is associated with disease. The dramatic and sometimes devastating clinical consequences of mutations in NLRP3 itself or in other inflammasome-related genes show how an overly active signaling pathway can lead to chronic pathology. Anti-IL-1 therapies have proven to be enormously successful in the treatment of auto-inflammatory diseases, which gives great hope that such therapies can also be effective in other inflammatory conditions that are linked to inflammasome activation. Future work should focus on deciphering the integral mechanisms that lead to NLRP3 inflammasome activation.

We believe that this volume written by several experts in the field will be useful for scientists and medicals doctors investigating the mechanisms of inflammation and the physiopathology of inflammatory diseases and developing new potential therapies. Since the first description in 2002, the scientific literature on inflammasome expanded exponentially with more than 600 scientific publications published. This comprehensive review on inflammasome provides a timely state of the art review of literature, describing the multiple involvements of inflammasome in Immunity.

Finally, we wished to thank the authors dedicating their precious time with expert contributions for this main volume in Inflammation Research.

Orleans, France I. Couillin
Lyon, France V. Pétrilli
Boston, MA, USA F. Martinon
March 2011

Contents

Contributors

Lionel Apetoh INSERM, U866, Dijon, France; INSERM AVENIR Team, Dijon, France; Centre Georges François Leclerc, Dijon, France, lionel.apetoh@yahoo.fr

Isabelle Couillin Molecular Immunology and Embryology, UMR6218, University of Orleans and CNRS, 3B rue de la Férollerie, 45071, Orleans, France; KeyObs SAS, Orleans, France, couillin@cnrs-orleans.fr

Olivier Gaide Department of Dermatology-Venereology, Geneva University, Rue Michel-Servet 1, 1211 Geneva, Switzerland; Department of Pathology-Immunology, Geneva University, Rue Michel-Servet 1, 1211 Geneva, Switzerland

Marco Gattorno 2nd Division of Pediatrics, "G. Gaslini" Scientific Institute, Largo G. Gaslini 5, 16147 Genova, Italy

Olaf Gross Département de Biochimie, Université de Lausanne, Chemin des Boveresses 155, 1066 Epalinges, Switzerland, ogross4926@lrz.tum.de

Juan-Carlos Hernandez Grupo Inmunovirología, Universidad de Antioquia, Medellín, Colombia

Thirumala-Devi Kanneganti Department of Immunology, St. Jude Children's Research Hospital, 262 Danny Thomas Place, MS 351, Memphis, TN 38105, USA, Thirumala-evi.Kanneganti@StJude.org

Katherine Labbé Department of Microbiology and Immunology, McGill University, Montreal, QC, Canada H3A 2B4

Helen J. Lachmann National Amyloidosis Centre, UCL Medical School, Royal Free Campus, Rowland Hill Street, London NW3 2PF, UK, h.lachmann@medsch.ucl.ac.uk

Eicke Latz Division of Infectious Diseases and Immunology, University of Massachusetts Medical School, Worcester, MA 01605, USA; Institute of Innate Immunity, University Hospitals, University of Bonn, 53127 Bonn, Germany, eicke.latz@unionn.de, eicke.latz@umassmed.edu

Laura E. Layland Institut für Medizinische Mikrobiologie, Immunologie und Hygiene, Technische Universität München, 81675 Munich, Germany; Institut für Medizinische Mikrobiologie, Immunologie und Parasitologie, Universität Bonn, 53127 Bonn, Germany

Christopher R. Lupfer Department of Immunology, St. Jude Children's Research Hospital, 262 Danny Thomas Place, MS 351, Memphis, TN 38105, USA

Fabio Martinon Department of Biochemistry, University of Lausanne, Chemin des Boveresses 155, 1066 Epalinges, Switzerland

Mihai G. Netea Department of Medicine, Radboud University Nijmegen Medical Center, Nijmegen, The Netherlands; Nijmegen Institute for Infection, Inflammation and Immunity (N4i), Nijmegen, The Netherlands

Virginie Pétrilli Université de Lyon, Université Lyon1, F-69000, Lyon, France; Inserm U1052, Centre de Recherche en Cancérologie de Lyon, F-69000, Lyon, France; CNRS5286, Centre de Recherche en Cancérologie de Lyon, F-69000, Lyon, France, petrilli@lyon.fnclcc.fr

Pablo Pelegrín Inflammation and Experimental Surgery Group, University Hospital "Virgen de la Arrixaca"-FFIS, 30120 Murcia, Spain, pablo.pelegrin@ffis.es

Fabio Re Department of Molecular Sciences, University of Tennessee Health Science Center, 858 Madison Avenue, MSB 501B, Memphis, TN 38163, USA, fre@uthsc.edu

Anna Rubartelli Division of Cell Biology, National Cancer Research Institute, Largo Rosanna Benzi 10, 16132 Genova, Italy, anna.rubartelli@istge.it

Bernhard Ryffel Molecular Immunology and Embryology, UMR6218, University of Orleans and CNRS, 3B rue de la Férollerie, 45071 Orleans, France

Maya Saleh Department of Microbiology and Immunology, McGill University, Montreal, QC, Canada H3A 2B4; Department of Medicine, McGill University, McGill Life Sciences Complex, Montreal, QC, Canada H3A 1A1; McGill Life Sciences Complex, Bellini Pavillion, Room 364, 3649 Promenade Sir-William Osler, Montreal, QC, Canada H3G 0B1, maya.saleh@mcgill.ca

Cherilyn M. Sirois Division of Infectious Diseases and Immunology, University of Massachusetts Medical School, Worcester, MA 01605, USA

Fayyaz S. Sutterwala Graduate Program in Immunology, University of Iowa, Iowa City, IA 52242, USA; Inflammation Program, University of Iowa, Iowa City, IA 52242, USA; Division of Infectious Diseases, Department of Internal Medicine, University of Iowa, Iowa City, IA 52242, USA; Veterans Affairs Medical Center, Iowa City, IA 52241, USA; University of Iowa, 2501 Crosspark Road, D156 MTF, Coralville, IA 52241, USA, fayyaz-utterwala@uiowa.edu

Christina J. Thomas Département de Biochimie, Université de Lausanne, Chemin des Boveresses 155, 1066 Epalinges, Switzerland, ogross4926@lrz.tum.de

Dieudonnée Togbe Molecular Immunology and Embryology, UMR6218, University of Orleans and CNRS, 3B rue de la Férollerie, 45071 Orleans, France; ArtImmune SAS, Orleans, France

Tyler K. Ulland Graduate Program in Immunology, University of Iowa, Iowa City, IA 52242, USA; Inflammation Program, University of Iowa, Iowa City, IA 52242, USA

Frank L. van de Veerdonk Department of Medicine, Radboud University Nijmegen Medical Center, Nijmegen, The Netherlands; Nijmegen Institute for Infection, Inflammation and Immunity (N4i), Nijmegen, The Netherlands, f.veerdonk@aig.umcn.nl

Molecular Definition of Inflammasomes

Virginie Pétrilli and Fabio Martinon

Abstract Inflammasomes are crucial actors of the innate immunity. They consist of cytoplasmic multiprotein complexes controlling the biological activities of the inflammatory cytokines IL-1β and IL-18. Inflammasome assembly depends on protein domain interactions. In this chapter, we focus on the biochemical aspects characterizing the components of the inflammasomes and their assembly into a high molecular weight proteolytic complex.

1 Introduction

Multicellular organisms face constant aggression from invasive pathogens. To protect themselves against aggression, hosts have evolved a system of defence called the immune system. In vertebrates, the immune system is subdivided in "innate" and "adaptive" immunity. The innate immune system, which is also the oldest in terms of evolution, constitutes the first line of defence against invasive pathogens. Because of the tremendous diversity of invaders, the host innate immune system has coevolved molecules that are able to detect essential constituent of the invaders such as bacterial wall or genetic material, giving rise to the notion of Pathogen-Associated Molecular

V. Pétrilli (✉)
Université de Lyon, Université Lyon1, F-69000, Lyon, France
and
Inserm U1052, Centre de Recherche en Cancérologie de Lyon, F-69000, Lyon, France
and
CNRS5286, Centre de Recherche en Cancérologie de Lyon, F-69000, Lyon, France
e-mail: petrilli@lyon.fnclcc.fr

F. Martinon
Department of Biochemistry, University of Lausanne, Chemin des Boveresses 155, 1066 Epalinges, Switzerland

I. Couillin et al. (eds.), *The Inflammasomes*, Progress in Inflammation Research,
DOI 10.1007/978-3-0348-0148-5_1,

Patterns (PAMPs). These molecules are designated as Pattern Recognition Receptors (PRRs) and are germ-line encoded; i.e. they do not need genomic rearrangement to be expressed. Such detection systems include C-type Lectin receptors, Toll-Like Receptors (TLRs) and pentraxins, which survey the extracellular milieu, and NOD (Nucleotide-Binding Oligomerization Domain)-Like Receptors (NLRs) and RIG-I-Like Receptors (RLRs), which survey the cytoplasm. Most of these PRR activate signalling cascades resulting in the triggering of an inflammatory response and activation of the adaptive immunity. Members of the NLR family are involved in the formation of structures named inflammasomes.

2 Definition of Inflammasomes

Inflammasomes are multiprotein complexes localized within the cytoplasm of the cell. These complexes are responsible for the activation of inflammatory caspase-1, resulting in processing of the proinflammatory cytokines pro-IL-1β and pro-IL-18 into active IL-1β and IL-18.

The first inflammasome complex reported in 2002 described the assembly of the PRR NLRP1 (NALP1) with the adaptor ASC (apoptosis-associated Speck-like protein containing a CARD), caspase-1 and caspase-5 resulting in caspase activation and processing of their substrate pro-IL-1β and pro-IL-18 [1]. Since then, different types of inflammasomes have been reported including NLRC4 (IPAF) and AIM2 inflammasomes [2–6].

Generally speaking, the inflammasome depends on the assembly of a sensor, for instance an NLRP, with an adaptor, ASC, allowing the recruitment and activation of an inflammatory caspase, caspase-1. Various microbial stimuli and danger signals can lead to inflammasome assembly. These aspects are covered within the different chapters of this book. In this chapter, we endeavour the biochemical description of these inflammasome complexes.

NLRP and NLRC4 proteins belong to the NLR family. Twenty-two NLRs have been reported in humans. These NLR proteins share a common structural organization with the disease Resistance (R) proteins found in plants [7]. R genes play an essential role in the defence against pathogens by activating the MAPKinase or/and cell death pathways. NLR proteins are characterized by three main structural domains: the N-terminus contains an effector domain that can be a Pyrin Domain (PYD), a Caspase-Recruitment Domain (CARD), or a baculovirus inhibitor of apoptosis protein repeats domain (BIR). This effector domain is followed by an intermediate domain named NOD (also known as NBS or NACHT for neuronal apoptosis inhibitor protein NAIP, MHC class II transcription activator CIITA, heterokaryon incompatibility factor HET-E and telomerase-associated protein TLP1), and finally the C-terminus of the protein contains several Leucine-Rich Repeats (LRRs).

3 Protein Domains Composing the Inflammasomes

For a good comprehension of inflammasome biochemistry, it is important to get some knowledge about the different structural domains that compose the proteins. Here, we introduce the major domains that are found on the different inflammasome components (Fig. 1).

3.1 Death Fold Domains

The Death Fold (DF) superfamily includes the Death Domain (DD), Death Effector Domain (DED), the Caspase-Recruitment Domain (CARD), and the Pyrin Domain (PYD also known as PAAD or DAPIN). The first DD was identified on the Fas receptor by Itoh and Nagata in 1993 [8].

In humans, 32 DDs have been described including in proteins involved in innate immunity and inflammation such as MyD88, and IRAK4. The human genome harbours 28 CARD-containing proteins. Many of these proteins are well-known regulators of immunity. For example, inflammatory caspases, caspase-1 and -5, the kinase RIP2, and the innate immune sensors NOD1 (NLRC1), NOD2 (NLRC2) and NLRP1 require a CARD domain to be fully functional. The DED is found in seven proteins mainly involved in apoptosis including the adaptor protein FADD and apoptotic caspases, caspase-8 and caspase-10. Finally, more than 20 PYDs have been identified mainly in proteins involved in inflammation and early stage development such as the adaptor ASC, the NLRPs and the DNA sensor AIM2.

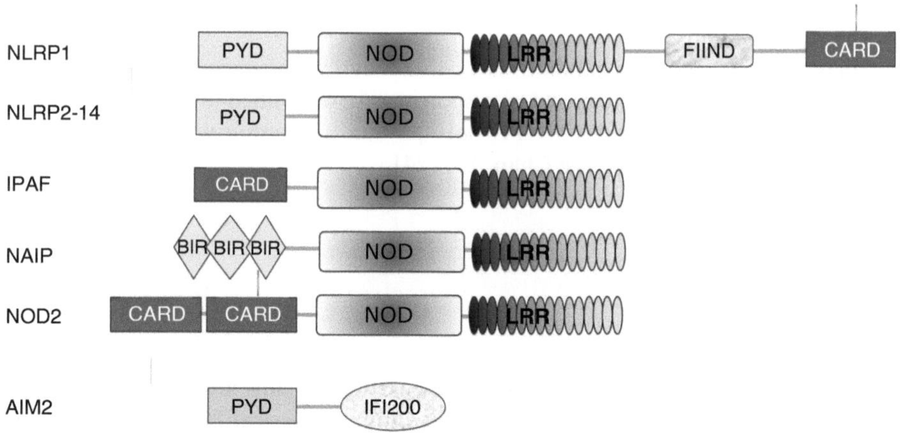

Fig. 1 Parenchymal

Although DFs share a low sequence homology, the common structural feature for this superfamily is the six-helix bundle fold as evidenced by the different domain structure resolutions [9–11].

By mediating homotypic interactions, these domains play a crucial role in the signalling cascades leading to cell death and inflammation [12]. For instance, DD interactions can give rise to oligomeric structures resulting in caspase activation or NF-κB signalization. DFs are found at the protein extremities at either N- or C-termini [13].

Inflammasome complexes are typically assembled around CARD- and PYD-containing proteins. NOD1, NOD2, ASC, inflammatory caspases, NLRP1 and NLRC4 harbour CARD domains, and NLRPs, ASC and IFI protein family harbour PYD domains.

CARDs display the typical death fold six-helix bundle, except for the first helix that tends to be bent or broken [14, 15]. In addition, CARD surfaces are polarized with basic and acidic residues, a feature that may be involved in CARD–CARD interactions. Although some DFs such as DD and DED are able to self-associate, little is known about CARD self-association among caspase prodomains and NLRs. While recent crystallography and NMR studies suggest that in certain conditions the CARD of NOD1 and ASC are able to homodimerize, a yeast-two-hybrid screen for CARD and PYD interactions showed that NOD2, RIP2, ASC and CARDINAL can homodimerize via their CARD [16–19].

PYD is the lastly identified DD superfamily member. PYD stands for PYrin Domain, highlighting the fact that Pyrin was among the first identified PYD-containing protein. In silico structural studies predicted that the PYD was a new member of the death fold superfamily, a prediction that was supported by NMR analysis of NLRP1- and ASC-PYD structures revealing a three-dimensional organization similar to death domain. Similarly to CARD, PYDs are highly bipolar molecules, suggesting a role for charges in domain interactions [11, 20–23]. Interestingly, NLRP1-PYD does not display the six strict α-helices as helix 3 is rather altered or absent [11, 12].

Interestingly, pathogen genomes have also coevolved to encode DF sequences in order to interfere with host stress response as exemplified by the M13L poxvirus protein, which encodes a PYD-only protein interfering with inflammasome assembly (for more details, see Chap. 15) [24].

3.2 BIR

Baculo-IAP (Inhibitor of Apoptosis) Repeat (BIR) domains are zinc-binding folds that are involved in protein–protein interactions. A classical BIR domain is composed of six α-helices and three-stranded β-sheets that coordinate one zinc ion with three conserved cysteines and one conserved histidine [25]. These domains allow interaction with caspases and IAP antagonists. BIR-containing proteins include

NAIP, XIAP and survivin. Several BIR motifs can be found in proteins; for instance, NAIP contains three BIRs at its N-terminus.

3.3 NOD/NACHT Domain

The NOD domain is essential for the oligomerization and the activation of the NLR. NOD domains are part of the P-loop NTPase class [26]. P-loop NTPase are characterized by very conserved sequence signature, the Walker A and Walker B motifs. These motifs bind respectively the β and γ phosphate moieties of bound NTP and a Mg^{2+} cation. P-loop NTPases display substrate preference for either ATP or GTP. NODs are generally involved in the formation of oligomeric structures. Until recently, it was not very clear whether NODs from NLR display a true ATPase activity. The studies by Ting and Reed groups suggested that NLRP1 and NLRP3 require ATP binding for their activity [27, 28].

NOD is crucial for the protein functions as hereditary mutations in NLRP3 NOD (encoded by exon 3) are responsible for three fever periodic syndromes: Muckle–Wells syndrome, familial cold urticaria and chronic infantile neurological cutaneous and articular syndrome [29–33].

3.4 LRR

Like TLR, NLR harbour an activator-sensing module made of several LRR. As it is the case for many recognition structures, LRRs are composed of several repeat motifs of short unit of 22–28 amino acid residues. NLRP LRRs are encoded by several exons of 171 nucleotides length with a conserved codon phase. Each exon codes for two half LRRs of a LRR flanking a full LRR [34]. Strikingly, this organization is conserved in most NLRP LRRs throughout evolution. While the reason for such conserved modular organization is unknown, it is tempting to speculate that it may promote extensive alternative splicing of the LRR module that can be shortened without affecting the three-dimensional structure. At the protein level, the LRR structure units consist in alternation between β-strands and α-helixes. The spatial organization of these secondary structures gives rise to a "horseshoe-shaped" molecule as found in many other LRR-containing proteins, a structure that was originally solved by crystallizing the ribonuclease inhibitor [35, 36].

These repetitive motifs form modules for which functions are not completely elucidated. LRR are thought to be the recognition modules of NLR. It is not clear yet if LRRs detect directly or indirectly their ligands/activators. NLR LRRs may also act as protein regulation modules, for instance, by controlling the activated state of the NLR. Indeed, different studies proposed that when at inactive state the LRR motifs of NLRP3, NOD1 and NOD2 fold back on the NOD, preventing any oligomerization and activation via the NOD. This inactive conformation is maintained by the interaction with the chaperones HSP90 and SGT1 [37, 38]. Furthermore, destabilization of the chaperones results in NLR degradation, a mechanism that is conserved from plants to mammals [39, 40].

4 NLR Repertoire Involved in Inflammasome Formation

4.1 NLRP

The human genome harbours 14 *NLRP* genes, and in mice 20 NLRPs have been identified. Interestingly, some of NLRPs such as NLRP1, NLRP4 and NLRP9 appear to have evolved in a relatively recent lineage-specific expansion during the vertebrate diversification leading to the generation of many paralogues [26] (Table 1). Recent gene duplication events are also observed in the human genome between the paralogues hNLRP2 and hNLRP7. NLRP structures are highly conserved and display the canonical structure of PYD-NOD-LRR (Fig. 1 and Table 1), with the exception of NLRP10 that lacks the LRR. Among NLRPs, NLRP1 and NLRP3 were shown to form effective inflammasomes.

NLRP1 is a special NLRP protein as it possesses two supplementary domains at the C-terminus: a Filamin Interactin Domain (FIIND) and a CARD. Intriguingly, both domains are found on an independent protein named CARDINAL/TUCAN/CARD8 [41, 42]. In humans, mutations within NLRP1 have been linked to Vitiligo, a autoimmune disorder causing depigmentation of the patient's skin [43]. In the mouse, the NLRP1 locus underwent duplication and contains three paralogues *NLRP1a*, *NLRP1b* and *NLRP1c*. It is of note that mouse NLRP1 proteins do not harbour a predicted functional PYD; therefore, the function of NLRP1 in mouse might slightly differ to that of human NLRP1. For instance, it could act more as an adaptor protein, which would oligomerize through its NOD. C-terminal mouse NLRP1s are the closest mouse sequences related to human CARDINAL; it is therefore possible that mouse NLRP1s are functional homologues of human CARDINAL. The first inflammasome reported was the NLRP1 inflammasome. Using an in vitro system activation, the NLRP1 inflammasome was shown to be composed of NLRP1, ASC, caspase-1 and caspase-5 [1]. In the mouse, *NLRP1* locus has been associated with strain resistance/sensitivity to anthrax Lethal Toxin (LT) [44].

NLRP3 is relatively conserved within species indicating a role in the recognition of conserved DAMP (Damage-Associated Molecular Patterns) and PAMP. NLRP3 is expressed by myeloid cells, keratinocytes, epithelial cells and osteoblasts; its expression is strongly inducible by TLR agonists ([45, 46]; V.P: personal observations). In humans, many missense hereditary mutations within the NOD have been reported to be responsible for periodic fever syndromes as described above [31–33]. They confer a gain-of-function mutation to the NLRP3 (for more details, see chapter "Mechanisms of NLRP3 inflammasome activation in CAPS patients"). The NLRP3 inflammasome has been extensively characterized so far. It can be

Table 1

NLR subfamily	Common nomenclature Human	Mouse	Chromosome localization	Structure
	NLRP1		17p13	PYD-NACHT-NAD-LRR-FIIND-CARD
		NLRP1a	11B4	NACHT-NAD-LRR-FIIND-CARD
		NLRP1a	11B4	NACHT-NAD-LRR-FIIND-CARD
		NLRP1a	11B4	NACHT-NAD-LRR-FIIND-CARD
	NLRP2		19q13.42	PYD-NACHT-NAD-LRR
		NLRP2	7A1	PYD-NACHT-NAD-LRR
	NLRP3		1q44	PYD-NACHT-NAD-LRR
		NLRP3	11B1.3	PYD-NACHT-NAD-LRR
	NLRP4		19q13.43	PYD-NACHT-NAD-LRR
		NLRP4a	7A1	PYD-NACHT-NAD-LRR
		NLRP4b	7A1	PYD-NACHT-NAD-LRR
		NLRP4c	7A1	PYD-NACHT-NAD-LRR
		NLRP4d	7A1	PYD-NACHT-NAD-LRR
		NLRP4e	7A2	PYD-NACHT-NAD-LRR
		NLRP4f	13B3	PYD-NACHT-NAD-LRR
		NLRP4g		PYD-NACHT-NAD-LRR
	NLRP5		19q13.42	PYD-NACHT-NAD-LRR
		NLRP5	7A2	NACHT-NAD-LRR
	NLRP6		11p15.5	PYD-NACHT-NAD-LRR
		NLRP6	7F4	PYD-NACHT-NAD-LRR
	NLRP7		19q13.42	PYD-NACHT-NAD-LRR
	NLRP8		19q13.42	PYD-NACHT-NAD-LRR
	NLRP9		19q13.42	PYD-NACHT-NAD-LRR
		NLRP9a	7A3	PYD-NACHT-NAD-LRR
		NLRP9b	7A2	PYD-NACHT-NAD-LRR
		NLRP9c	7A3	PYD-NACHT-NAD-LRR
	NLRP10		11p15.4	PYD-NACHT-NAD
		NLRP10	7E3	PYD-NACHT-NAD
	NLRP11		19q13.42	PYD-NACHT-NAD-LRR
	NLRP12		19q13.42	PYD-NACHT-NAD-LRR
		NLRP12	7A1	PYD-NACHT-NAD-LRR
	NLRP13		19q13.42	PYD-NACHT-NAD-LRR
	NLRP14		11p15.4	PYD-NACHT-NAD-LRR
NLRP/NALP		NLRP14	7E3	PYD-NACHT-NAD-LRR
	NLRC4		2p22-p21	CARD-NACHT-LRR
NLRC4/IPAF		NLRC4	17 E3	CARD-NACHT-LRR
	NAIP		5q13.1	BIR3x-NACHT-LRR
		NAIPa	13D1	BIR3x-NACHT-LRR
		NAIPb	13D1	BIR3x-NACHT-LRR
		NAIPc	13D1	BIR3x-NACHT-LRR
		NAIPd	13D1	BIR3x-NACHT-LRR
		NAIPe	13D1	BIR3x-NACHT-LRR
		NAIPf	13D1	BIR3x-NACHT-LRR
NAIP		NAIPg	13D1	BIR3x-NACHT-LRR

activated by a wide range of stimuli including ATP, Monosodium Urate crystals (MSU) and Gram-positive bacteria (fore more details, see chapters "The Role of Inflammasomes in Viral Infection", "Inflammasome Activation in Response to Eukaryotic Pathogens", "Inflammasome Activation by Pathogenic Crystals and Particles", "Inflammasome Activation by Danger Signals" and "Toll-Like Receptors and Inflammasomes"). Further studies are required to investigate whether the other NLRP members are capable of forming inflammasomes.

4.2 *NLRC4/IPAF*

NRLC4 harbours a N-terminus CARD domain and is therefore believed to directly interact with caspase-1 through CARD–CARD homotypic interaction (Fig. 1). However, ASC has been shown to contribute to caspase-1 activation and IL-1β maturation within the NLRC4 inflammasome [2]. NLRC4 is the sensor of Gram-negative bacteria *Salmonella typhimurium*, *Shigella flexneri* or *Legionella pneumophila* [47]. The function of NLRC4 is detailed in chapter "Activation of the Inflammasome by Bacterial Pathogens".

4.3 *NAIP*

NAIP contains three BIR domains at the N-terminus, followed by NACHT and LRR domains displaying the highest sequence similarity with NLRC4. In humans, mutations within *NAIP* gene are associated with spinal muscular dystrophy. NAIP is expressed in macrophages, however, its function is still unclear. In the mouse, seven NAIP gene paralogues have been identified: *NAIP1* to *NAIP7*. Although BIR domains are supposed to be caspase inhibitory domains, NAIP may form an inflammasome together with NLRC4. NAIP5 has also been shown to be a critical factor to restrict *Legionella* growth [48] (see chapter "Activation of the Inflammasome by Bacterial Pathogens").

4.4 *NOD2*

NOD2/NLRC2 displays two CARDs at its N-terminus. Mutations in the *NOD2* gene are detected in patients suffering from Blau syndrome (mutations in the NOD) and Crohn's disease. NOD2 per se has not been shown to form an inflammasome, but was found in certain conditions to modulate NLRP3 inflammasome activity in response to MDP [49].

5 Non-NLR Inflammasome, AIM2

Recently, the protein AIM2 has been identified as the cytoplasmic DNA sensor capable of interacting with ASC to trigger caspase-1 activation in macrophages (reviewed in detail in chapter "The Role of Inflammasomes in Viral Infection" [50]). It is the first non-NLR inflammasome described. AIM2 belongs to the protein family HIN or IFI-200. The first member of this family was described in mouse. There are several murine HIN family members, including p202a, p202b, p203, p204 and p205, while there are four human HIN proteins known as IFI16 (Interferon-inducible protein 16), AIM2 (Absent In Melanoma 2), IFIX (IFN-Inducible protein X) and MNDA (Myeloid Nuclear Differentiation Antigen). These proteins contain both an "IFI200" domain and a PYD except for mouse p202 that lacks PYD (Fig. 1) [50, 51].

6 ASC

ASC, also known as CARD5, PYCARD or TMS1, is an ubiquitous protein of 22 kDa. It is an adaptor containing two DFs: a PYD at the N-terminus and a CARD at the C-terminus. ASC plays a crucial role by mediating the interaction between the scaffold receptor or PRR and the caspase-1. Invalidating ASC in the mouse results in ablation of caspase-1 activation in response to NLRP3 and AIM2 activators and reduces NLRC4 ability to activate caspase-1 [2, 52].

7 Inflammatory Caspases

Caspases are cysteine proteases that have major functions in cell death and inflammation. They cleave their substrates following a very conserved tetrapeptide sequence containing at the C-terminus an aspartate residue. Thirteen caspases have been described in humans and mice. Caspases can be subdivided into inflammatory and apoptotic caspases but also into initiator and effector caspases. Initiator caspases possess a large prodomain. This prodomain belongs to the DF family and is essential for the interaction of the initiator caspase with the signalling machinery that triggers caspase activation. In mammals, initiator caspases caspase-8, -10, -2 and -9 are involved in the apoptosis pathway, and caspase-1, -4, -5, -11 and -12 in the inflammatory pathway. Executioner caspases such as caspase-3, -6 and -7 have a short prodomain and are activated by the initiator caspases. Executioner caspases are responsible for the cleavage of a tremendous number of cellular substrates that results in the morphological changes of apoptosis.

Inflammatory caspases are found only in vertebrates. In mammals, they are five inflammatory caspases that all contain a CARD prodomain. Interestingly, in humans, chimps and mice, inflammatory caspases cluster on a single locus and share significant similarities. Mouse inflammatory caspases include caspase-1, -11 and -12 located on chr9A1. Human inflammatory caspases consist of caspase-1, -4, -5 and -12, but caspase-12 does not encode an active enzyme. Phylogenic analysis suggests that caspase-4 and -5 arise from a gene duplication of caspase-11 [34]. These caspases are clustered on chr11q22.

7.1 Human and Mouse Caspase-1

Caspase-1, also called Interleukin-1 Converting Enzyme (ICE), is the first caspase discovered in mammals, but also the best characterized inflammatory caspase [53, 54]. Caspase-1 is responsible for the activation of proinflammatory cytokines that are synthesized as inactive precursors: pro-IL-1β and pro-IL-18. Pro-IL-1β cleavage occurs at the site Tyr-Val-Ala-Asp116/-Ala117, giving rise to a 17 kDa active cytokine (C-terminus part) that is secreted by the cell [55, 56]. Evidence for the indispensable role of caspase-1 in IL-1β activity was definitively settled by the generation of caspase-1-deficient mice [57, 58]. These mice are unable to produce active IL-1β and are more sensitive to infections by *Shigella*, *Francisella* or *Salmonella* [59–61]. IL-1β is a plethoric cytokine that, for instance, induces T-cell proliferation through the induction of IL-2 and neutrophil recruitment. It can act on the CNS by inducing fever and fatigue, but also promotes the expression of angiogenic factors [62]. Similarly, caspase-1 cleaves pro-IL-18 after Asp37 to give rise to an active IL-18 (previously named interferon-γ inducing factor). Indeed, IL-18 is also a proinflammatory cytokine participating, for instance, in the TH1 polarization and NK cell activation [63, 64]. Strikingly, IL-18 participates in the control of the certain central nervous system functions such as the regulation of energy uptake and stress [65, 66]. Recent works reported that caspase-1 processes IL-33, another member of IL-1 family, although the outcome is most likely the inactivation of the cytokine [67, 68]. Besides the maturation of cytokines, the group of Maya Saleh reported a panel of new caspase-1 substrates including chaperone, cytoskeletal proteins and glycolytic enzymes such as α-enolase or the pyruvate kinase [69]. Finally, activation of caspase-1 can also result in a particular cell death named the pyroptosis, "for more details see chapter "Pyroptosis: A Caspase-1-Dependent Programmed Cell Death and a Barrier to Infection, this volume.""

7.2 Human Caspase-4 and Caspase-5 and Murine Caspase-11

Despite the generation of caspase-11-deficient mice, the functions of caspase-11 are still poorly understood. Unlike caspase-1, caspase-11 expression is inducible (e.g. by LPS). Caspase-11−/− mice are more resistant to endotoxic (LPS) shock and fail to produce mature IL-1β, suggesting a role for caspase-11 in caspase-1 activation [70]. Recently, a few studies described a role for caspase-11 in brain ischemia-induced injury, ERS-mediated astrocyte cell death, Lethal Toxin-induced cell death or interferon-β-induced cell death of dendritic cells [71–74]. Strikingly, Li and co-workers demonstrated an unexpected role for caspase-11 in cell migration [75]. By interacting with Actin-Interacting Protein-1 (AIP1) caspase-11 controls a cofilin-mediated actin depolimerization, a function that appears to be independent of its enzymatic activity.

No defined roles and substrates have been reported for human caspase-4 and -5. Caspase-5 was shown to be part of the NLRP1 inflammasome [1], whereas caspase-4 is proposed to be involved in ERS-mediated cell death, although these data are matter of debate [76–78]. Finally, a recent genetic analysis on cancer microdissections reports an association with caspase-5 mutation (frameshift mutations mainly in the CARD) and gastric carcinoma [79].

7.3 Human and Mouse Caspase-12

A first study using mice deficient for caspase-12 proposed a role for this caspase in ERS-mediated cell death [80, 81]. However, other studies have contradicted these findings and excluded caspase-12 as an ERS actor in cell death. Particularly, the generation of another caspase-12-deficient mouse strain demonstrated rather a role for caspase-12 in bacterial infection by dampening the caspase-1 activity [78, 82, 83]. More recently, the same group showed that caspase-12 could be a negative regulator of the NOD signalling pathway [84]. Caspase-12 is likely to act as a negative regulator of the inflammatory response and positively on anti-viral response [85, 86]. Finally, murine caspase-12 enzymatic activity seems to be restricted to autocatalytic activity [87].

Human caspase-12 gene displays a single nucleotide polymorphism in exon 4 that results either in a truncated protein (CARD only) or in full-length variant molecule putatively enzymatically inactive. The full-length allele is the less frequent among human population and is specifically expressed by the African descendant population [88]. Interestingly, this variant is also present in the caspase-12 gene of non-human primates, suggesting a selection advantage for these populations. The FL caspase-12 linked to African descendants may confer a hyporesponsiveness to endotoxic shock. Finally, a "humanized" mouse strain expressing the hcasp12L transgene in a caspase-12-deficient background revealed that in females, estrogens repress hcaspase-12L expression resulting in better protection from *L. monocytogenes* [89].

8 Conclusions

Based on the modular structure of inflammasome components, and initial biochemical evidence, a model has emerged suggesting that inflammatory caspases get activated upon inflammasome formation. Upon sensing of DAMP and PAMP, NLR and AIM2 proteins induce the recruitment of ASC and caspase-1 within a multimolecular platform resulting in caspase-1 activation. As detailed in the following chapters, recent years have provided a plethora of examples of conditions and pathologies involving inflammasome components including NLRP3, ASC and IL-1β. However, the biochemical and molecular mechanisms of how the inflammasome assembles are largely unknown. For instance, whether the NLRP3s sense the PAMP/DAMP directly or not is not yet well established. Similarly, the exact composition of the various inflammasomes and its overall structure are still lacking. This is still an emerging question and future studies will undoubtedly shed more light on the diversity and respective roles of various inflammasome components in the assembly and activation of inflammasomes.

References

1. Martinon F, Burns K, Tschopp J (2002) The inflammasome: a molecular platform triggering activation of inflammatory caspases and processing of proIL-beta. Mol Cell 10:417–426
2. Mariathasan S, Newton K, Monack DM, Vucic D, French DM, Lee WP, Roose-Girma M, Erickson S, Dixit VM (2004) Differential activation of the inflammasome by caspase-1 adaptors ASC and Ipaf. Nature 430:213–218
3. Schroder K, Muruve DA, Tschopp J (2009) Innate immunity: cytoplasmic DNA sensing by the AIM2 inflammasome. Curr Biol 19:R262–R265
4. Burckstummer T, Baumann C, Bluml S, Dixit E, Durnberger G, Jahn H, Planyavsky M, Bilban M, Colinge J, Bennett KL et al (2009) An orthogonal proteomic-genomic screen identifies AIM2 as a cytoplasmic DNA sensor for the inflammasome. Nat Immunol 10:266–272
5. Fernandes-Alnemri T, Yu JW, Datta P, Wu J, Alnemri ES (2009) AIM2 activates the inflammasome and cell death in response to cytoplasmic DNA. Nature 458:509–513
6. Hornung V, Ablasser A, Charrel-Dennis M, Bauernfeind F, Horvath G, Caffrey DR, Latz E, Fitzgerald KA (2009) AIM2 recognizes cytosolic dsDNA and forms a caspase-1-activating inflammasome with ASC. Nature 458:514–518
7. Meyers BC, Morgante M, Michelmore RW (2002) TIR-X and TIR-NBS proteins: two new families related to disease resistance TIR-NBS-LRR proteins encoded in Arabidopsis and other plant genomes. Plant J 32:77–92
8. Itoh N, Nagata S (1993) A novel protein domain required for apoptosis. Mutational analysis of human Fas antigen. J Biol Chem 268:10932–10937
9. Chou JJ, Matsuo H, Duan H, Wagner G (1998) Solution structure of the RAIDD CARD and model for CARD/CARD interaction in caspase-2 and caspase-9 recruitment. Cell 94:171–180
10. Huang B, Eberstadt M, Olejniczak ET, Meadows RP, Fesik SW (1996) NMR structure and mutagenesis of the Fas (APO-1/CD95) death domain. Nature 384:638–641
11. Hiller S, Kohl A, Fiorito F, Herrmann T, Wider G, Tschopp J, Grutter MG, Wuthrich K (2003) NMR structure of the apoptosis- and inflammation-related NALP1 pyrin domain. Structure 11:1199–1205
12. Park HH, Lo YC, Lin SC, Wang L, Yang JK, Wu H (2007) The death domain superfamily in intracellular signaling of apoptosis and inflammation. Annu Rev Immunol 25:561–586
13. Hofmann K, Tschopp J (1995) The death domain motif found in Fas (Apo-1) and TNF receptor is present in proteins involved in apoptosis and axonal guidance. FEBS Lett 371:321–323
14. Day CL, Dupont C, Lackmann M, Vaux DL, Hinds MG (1999) Solution structure and mutagenesis of the caspase recruitment domain (CARD) from Apaf-1. Cell Death Differ 6:1125–1132
15. Zhou P, Chou J, Olea RS, Yuan J, Wagner G (1999) Solution structure of Apaf-1 CARD and its interaction with caspase-9 CARD: a structural basis for specific adaptor/caspase interaction. Proc Natl Acad Sci USA 96:11265–11270

16. de Alba E (2009) Structure and interdomain dynamics of apoptosis-associated speck-like protein containing a CARD (ASC). J Biol Chem 284:32932–32941
17. Srimathi T, Robbins SL, Dubas RL, Hasegawa M, Inohara N, Park YC (2008) Monomer/dimer transition of the caspase-recruitment domain of human Nod1. Biochemistry 47:1319–1325
18. Coussens NP, Mowers JC, McDonald C, Nunez G, Ramaswamy S (2007) Crystal structure of the Nod1 caspase activation and recruitment domain. Biochem Biophys Res Commun 353:1–5
19. Wagner RN, Proell M, Kufer TA, Schwarzenbacher R (2009) Evaluation of Nod-like receptor (NLR) effector domain interactions. PLoS One 4:e4931
20. Liepinsh E, Barbals R, Dahl E, Sharipo A, Staub E, Otting G (2003) The death-domain fold of the ASC PYRIN domain, presenting a basis for PYRIN/PYRIN recognition. J Mol Biol 332:1155–1163
21. Martinon F, Hofmann K, Tschopp J (2001) The pyrin domain: a possible member of the death domain-fold family implicated in apoptosis and inflammation. Curr Biol 11:R118–R120
22. Bertin J, DiStefano PS (2000) The PYRIN domain: a novel motif found in apoptosis and inflammation proteins. Cell Death Differ 7:1273–1274
23. Pawlowski K, Pio F, Chu Z, Reed JC, Godzik A (2001) PAAD – a new protein domain associated with apoptosis, cancer and autoimmune diseases. Trends Biochem Sci 26:85–87
24. Johnston JB, Barrett JW, Nazarian SH, Goodwin M, Ricciuto D, Wang G, McFadden G (2005) A poxvirus-encoded pyrin domain protein interacts with ASC-1 to inhibit host inflammatory and apoptotic responses to infection. Immunity 23:587–598
25. Verhagen AM, Coulson EJ, Vaux DL (2001) Inhibitor of apoptosis proteins and their relatives: IAPs and other BIRPs. Genome Biol 2:reviews3009–reviews3009.10
26. Leipe DD, Koonin EV, Aravind L (2004) STAND, a class of P-loop NTPases including animal and plant regulators of programmed cell death: multiple, complex domain architectures, unusual phyletic patterns, and evolution by horizontal gene transfer. J Mol Biol 343:1–28
27. Duncan JA, Bergstralh DT, Wang Y, Willingham SB, Ye Z, Zimmermann AG, Ting JP (2007) Cryopyrin/NALP3 binds ATP/dATP, is an ATPase, and requires ATP binding to mediate inflammatory signaling. Proc Natl Acad Sci USA 104:8041–8046
28. Faustin B, Lartigue L, Bruey JM, Luciano F, Sergienko E, Bailly-Maitre B, Volkmann N, Hanein D, Rouiller I, Reed JC (2007) Reconstituted NALP1 inflammasome reveals two-step mechanism of caspase-1 activation. Mol Cell 25:713–724
29. Hawkins PN, Lachmann HJ, Aganna E, McDermott MF (2004) Spectrum of clinical features in Muckle-Wells syndrome and response to anakinra. Arthritis Rheum 50:607–612
30. McDermott MF, Aksentijevich I (2002) The autoinflammatory syndromes. Curr Opin Allergy Clin Immunol 2:511–516
31. Aganna E, Martinon F, Hawkins PN, Ross JB, Swan DC, Booth DR, Lachmann HJ, Bybee A, Gaudet R, Woo P et al (2002) Association of mutations in the NALP3/CIAS1/PYPAF1 gene with a broad phenotype including recurrent fever, cold sensitivity, sensorineural deafness, and AA amyloidosis. Arthritis Rheum 46:2445–2452
32. Hoffman HM, Gregory SG, Mueller JL, Tresierras M, Broide DH, Wanderer AA, Kolodner RD (2003) Fine structure mapping of CIAS1: identification of an ancestral haplotype and a common FCAS mutation, L353P. Hum Genet 112:209–216
33. Feldmann J, Prieur AM, Quartier P, Berquin P, Certain S, Cortis E, Teillac-Hamel D, Fischer A, de Saint BG (2002) Chronic infantile neurological cutaneous and articular syndrome is caused by mutations in CIAS1, a gene highly expressed in polymorphonuclear cells and chondrocytes. Am J Hum Genet 71:198–203
34. Martinon F, Tschopp J (2007) Inflammatory caspases and inflammasomes: master switches of inflammation. Cell Death Differ 14:10–22
35. Kobe B, Deisenhofer J (1993) Crystal structure of porcine ribonuclease inhibitor, a protein with leucine-rich repeats. Nature 366:751–756

36. Istomin AY, Godzik A (2009) Understanding diversity of human innate immunity receptors: analysis of surface features of leucine-rich repeat domains in NLRs and TLRs. BMC Immunol 10:48
37. Mayor A, Martinon F, De Smedt T, Petrilli V, Tschopp J (2007) A crucial function of SGT1 and HSP90 in inflammasome activity links mammalian and plant innate immune responses. Nat Immunol 8:497–503
38. da Silva CJ, Miranda Y, Leonard N, Ulevitch R (2007) SGT1 is essential for Nod1 activation. Proc Natl Acad Sci USA 104:6764–6769
39. Azevedo C, Betsuyaku S, Peart J, Takahashi A, Noel L, Sadanandom A, Casais C, Parker J, Shirasu K (2006) Role of SGT1 in resistance protein accumulation in plant immunity. EMBO J 25:2007–2016
40. Liu Y, Burch-Smith T, Schiff M, Feng S, Dinesh-Kumar SP (2004) Molecular chaperone Hsp90 associates with resistance protein N and its signaling proteins SGT1 and Rar1 to modulate an innate immune response in plants. J Biol Chem 279:2101–2108
41. Bouchier-Hayes L, Conroy H, Egan H, Adrain C, Creagh EM, MacFarlane M, Martin SJ (2001) CARDINAL, a novel caspase recruitment domain protein, is an inhibitor of multiple NF-kappa B activation pathways. J Biol Chem 276:44069–44077
42. Razmara M, Srinivasula SM, Wang L, Poyet JL, Geddes BJ, DiStefano PS, Bertin J, Alnemri ES (2002) CARD-8 protein, a new CARD family member that regulates caspase-1 activation and apoptosis. J Biol Chem 277:13952–13958
43. Jin Y, Mailloux CM, Gowan K, Riccardi SL, LaBerge G, Bennett DC, Fain PR, Spritz RA (2007) NALP1 in vitiligo-associated multiple autoimmune disease. N Engl J Med 356:1216–1225
44. Boyden ED, Dietrich WF (2006) Nalp1b controls mouse macrophage susceptibility to anthrax lethal toxin. Nat Genet 38:240–244
45. Bauernfeind FG, Horvath G, Stutz A, Alnemri ES, MacDonald K, Speert D, Fernandes-Alnemri T, Wu J, Monks BG, Fitzgerald KA et al (2009) Cutting edge: NF-kappaB activating pattern recognition and cytokine receptors license NLRP3 inflammasome activation by regulating NLRP3 expression. J Immunol 183:787–791
46. McCall SH, Sahraei M, Young AB, Worley CS, Duncan JA, Ting JP, Marriott I (2008) Osteoblasts express NLRP3, a nucleotide-binding domain and leucine-rich repeat region containing receptor implicated in bacterially induced cell death. J Bone Miner Res 23:30–40
47. Franchi L, Amer A, Body-Malapel M, Kanneganti TD, Ozoren N, Jagirdar R, Inohara N, Vandenabeele P, Bertin J, Coyle A et al (2006) Cytosolic flagellin requires Ipaf for activation of caspase-1 and interleukin 1beta in Salmonella-infected macrophages. Nat Immunol 7: 576–582
48. Zamboni DS, Kobayashi KS, Kohlsdorf T, Ogura Y, Long EM, Vance RE, Kuida K, Mariathasan S, Dixit VM, Flavell RA et al (2006) The Birc1e cytosolic pattern-recognition receptor contributes to the detection and control of Legionella pneumophila infection. Nat Immunol 7:318–325
49. Pan Q, Mathison J, Fearns C, Kravchenko VV, Da Silva CJ, Hoffman HM, Kobayashi KS, Bertin J, Grant EP, Coyle AJ et al (2007) MDP-induced interleukin-1beta processing requires Nod2 and CIAS1/NALP3. J Leukoc Biol 82:177–183
50. Choubey D, Duan X, Dickerson E, Ponomareva L, Panchanathan R, Shen H, Srivastava R (2010) Interferon-inducible p200-family proteins as novel sensors of cytoplasmic DNA: role in inflammation and autoimmunity. J Interferon Cytokine Res 30:371–380
51. Choubey D, Snoddy J, Chaturvedi V, Toniato E, Opdenakker G, Thakur A, Samanta H, Engel DA, Lengyel P (1989) Interferons as gene activators. Indications for repeated gene duplication during the evolution of a cluster of interferon-activatable genes on murine chromosome 1. J Biol Chem 264:17182–17189
52. Martinon F, Petrilli V, Mayor A, Tardivel A, Tschopp J (2006) Gout-associated uric acid crystals activate the NALP3 inflammasome. Nature 440:237–241

53. Cerretti DP, Kozlosky CJ, Mosley B, Nelson N, Van Ness K, Greenstreet TA, March CJ, Kronheim SR, Druck T, Cannizzaro LA et al (1992) Molecular cloning of the interleukin-1 beta converting enzyme. Science 256:97–100
54. Thornberry NA, Bull HG, Calaycay JR, Chapman KT, Howard AD, Kostura MJ, Miller DK, Molineaux SM, Weidner JR, Aunins J et al (1992) A novel heterodimeric cysteine protease is required for interleukin-1 beta processing in monocytes. Nature 356:768–774
55. March CJ, Mosley B, Larsen A, Cerretti DP, Braedt G, Price V, Gillis S, Henney CS, Kronheim SR, Grabstein K et al (1985) Cloning, sequence and expression of two distinct human interleukin-1 complementary DNAs. Nature 315:641–647
56. Cameron P, Limjuco G, Rodkey J, Bennett C, Schmidt JA (1985) Amino acid sequence analysis of human interleukin 1 (IL-1). Evidence for biochemically distinct forms of IL-1. J Exp Med 162:790–801
57. Li P, Allen H, Banerjee S, Franklin S, Herzog L, Johnston C, McDowell J, Paskind M, Rodman L, Salfeld J et al (1995) Mice deficient in IL-1 beta-converting enzyme are defective in production of mature IL-1 beta and resistant to endotoxic shock. Cell 80:401–411
58. Kuida K, Lippke JA, Ku G, Harding MW, Livingston DJ, Su MS, Flavell RA (1995) Altered cytokine export and apoptosis in mice deficient in interleukin-1 beta converting enzyme. Science 267:2000–2003
59. Sansonetti PJ, Phalipon A, Arondel J, Thirumalai K, Banerjee S, Akira S, Takeda K, Zychlinsky A (2000) Caspase-1 activation of IL-1beta and IL-18 are essential for Shigella flexneri-induced inflammation. Immunity 12:581–590
60. Mariathasan S, Weiss DS, Dixit VM, Monack DM (2005) Innate immunity against Francisella tularensis is dependent on the ASC/caspase-1 axis. J Exp Med 202:1043–1049
61. Lara-Tejero M, Sutterwala FS, Ogura Y, Grant EP, Bertin J, Coyle AJ, Flavell RA, Galan JE (2006) Role of the caspase-1 inflammasome in Salmonella typhimurium pathogenesis. J Exp Med 203:1407–1412
62. Dinarello CA (2009) Immunological and inflammatory functions of the interleukin-1 family. Annu Rev Immunol 27:519–550
63. Hunter CA, Timans J, Pisacane P, Menon S, Cai G, Walker W, Aste-Amezaga M, Chizzonite R, Bazan JF, Kastelein RA (1997) Comparison of the effects of interleukin-1 alpha, interleukin-1 beta and interferon-gamma-inducing factor on the production of interferon-gamma by natural killer. Eur J Immunol 27:2787–2792
64. Kohno K, Kataoka J, Ohtsuki T, Suemoto Y, Okamoto I, Usui M, Ikeda M, Kurimoto M (1997) IFN-gamma-inducing factor (IGIF) is a costimulatory factor on the activation of Th1 but not Th2 cells and exerts its effect independently of IL-12. J Immunol 158:1541–1550
65. Netea MG, Joosten LA, Lewis E, Jensen DR, Voshol PJ, Kullberg BJ, Tack CJ, van Krieken H, Kim SH, Stalenhoef AF et al (2006) Deficiency of interleukin-18 in mice leads to hyperphagia, obesity and insulin resistance. Nat Med 12:650–656
66. Sugama S, Conti B (2008) Interleukin-18 and stress. Brain Res Rev 58:85–95
67. Liew FY, Pitman NI, McInnes IB (2010) Disease-associated functions of IL-33: the new kid in the IL-1 family. Nat Rev Immunol 10:103–110
68. Sims JE, Smith DE (2010) The IL-1 family: regulators of immunity. Nat Rev Immunol 10:89–102
69. Shao W, Yeretssian G, Doiron K, Hussain SN, Saleh M (2007) The caspase-1 digestome identifies the glycolysis pathway as a target during infection and septic shock. J Biol Chem 282:36321–36329
70. Wang S, Miura M, Jung YK, Zhu H, Li E, Yuan J (1998) Murine caspase-11, an ICE-interacting protease, is essential for the activation of ICE. Cell 92:501–509
71. Nour AM, Yeung YG, Santambrogio L, Boyden ED, Stanley ER, Brojatsch J (2009) Anthrax lethal toxin triggers the formation of a membrane-associated inflammasome complex in murine macrophages. Infect Immun 77:1262–1271
72. Yen JH, Ganea D (2009) Interferon beta induces mature dendritic cell apoptosis through caspase-11/caspase-3 activation. Blood 114:1344–1354

73. Hisahara S, Yuan J, Momoi T, Okano H, Miura M (2001) Caspase-11 mediates oligodendrocyte cell death and pathogenesis of autoimmune-mediated demyelination. J Exp Med 193:111–122
74. Fradejas N, Pastor MD, Burgos M, Beyaert R, Tranque P, Calvo S (2010) Caspase-11 mediates ischemia-induced astrocyte death: involvement of endoplasmic reticulum stress and C/EBP homologous protein. J Neurosci Res 88:1094–1105
75. Li J, Brieher WM, Scimone ML, Kang SJ, Zhu H, Yin H, von Andrian UH, Mitchison T, Yuan J (2007) Caspase-11 regulates cell migration by promoting Aip1-Cofilin-mediated actin depolymerization. Nat Cell Biol 9:276–286
76. Pelletier N, Casamayor-Palleja M, De Luca K, Mondiere P, Saltel F, Jurdic P, Bella C, Genestier L, Defrance T (2006) The endoplasmic reticulum is a key component of the plasma cell death pathway. J Immunol 176:1340–1347
77. Hitomi J, Katayama T, Eguchi Y, Kudo T, Taniguchi M, Koyama Y, Manabe T, Yamagishi S, Bando Y, Imaizumi K et al (2004) Involvement of caspase-4 in endoplasmic reticulum stress-induced apoptosis and Abeta-induced cell death. J Cell Biol 165:347–356
78. Obeng EA, Boise LH (2005) Caspase-12 and caspase-4 are not required for caspase-dependent endoplasmic reticulum stress-induced apoptosis. J Biol Chem 280:29578–29587
79. Soung YH, Jeong EG, Ahn CH, Kim SS, Song SY, Yoo NJ, Lee SH (2008) Mutational analysis of caspase 1, 4, and 5 genes in common human cancers. Hum Pathol 39:895–900
80. Nakagawa T, Zhu H, Morishima N, Li E, Xu J, Yankner BA, Yuan J (2000) Caspase-12 mediates endoplasmic-reticulum-specific apoptosis and cytotoxicity by amyloid-beta. Nature 403:98–103
81. Morishima N, Nakanishi K, Takenouchi H, Shibata T, Yasuhiko Y (2002) An endoplasmic reticulum stress-specific caspase cascade in apoptosis. Cytochrome c-independent activation of caspase-9 by caspase-12. J Biol Chem 277:34287–34294
82. Saleh M, Mathison JC, Wolinski MK, Bensinger SJ, Fitzgerald P, Droin N, Ulevitch RJ, Green DR, Nicholson DW (2006) Enhanced bacterial clearance and sepsis resistance in caspase-12-deficient mice. Nature 440:1064–1068
83. Kalai M, Lamkanfi M, Denecker G, Boogmans M, Lippens S, Meeus A, Declercq W, Vandenabeele P (2003) Regulation of the expression and processing of caspase-12. J Cell Biol 162:457–467
84. LeBlanc PM, Yeretssian G, Rutherford N, Doiron K, Nadiri A, Zhu L, Green DR, Gruenheid S, Saleh M (2008) Caspase-12 modulates NOD signaling and regulates antimicrobial peptide production and mucosal immunity. Cell Host Microbe 3:146–157
85. Miu J, Saleh M, Stevenson MM (2010) Caspase-12 deficiency enhances cytokine responses but does not protect against lethal Plasmodium yoelii 17XL infection. Parasite Immunol 32: 773–778
86. Wang P, Arjona A, Zhang Y, Sultana H, Dai J, Yang L, LeBlanc PM, Doiron K, Saleh M, Fikrig E (2010) Caspase-12 controls West Nile virus infection via the viral RNA receptor RIG-I. Nat Immunol 11:912–919
87. Roy S, Sharom JR, Houde C, Loisel TP, Vaillancourt JP, Shao W, Saleh M, Nicholson DW (2008) Confinement of caspase-12 proteolytic activity to autoprocessing. Proc Natl Acad Sci USA 105:4133–4138
88. Saleh M, Vaillancourt JP, Graham RK, Huyck M, Srinivasula SM, Alnemri ES, Steinberg MH, Nolan V, Baldwin CT, Hotchkiss RS et al (2004) Differential modulation of endotoxin responsiveness by human caspase-12 polymorphisms. Nature 429:75–79
89. Yeretssian G, Doiron K, Shao W, Leavitt BR, Hayden MR, Nicholson DW, Saleh M (2009) Gender differences in expression of the human caspase-12 long variant determines susceptibility to Listeria monocytogenes infection. Proc Natl Acad Sci USA 106:9016–9020

Pyroptosis: A Caspase-1-Dependent Programmed Cell Death and a Barrier to Infection

Katherine Labbé and Maya Saleh

Abstract Infection elicits a number of innate protective responses in the host that cooperate to promote effective pathogen clearance. Increasingly, the inflammatory response to infection appears to be coupled to cell death as an important mediator of host defence. In this chapter we review the modalities of "pyroptosis", a highly inflammatory form of cell death mediated by the inflammasome and caspase-1 activation. Occurring in the context of infection, pyroptosis is morphologically, mechanistically and physiologically distinct from other forms of cell death. The pathogenic factors that initiate pyroptosis and the cellular mechanisms and signalling pathways responsible for its execution are examined, with a focus on the role of the inflammasome in these processes. Finally, we discuss the possible physiological significance of this unique form of cell death during infection, that is, how pyroptosis can favour pathogen elimination on one hand, while contributing to the pathophysiology of disease on the other.

K. Labbé
Department of Microbiology and Immunology, McGill University, Montreal, QC, Canada H3A 2B4

M. Saleh (✉)
Department of Microbiology and Immunology, McGill University, Montreal, QC, Canada H3A 2B4
and
Department of Medicine, McGill University, McGill Life Sciences Complex, Montreal, QC, Canada H3A 1A1
and
McGill Life Sciences Complex, Bellini Pavillion, Room 364, 3649 Promenade Sir-William Osler, Montreal, QC, Canada H3G 0B1
e-mail: maya.saleh@mcgill.ca

I. Couillin et al. (eds.), *The Inflammasomes*, Progress in Inflammation Research, DOI 10.1007/978-3-0348-0148-5_2, © Springer Basel AG 2011

1 Introduction

One of the most primitive antimicrobial responses consists of the elimination of the infected cell by programmed cell death, a response found in all metazoan phyla, including plants [1]. Although mammals have developed several additional layers of immune defence, cell death remains a key component of the host response against infection. Eliminating an infected cell results in the death of the infectious agent, promoting effective pathogen clearance and the elimination of a pathogenic niche. Intracellular pathogens require a viable host cell within which to replicate and bacteria, such as *Mycobacterium tuberculosis* [2], *Rickettsia rickettsii* [3] and *Chlamydia* spp. [4], have evolved mechanisms to prevent host cell death to assure their own survival. Conversely, certain pathogens have devised strategies to use cell death to their advantage, to subvert normal host defence mechanisms or as a way to penetrate the epithelial barrier and reach deeper layers of tissue or the blood stream. *Bacillus anthracis* [5], *Actinobacillus actinomycetemcomitans* [6, 7] and *Pseudomonas aeruginosa* [8, 9] secrete cytotoxic exotoxins to kill macrophages before they themselves are phagocytosed and destroyed. *Bordetella pertussis* adenylate cyclase haemolysin promotes successful colonization of alveolar tissue by eliminating the local monocyte population [10]. Cell death therefore plays a major role in determining the outcome of host–pathogen interactions.

During infection, recognition by the innate immune system is achieved through a number of pattern-recognition receptors including Toll-like, Nod-like and RIG-I-like receptors. Activation of these receptors initiates an array of signalling networks that culminate in the mounting of a proinflammatory immune response [11]. These innate mechanisms are essential for primary pathogen clearance as well as the development of an adequate adaptive response to the infection. Increasingly, the inflammatory response to infection appears to be coupled to the induction of cell death as an important mediator of host defence. Understanding the modalities of cell death is therefore critical to the elucidation of pathogenic mechanisms. Here, we review the current knowledge on the mechanisms and functions of "pyroptosis", an inflammatory form of cell death initiated by infection and mediated by the activities of the inflammasome and caspase-1.

2 Cell Death Pathways

Cell death is generally described dichotomously as either programmed or passive. The former requires metabolic energy and is mediated by specific cellular pathways and effector molecules, while the latter occurs uncontrollably due to extracellular stresses. Programmed cell death can be further classified as either apoptosis (type I) or autophagic cell death (type II), each with a unique set of cellular mechanisms and morphologies (Fig. 1). Apoptosis is the best described form of cell death and is mediated by the apoptotic caspase enzymes, a family of cysteinyl aspartate-specific

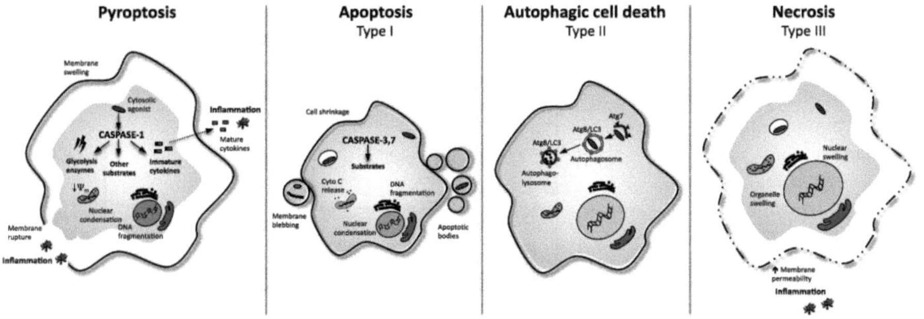

Fig. 1 Cell death pathways. Cell death is generally described as either programmed (apoptosis, or type I and autophagic cell death, or type II) or passive (necrosis or type III). Pyroptosis is categorized as programmed cell death, as it requires metabolic energy and is mediated by specific cellular pathways, namely, the inflammasome and caspase-1. Pyroptotic cells display a distinct set of morphological and biochemical characteristics, some of which are shared with apoptosis and necrosis. Unlike apoptosis and autophagic cell death, which do not induce inflammation, cytokine release and escape of cytoplasmic content during pyroptosis are highly inflammatory events

proteases. Eleven human caspases have been identified and are grouped into two major subfamilies according to their function in either apoptosis or cytokine maturation [12]. There are two major apoptosis pathways, extrinsic or intrinsic, and apoptotic caspases are classified as either upstream initiators (caspase-2, -8, -9 and -10) or executioners (caspase-3, -6 and -7) of these pathways. Caspases are synthesized as inactive precursors but in response to apoptotic signals become enzymatically active through processing or by a conformational change induced by oligomerization [13]. The intrinsic pathway is initiated when intracellular stresses induce the activation of Bcl-2 homology 3 (BH3)-only proteins, which leads to the oligomerization of the pro-apoptotic Bcl-2 family proteins BAX and BAK that form pores in the outer mitochondrial membrane [14]. This mitochondrial outer membrane permeabilization (MOMP) causes the release of apoptogenic factors, including Smac/DIABLO and cytochrome *c*, into the cytosol [15]. Cytochrome *c* associates with the apoptosis protease activating factor-1 (Apaf-1) to recruit and activate caspase-9 in a protein complex termed the "apoptosome" [16]. Active caspase-9 cleaves and activates the executioner caspases that in turn process cellular substrates to ultimately kill the cell. The extrinsic apoptosis pathway is induced by stimulation of death receptors of the TNF receptor family. At the active receptor, the adaptor proteins Fas-associated via death domain (FADD) and TNFR1-associated death domain (TRADD) recruit caspases-8 and -10 to form the death inducing signalling complex (DISC) [17]. Active caspase-8 cleaves and activates the executioner caspases, and, in certain cells, amplifies the cell death signal by cleaving the BH3-only protein Bid to induce MOMP and caspase-9 activation. Apoptotic cells are characterized by DNA fragmentation and chromatin condensation, nuclear fragmentation, cell shrinkage, loss of membrane asymmetry and the formation of cytoplasmic blebs and apoptotic bodies (Table 1).

Table 1 Pyroptosis is distinct from apoptosis. Despite sharing characteristics with apoptosis, the morphological, mechanistic and physiological features of pyroptosis make it a distinct form of programmed cell death

	Characteristics	Apoptosis	Pyroptosis	References
Morphology	Cell lysis	✖	✔	35, 38, 51
	Cell swelling	✖	✔	44, 51, 53
	Pore formation	✖	✔	35, 42, 44, 53
	Membrane blebbing	✔	✖	51, 52
	Nuclear condensation	✔	✔	40, 51
	DNA fragmentation (TUNEL positive)	✔	✔	42, 45, 47, 53
Mechanisms	Caspase-1 activation	✖	✔	26, 27
	Caspase-3 activation	✔	✖	34
	Caspase-7 activation	✔	✔	58
	MOMP/Cytochrome C release	✔	✖	38, 39, 53
	ICAD cleavage	✔	✖	53
	PARP cleavage	✔	✖	53
	Glycolysis enzyme inactivation	✔	✔	57
Out-come	Inflammation	✖	✔	112

Autophagy is also a tightly regulated process. Orchestrated by the ATG/Beclin proteins, autophagy results in lysosomal enzyme degradation of intracellular components captured within a double-membraned vacuole termed the autophagosome [18]. This catabolic process is essential during starvation conditions to maintain energy homeostasis and cell survival. Excessive autophagy has, however, been associated with a form of "autophagic cell death" characterized by massive accumulation of autophagic vacuoles in the cytoplasm in the absence of chromatin condensation [19]. Cytoplasmic content is not spilled into the extracellular space and thus, like apoptosis, autophagic cell death is non-inflammatory.

Passive cell death, or necrosis (type III), is often identified in negative terms for cells that do not have the markers of type I or type II programmed cell death pathways. Necrosis is thought to occur accidentally and uncontrollably as a result of environmental perturbations. Cells undergo cytoplasmic and organellar swelling, resulting in plasma membrane lysis and release of intracellular content, which is a highly inflammatory event [19]. The causative agents of necrosis are still unclear, as are the cellular events they initiate. Mitochondrial dysfunction such as production of reactive oxygen species and membrane permeabilization has been implicated, as have ATP depletion, loss of Ca^{2+} homeostasis, protease activation and lysosomal rupture [20]. Increasingly, there is evidence that necrotic cell death may be regulated by signal transduction and catabolic mechanisms and is not a completely passive event [21].

Recently, a novel form of programmed cell death has been described that occurs specifically in the context of infection. Like apoptosis, it is mediated by caspases, but rather than depending on the action of classical apoptotic caspases, it is an inflammatory caspase, caspase-1, that is critical. Caspase-1 was the first caspase to be described as the enzyme responsible for the cleavage of pro-IL-1β and was thus initially named Interleukin-1β converting enzyme, or ICE [22]. Caspase-1 is the prototypical member of the inflammatory caspase family, which includes caspases-4, -5 and -12 (and -11 in rodents). Caspase-1 activation is triggered by the formation of a cytosolic complex termed the "inflammasome" (see previous chapters for a detailed description). The 44-kDa pro-caspase-1 consists of a 10-kDa CARD domain (caspase-activation and recruitment domain), a large subunit (p20) and a small subunit (p10). Proximity-induced oligomerization of caspase-1 [13] results in auto-processing, release of the CARD domain and tetramerization of two small and two large subunits to form the active enzyme [22, 23]. Residues from both the p10 and the p20 subunits form the active site of the enzyme. The catalytic cysteine, Cys^{285}, and histidine, His^{237}, are found in the p20, while substrate specificity is determined by residues of the p10 [24]. Like all caspases, caspase-1 has an absolute requirement for Asp in the P1 position of its substrates, immediately N-terminal of the scissile bond. The optimal caspase-1 amino acid recognition sequence is $Tyr_{(P4)}$–$Val_{(P3)}$–$Ala_{(P2)}$–$Asp_{(P1)}$, though it can tolerate conservative substitutions at P2 and P3, and has a preference for hydrophobic amino acids at P4 [25]. Substrate cleavage by caspase-1 does not contribute to classical apoptosis pathways [26, 27]. Rather, activation of caspase-1 results in the cleavage of a unique array of proteins, including the preferred substrates pro-IL-1β and pro-Il-18, that are converted into their secreted, biologically active forms. Both cytokines are highly inflammatory and play important roles in the immune response by recruiting and activating immune cells [28]. Indeed, caspase-1 activation is essential for the mounting of an efficient immune response to a number of infectious pathogens [29].

There are striking parallels between the pathways controlling intrinsic apoptosis and those that activate caspase-1. Caspase-1-dependent cell death is initiated by infection, while apoptosis is can be induced by the mitochondria, an organelle reminiscent of bacteria. Both release stimulatory products (PAMPs and cytochrome *c*) into the cytosol to activate sensors (NLR/PYHIN and Apaf-1) that undergo oligomerization to form an activation platform (inflammasome and apoptosome). CARD-containing caspases are subsequently recruited (caspase-1 and caspase-9) where they are activated by proximity-induced catalysis, resulting in substrate recognition and cleavage. In the case of caspase-9, this leads to apoptosis, a tolerogenic form of cell death. For caspase-1, the outcome is a form of cell death tightly coupled to inflammation and the anti-microbial response. Such similarities may be the result of the co-evolution of these pathways under the pressures imposed by infection. This homology also provides important insight into the mechanisms of cell death and inflammation, furthering our understanding of the roles played by both in the host response to infection.

3 Caspase-1-Dependent Cell Death: Pyroptosis

The first report of a caspase-1-dependent cell death was in mouse macrophages infected with the gram-negative bacteria *Shigella flexneri* [30], the etiological agent of bacillary dysentery. In its human host, *Shigella* invades the colonic mucosa, where it encounters and infects the phagocytes of the lamina propria, resulting in extensive macrophage death and abscess formation [31]. *Shigella* was the first invasive bacteria reported to induce host cell death, which was initially described as apoptosis [32]. Further mechanistic studies uncovered a cell death pathway occurring independently of the apoptotic effector caspase-3, but contingent on the activity of the inflammatory caspase-1 [30]. Pharmacological inhibition of caspase-1 by Ac-YVAD-CHO [30, 33] or genetic ablation in *casp1*$^{-/-}$ mice [34] rendered macrophages fully resistant to *Shigella*-induced cytotoxicity while caspase-3, caspase-11 or p53 deficiency did not [34].

The findings in *Shigella*-infected macrophages were further corroborated by reports of caspase-1-dependent cell death induction by *Salmonella typhimurium* [35–37]. Macrophages are killed within minutes of *Salmonella* infection but are rescued by YVAD treatment [35] or if derived from *casp1*$^{-/-}$ mice [37]. Furthermore, caspases-3, -6 and -7 remain inactive in these cells [35, 38] and cytochrome *c* release does not occur [38]. Together, these findings firmly established the existence of a caspase-1-mediated cell death pathway distinct from apoptosis. Caspase-1-dependent cell death has since been reported in macrophages infected with a number of pathogens including *Listeria monocytogenes* [39], *Legionella pneumophila* [40, 41], *Yersinia pseudotuberculosis* [42], *P. aeruginosa* [43], *Burkholderia pseudomallei* [44] and *Francisella tularensis* [45], though the possible contribution of other inflammatory caspases to the cell death induced by these bacteria has not been fully investigated.

Concomitant with cell death are the inflammatory consequences of caspase-1 activation. Secretion of IL-1β promotes inflammatory cell recruitment and further production of proinflammatory mediators, resulting in important and sometimes severe physiological consequences such as fever, hypotension and metabolic derangements [46]. The release of mature IL-18 further amplifies the inflammatory response by stimulating immune cell activation and cytokine secretion [28]. The cleavage and release of IL-1β and IL-18 are not, however, required for caspase-1-mediated cell death [47]. Indeed, the action of caspase-1 is not limited to the processing of these cytokines, as demonstrated by the resistance of *casp1*$^{-/-}$ mice [27] but not *IL-1β*$^{-/-}$ [48] or IL-1β/IL-18 double knockout animals [49] to endotoxemia and septic shock. Although neither IL-1β nor IL-18 is required for its execution [47], caspase-1-dependent cell death is a highly inflammatory event, a key feature distinguishing it from apoptosis. The term "pyroptosis" has been proposed to describe this unique form of programmed cell death [50] as "pyro" or fire, denotes the release of proinflammatory mediators, while "ptosis" denotes falling, a term commonly used to describe cell death.

4 Execution of Pyroptosis

Pyroptosis is a distinct form of cell death [19] with a unique combination of morphological and mechanistic features (Table 1).

4.1 Morphology

Morphologically, pyroptosis is most notably characterized by loss of plasma membrane integrity and release of cytoplasmic content into the extracellular milieu [35, 38, 51]. This feature is shared with necrosis but not with apoptosis, in which cytosolic content is contained within cytoplasmic blebs and apoptotic bodies [52]. Microscopically, the pyroptotic plasma membrane appears to rupture, then rapidly reseal and swell, forming a "balloon-shaped" vesicle around the nucleus [51] (Fig. 2). Indeed, during pyroptosis, cells undergo a measurable size increase [44, 53]. As the membrane swells, the nucleus also undergoes rounding and condensation [40, 51], but, unlike apoptosis, nuclear integrity is maintained. Pyroptotic cells undergo DNA fragmentation and, like apoptotic cells, show positive TUNEL staining [42, 45, 47, 53] (Table 1). The TUNEL positivity of *S. flexneri* and *S. typhimurium* infected cells initially led to the assumption that cell death induced by these pathogens was apoptotic.

4.2 Mechanisms

The cellular mechanisms that mediate caspase-1-dependent cell death are still largely unknown, but are distinct from the classical apoptotic pathways. The apoptotic caspases, including caspases-3, -6 and -7, are not involved in pyroptosis [34, 35, 38]. During apoptosis, the poly-ADP ribose polymerase (PARP) is cleaved by executioner caspases in an attempt to preserve cellular ATP energy stores [54]. PARP is also cleaved by caspase-1 [55] but studies using PARP inhibitors [53] and *PARP1*$^{-/-}$ macrophages [56] suggest that PARP activity does not significantly

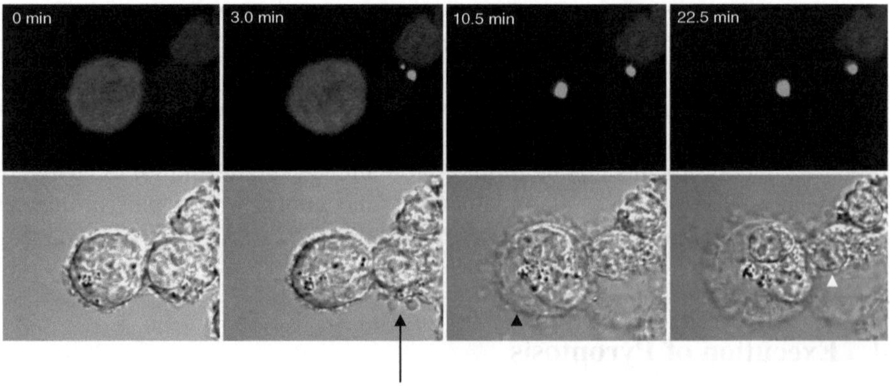

Fig. 2 Imaging of pyroptosis by time-lapse confocal microscopy. Pyroptosis was induced in differentiated THP-1-ASC-GFP cells with crude LPS. *Top row*; fluorescence images depicting the formation of the cytoplasmic ASC-GFP pyroptosome. *Bottom row*; plasma membrane rupture was observed rapidly (*black arrow*), followed by membrane re-sealing and swelling (*black arrowhead*) and nuclear condensation (*white arrowhead*) (these images are courtesy of Dr. E. Alnemri; reprinted with the permission of Nature Publishing Group)

impact pyroptosis. During apoptosis, DNA is degraded by the caspase activated DNase (CAD). The activity of CAD is controlled by its inhibitor, ICAD, a caspase-3 substrate [57]. During pyroptosis, DNA fragmentation depends on nuclease activity, but the lack of detectable ICAD processing suggests that this is a CAD-independent process [53]. Finally, a crucial step in apoptosis is the induction of MOMP, which causes the release of cytochrome c and Smac/DIABLO into the cytosol [14]. In pyroptosis, mitochondrial integrity is maintained and cytochrome c is not released [38, 39, 53].

Although distinct from apoptotic pathways, the question remains, what are the pro-death mechanisms occurring downstream of caspase-1 (Fig. 3)? The finding that mice genetically deficient in both IL-1β and IL-18 do not phenocopy caspase-1-deficient animals in septic shock models prompted investigation into the discovery of novel caspase-1 substrates and the description of novel caspase-1 mediated signalling pathways. Some insight into these signals has been gained with the report of the caspase-1 digestome using the diagonal gel approach on the human monocytic THP-1 cell line [55]. Caspase-1 was shown to target substrates involved in cellular functions as diverse as maintenance of the cytoskeleton, ATP metabolism, detoxification, trafficking, RNA/protein synthesis and degradation, signal transduction and cytokine production. Caspase-1 was further shown to cleave and inactivate a number of glycolysis enzymes, linking inactivation of bioenergetic pathways to cell death. An interesting conclusion of this study is the seemingly dual function of caspase-1 as both an initiator and executioner caspase.

Another report of caspase-1 substrates used the gel-free COFRADIC peptide sorting methodology on the mouse Mf4/4 macrophage cell line [58]. This study identified caspase-7 as a specific caspase-1 substrate. Caspase-7 cleavage was

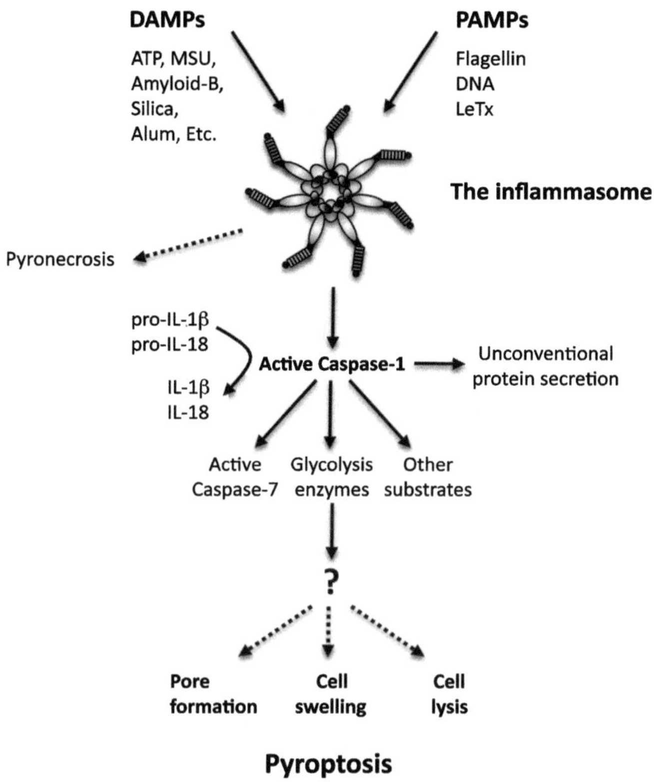

Fig. 3 Cellular mechanisms of pyroptosis. Caspase-1 is activated within the "inflammasome" a cytosolic multiprotein complex assembled in response to cytosolic PAMPs or host-derived DAMPs. Caspase-1 activation results in proinflammatory cytokine processing, as well as the cleavage of a number of additional substrates such as pro-caspase-7 and glycolysis enzymes. The nature of the pathways downstream of caspase-1 that ultimately result in cell lysis and death are still unknown

identified downstream of a number of caspase-1 activators, but *casp7*$^{-/-}$ cells were not deficient in IL-1β or IL-18 production and were not protected from death during *S. typhimurium* infection. A follow-up study using *L. pneumophila*-infected macrophages did demonstrate that caspase-7 was required for efficient bacterial clearance by promoting phagosome fusion with the lysosome [58]. Both studies reported caspase-3 processing in pyroptotic macrophages, but this occurred independently of caspase-1 activity and caspase-3-deficiency did not impair cytokine processing or cell death, leaving the function of this processing undetermined.

A downstream mechanism proposed to mediate pyroptotic cell lysis during *S. typhimurium* infection is the formation of small-ion permeable pores in the plasma membrane [35, 42, 44, 53]. Through size-exclusion studies and flow cytometry, the pore size has been estimated to be 1.1–2.4 nm in diameter [53]. Addition of extracellular glycine or osmoprotectants of 2.4 nm or greater to pyroptotic macrophages prevented cell swelling and lysis [35, 53]. Thus, it is suggested that plasma-membrane pores dissipate the cellular ionic gradient, producing an increase in osmotic pressure that results in water influx, cell swelling and membrane lysis [53]. Whether these events are downstream of substrate processing by caspase-1 remains to be examined.

5 The Role of the Inflammasome

The inflammasome is activated by bacterial [60], viral [61] and parasitic infection [62], as well as by host-derived danger-associated molecular patterns (DAMPs) [63]. The generation of knock-out mice for the various components of the inflammasome has allowed the investigation of which pathways are required for caspase-1 activation and pyroptosis.

5.1 NLRC4 and NAIP5

The NLRC4 inflammasome is most closely associated with the induction of pyroptosis. It is activated by the flagellin [64] and the rod proteins of the type-three secretion system [65] of a number of bacterial pathogens. Expression of NLRC4 is essential for pyroptosis induced by *S. typhimurium* [66], *P. aeruginosa* [43, 67, 68] and *S. flexneri* [69]. The role of ASC in mediating NLRC4 signalling differs between cytokine production and cell death. Indeed, macrophages infected with *S. typhimurium* require both NLRC4 and ASC for IL-1β production, but only *Nlrc4*$^{-/-}$ cells are resistant to pyroptosis [64, 66, 70, 71]. The same conditions are also true for *S. flexneri* and *P. aeruginosa*-infected macrophages [43, 67–69].

The facultative intracellular bacteria *L. pneumophila* is distinct in that it has an absolute requirement for a functional NAIP5 to stimulate caspase-1 activity. The action of NAIP5 is coupled to that of NLRC4 as both mediate cytotoxicity but require the additional presence of ASC to permit IL-1β secretion [40, 41, 72, 74]. NAIP5 was identified as the effector gene of the *Lgn* locus that controls macrophage permissiveness to *Legionella* replication [75]. Cells derived from A/J mice express a mutant NAIP5 and are resistant to pyroptosis, whereas those derived from C57BL/6 mice, which carry the functional *Lgn* allele, are susceptible. To better understand the function of NAIP5, complete *Naip5*$^{-/-}$ mice were recently generated on a C57BL/6 background [73]. Infectious studies in macrophages derived from these animals have now demonstrated a partial dependence on NAIP5 expression for *S. typhimurium* and *P. aeruginosa* induced pyroptosis [73].

5.2 *NLRP1*

A critical factor in the virulence of *B. anthracis* is the production of anthrax lethal toxin (LeTx). LeTx is produced during infection and is secreted as two proteinaceous subunits, protective antigen (PA) and lethal factor (LF). When PA-LF complexes are endocytosed and trafficked to acidic vesicles, conformational changes in PA allow LF to translocate to the cytosol [76]. LF, a zinc-dependent metalloprotease, specifically cleaves MAPKKs, disrupting MAPK signalling pathways [77]. Although MAPKK cleavage occurs in all infected mammalian cells, LeTx is not universally lethal. In mouse macrophages, cell death is strain dependent, where C57BL/6-derived cells are resistant and 129S1 or BALB/c cells are susceptible. Genetic studies have mapped susceptibility to the *Ltxs1* locus and subsequently shown that *Nlrp1b* mediates strain susceptibility. NLRP1b is essential for caspase-1 activation by lethal toxin and is required for both IL-1β production and cytotoxicity, suggesting that LeTx kills cells by pyroptosis [78]. Macrophages derived from 129S1 or BALBc mice express a functional NLRP1, and are susceptible to LeTx pyroptosis, whereas the C57Bl/6 strain *Nlrp1b* gene is mutated and non-functional, conferring LeTx-resistance to macrophages derived from these mice. *B. anthracis* spores activate caspase-1 and promote IL-1β release by macrophages, but do not induce cell death [79]. It is not yet known whether live *B. anthracis* infection activates caspase-1.

5.3 *NLRP3*

The list of activators of the NLRP3 inflammasome is increasingly long and includes pathogen-derived signals (viral [80], fungal [81, 82] and bacterial infection [83], pore-forming toxins [71]), environment-derived factors (silica [84], asbestos [85], alum [84, 86]) and host-derived danger signals (ATP [71], uric acid [87], hyaluronan [88], amyloid-β [89]). Yet despite this list of agonists, NLRP3 is not known to mediate caspase-1-dependent cell death. Mutant, overactive forms of NLRP3 are capable of driving an inflammatory form of cell death during infection, distinct from pyroptosis, termed "pyronecrosis". It is entirely caspase independent but requires the activity of cathepsin B [90]. Pyronecrosis promotes and the release of HMGB1, a cellular DAMP [91], and is induced by high multiplicity of infection with *S. flexneri* [90], *K. pneumoniae* [92] and *N. gonorrhoeae* [83].

5.4 *AIM2*

During infection by intracellular pathogens, the presence of foreign double-stranded DNA (dsDNA) leads to robust caspase-1 activation. The inflammasome component involved in this case is AIM2, a cytosolic member of the HIN-200 protein family that specifically binds dsDNA and associates with ASC to activate caspase-1 [93–96]. Transfection of the synthetic dsDNA poly(dA:dT) results in AIM2-dependent IL-1β processing and macrophage death, both of which require the presence of ASC [95]. Pyroptosis of macrophages infected with *F. tularensis* or *L. monocytogenes* was long known to depend on ASC but the PRR involved was unknown until AIM2 was identified as the activated inflammasome [97, 98]. Cytosolic release of dsDNA by these intracellular Gram-negative bacteria is directly sensed by AIM2, resulting in caspase-1 activation, IL-1β secretion and pyroptosis.

5.5 ASC

The role of the adaptor protein ASC in mediating pyroptosis is still unclear. As described above, NLRC4/NAIP5 inflammasome-dependent pyroptosis does not require ASC whereas AIM2-mediated cell death does. A recent study [99] determined that it is the presence of a CARD domain that determines whether an inflammasome requires ASC to induce pyroptosis. It was further reported that ASC-independent pyroptosis does not induce caspase-1 auto-proteolysis as does ASC-dependent pro-IL-1β processing, suggesting the existence of two different caspase-1 activation pathways.

ASC has been implicated in other pyroptotic conditions. For instance, ASC has been shown to induce pyroptosis in a human THP-1 monocytic cell line engineered to stably express an ASC-GFP fusion protein. These cells were used to demonstrate the formation of a large supramolecular ASC complex termed the "pyroptosome" in response to a number of inflammasome agonists [51, 94]. Upon stimulation, a single ASC pyroptosome is formed in the cell and is required for caspase-1 activation and the induction of pyroptosis. ASC oligomers were also shown to form in primary macrophages from both wild-type and *casp1*$^{-/-}$ mice, though only the former succumbed to cell death [51].

The observation that caspase-1 activation does not always result in cell death is indicative of the existence of a unique, pro-death inflammasome signalling pathway. A certain level of signalling specificity is achieved by the type of inflammasome activated by an agonist; while NLRC4 activation is closely associated with pyroptosis, NLRP3 responses are predominantly cytokine based. The types of inflammasome agonists that induce pyroptosis also reflect this specificity. NLRC4 is primarily activated by bacteria in the context of a live infection. In these conditions, pyroptosis would result in the elimination of the infected cell and promote a strong inflammatory response at the site of infection. NLRP3, on the other hand, is mainly activated by danger signals derived from the environment or from other cells. In this case, a cytokine-only response would promote tissue repair without causing cell death.

6 Physiological Significance

During infection, pyroptosis can be a beneficial event for the host. The compromised cell is eliminated, effectively destroying a protective milieu in which infectious agents can thrive. For instance, macrophages deficient in caspase-1, NLRC4 or NAIP5 are protected from pyroptosis during *L. pneumophila* infection and support higher intracellular bacterial loads [41, 73]. Pyroptosis also promotes pathogen clearance by acting as an alarm signal that recruits immune cells to the site of infection. The production of IL-1β and IL-18 promotes leukocyte infiltration and activation [46] and cell lysis releases immuno-stimulatory factors into the extracellular milieu. Several cytosolic products are potent DAMPs, such as HMGB1 [100], ATP [71], uric acid [87], heat-shock proteins [101] and DNA–chromatin complexes [102], that promote proinflammatory cytokine production through the activation of pattern-recognition receptors.

The hypothesis that pyroptosis is detrimental to the pathogen is also supported by the description of several microbial inflammasome-evasion mechanisms. One such strategy is to reduce the presence of pyroptosis-inducing agonists, as does *S. typhimurium*. By downregulating flagellin expression during *in vivo* infection, *Salmonella* evades detection by the inflammasome, thus rendering *Nlrc4*$^{-/-}$ mice as susceptible to infection as wild-type animals [37, 64]. Other pathogens have evolved effector proteins capable of directly inhibiting inflammasome activation, such as the *Y. pseudotuberculosis* T3SS Rho-GTPase activating protein YopE [103], the *M. tuberculosis* zinc metalloprotease Zmp1 [104], the poxvirus M13L [105] or the influenza virus NS1 protein [106]. Recently, a screen for *F. tularensis* virulence genes identified two targets involved in the delay of IL-1β release and cell death in infected cells [107]. Similarly, virulent *Pseudomonas* strains express the effector protein ExoU, which blocks caspase-1-dependent cell death and promotes necrosis [43]. The evolution of such mechanisms would suggest that preventing inflammasome activation contributes to pathogen fitness, and that the activity of caspase-1 leading to pyroptosis represents an important selective pressure that prevents replication and spread of the pathogen to neighbouring cells.

Yet, as with any physiological process, excessive pyroptosis is detrimental to the host and may contribute to histopathology and disease. Macrophage cell death prevents appropriate elimination of the infectious agent and pyroptosis of dendritic cells [108, 109] leads to immunosuppression by impairing cytokine production and antigen presentation. Excessive pyroptosis also leads to the development of severe sepsis and septic shock, through the release of the alarmin HMGB1. When released into the circulation, HMGB1 acts as a critical mediator of severe sepsis [110] and administration of HMGB1 neutralizing antibodies confers resistance from lethality [111]. During apoptosis, oxidation of HMGB1 neutralizes its stimulatory activity [112].

Finally, it is difficult to determine the immunological function of pyroptosis, as caspase-1 activities are numerous and extend beyond cell death. Furthermore, there is enormous redundancy in cell death pathways, and cells in which pyroptosis is blocked will often die by alternative pathways. Thus, although *casp1*$^{-/-}$ macrophages are protected from pyroptosis during *S. typhimurium* infection, they nonetheless succumb to a delayed form of cell death [38]. In *S. flexneri*-infected macrophages, deficiency in either caspase-1 or NLRC4 results in increased activation of autophagy [69]. The induction of alternate cell death pathways could provide a backup mechanism for the infected cell in cases where the inflammasome is not activated or actively is inhibited by the pathogen in an attempt to clear the infection. On the other hand, these alternate routes may provide time for the pathogen to replicate before the cell is destroyed. One study [113] sought to tease out the contribution of pyroptosis to the host response by using a strain of *S. typhimurium* engineered to persistently express flagellin. Enhanced clearance of these bacteria did not depend on cytokine production, but on the pyroptotic release of the pathogen from macrophages, where it was subsequently killed by neutrophil ROS production. A similar result was obtained for *L. pneumophilia* and *B. thailandensis*, illustrating that pyroptosis is an important innate effector mechanisms protecting from bacterial infection *in vivo*.

7 Conclusion

Cell death is an important factor in host–pathogen interactions. The elimination of an infected cell can be beneficial or detrimental to both parties, and each utilizes a number of strategies to regulate the outcome in its favour. Both host and pathogen responses to infection also rely on the modulation of the proinflammatory response to promote own survival. It should come as no surprise, then, that the pathways responsible for mounting the inflammatory response are also involved in the regulation of certain forms of cell death. Elucidation of cell death mechanisms will undoubtedly reveal information on inflammatory processes, and vice versa. Thus, a better understanding of pyroptosis will provide information on the function and regulation of the inflammasome, as well as important insights into the role of cell death during infection.

Acknowledgements We thank Dr. Emad Alnemri, Kimmel Cancer Institute, Thomas Jefferson University for providing us with the images in Fig. 2, reproduced with permission from Nature Publishing Group, Macmillan Publishers Ltd: Labbé K, Saleh M (2008) Cell death in the host response to infection. Cell Death Differ 15(9):1339–1349, copyright 2008.

References

1. Zitvogel L, Kroemer G (2008) The immune response against dying tumor cells: avoid disaster, achieve cure. Cell Death Differ 15:1–2
2. Fairbairn IP (2004) Macrophage apoptosis in host immunity to mycobacterial infections. Biochem Soc Trans 32:496–498

3. Clifton DR, Goss RA, Sahni SK, van Antwerp D, Baggs RB, Marder VJ, Silverman DJ, Sporn LA (1998) NF-kappa B-dependent inhibition of apoptosis is essential for host cellsurvival during Rickettsia rickettsii infection. Proc Natl Acad Sci U S A 95:4646–4651
4. Fan T, Lu H, Hu H, Shi L, McClarty GA, Nance DM, Greenberg AH, Zhong G (1998) Inhibition of apoptosis in chlamydia-infected cells: blockade of mitochondrial cytochrome c release and caspase activation. J Exp Med 187:487–496
5. Hanna PC, Acosta D, Collier RJ (1993) On the role of macrophages in anthrax. Proc Natl Acad Sci U S A 90:10198–10201
6. Kato S, Muro M, Akifusa S, Hanada N, Semba I, Fujii T, Kowashi Y, Nishihara T (1995) Evidence for apoptosis of murine macrophages by Actinobacillus actinomycetemcomitans infection. Infect Immun 63:3914–3919
7. Taichman NS, Dean RT, Sanderson CJ (1980) Biochemical and morphological characterization of the killing of human monocytes by a leukotoxin derived from Actinobacillus actinomycetemcomitans. Infect Immun 28:258–268
8. Kochi SK, Collier RJ (1993) DNA fragmentation and cytolysis in U937 cells treated with diphtheria toxin or other inhibitors of protein synthesis. Exp Cell Res 208:296–302
9. Morimoto H, Bonavida B (1992) Diphtheria toxin- and Pseudomonas A toxin-mediated apoptosis. ADP ribosylation of elongation factor-2 is required for DNA fragmentation and cell lysis and synergy with tumor necrosis factor-alpha. J Immunol 149:2089–2094
10. Khelef N, Zychlinsky A, Guiso N (1993) Bordetella pertussis induces apoptosis in macrophages: role of adenylate cyclase-hemolysin. Infect Immun 61:4064–4071
11. Lee MS, Kim YJ (2007) Signaling pathways downstream of pattern-recognition receptors and their cross talk. Annu Rev Biochem 76:447–480
12. Lamkanfi M, Declercq W, Kalai M, Saelens X, Vandenabeele P (2002) Alice in caspase land. A phylogenetic analysis of caspases from worm to man. Cell Death Differ 9:358–361
13. Boatright KM, Renatus M, Scott FL, Sperandio S, Shin H, Pedersen IM, Ricci JE, Edris WA, Sutherlin DP, Green DR et al (2003) A unified model for apical caspase activation. Mol Cell 11:529–541
14. Green DR, Kroemer G (2004) The pathophysiology of mitochondrial cell death. Science 305:626–629
15. Ekert PG, Vaux DL (2005) The mitochondrial death squad: hardened killers or innocent bystanders? Curr Opin Cell Biol 17:626–630
16. Riedl SJ, Salvesen GS (2007) The apoptosome: signalling platform of cell death. Nat Rev Mol Cell Biol 8:405–413
17. Peter ME, Krammer PH (2003) The CD95(APO-1/Fas) DISC and beyond. Cell Death Differ 10:26–35
18. Levine B, Klionsky DJ (2004) Development by self-digestion: molecular mechanisms and biological functions of autophagy. Dev Cell 6:463–477
19. Kroemer G, Galluzzi L, Vandenabeele P, Abrams J, Alnemri ES, Baehrecke EH, Blagosklonny MV, El-Deiry WS, Golstein P, Green DR et al (2009) Classification of cell death: recommendations of the Nomenclature Committee on Cell Death 2009. Cell Death Differ 16:3–11
20. Golstein P, Kroemer G (2007) Cell death by necrosis: towards a molecular definition. Trends Biochem Sci 32:37–43
21. Vandenabeele P, Galluzzi L, Vanden Berghe T, Kroemer G. Molecular mechanisms of necroptosis: an ordered cellular explosion. Nat Rev Mol Cell Biol 11:700–714
22. Thornberry NA, Bull HG, Calaycay JR, Chapman KT, Howard AD, Kostura MJ, Miller DK, Molineaux SM, Weidner JR, Aunins J et al (1992) A novel heterodimeric cysteine protease is required for interleukin-1 beta processing in monocytes. Nature 356:768–774
23. Ayala JM, Yamin TT, Egger LA, Chin J, Kostura MJ, Miller DK (1994) IL-1 beta-converting enzyme is present in monocytic cells as an inactive 45-kDa precursor. J Immunol 153:2592–2599

24. Thornberry NA (1994) Interleukin-1 beta converting enzyme. Methods Enzymol 244: 615–631
25. Sleath PR, Hendrickson RC, Kronheim SR, March CJ, Black RA (1990) Substrate specificity of the protease that processes human interleukin-1 beta. J Biol Chem 265:14526–14528
26. Kuida K, Lippke JA, Ku G, Harding MW, Livingston DJ, Su MS, Flavell RA (1995) Altered cytokine export and apoptosis in mice deficient in interleukin-1 beta converting enzyme. Science 267:2000–2003
27. Li P, Allen H, Banerjee S, Franklin S, Herzog L, Johnston C, McDowell J, Paskind M, Rodman L, Salfeld J et al (1995) Mice deficient in IL-1 beta-converting enzyme are defective in production of mature IL-1 beta and resistant to endotoxic shock. Cell 80:401–411
28. Arend WP, Palmer G, Gabay C (2008) IL-1, IL-18, and IL-33 families of cytokines. Immunol Rev 223:20–38
29. Mariathasan S, Monack DM (2007) Inflammasome adaptors and sensors: intracellular regulators of infection and inflammation. Nat Rev Immunol 7:31–40
30. Chen Y, Smith MR, Thirumalai K, Zychlinsky A (1996) A bacterial invasin induces macrophage apoptosis by binding directly to ICE. Embo J 15:3853–3860
31. Mathan MM, Mathan VI (1991) Morphology of rectal mucosa of patients with shigellosis. Rev Infect Dis 13 Suppl 4: S314-318
32. Zychlinsky A, Prevost MC, Sansonetti PJ (1992) Shigella flexneri induces apoptosis in infected macrophages. Nature 358:167–169
33. Hilbi H, Chen Y, Thirumalai K, Zychlinsky A (1997) The interleukin 1beta-converting enzyme, caspase 1, is activated during Shigella flexneri-induced apoptosis in human monocyte-derived macrophages. Infect Immun 65:5165–5170
34. Hilbi H, Moss JE, Hersh D, Chen Y, Arondel J, Banerjee S, Flavell RA, Yuan J, Sansonetti PJ, Zychlinsky A (1998) Shigella-induced apoptosis is dependent on caspase-1 which binds to IpaB. J Biol Chem 273:32895–32900
35. Brennan MA, Cookson BT (2000) Salmonella induces macrophage death by caspase-1-dependent necrosis. Mol Microbiol 38:31–40
36. Hersh D, Monack DM, Smith MR, Ghori N, Falkow S, Zychlinsky A (1999) The Salmonella invasin SipB induces macrophage apoptosis by binding to caspase-1. Proc Natl Acad Sci U S A 96:2396–2401
37. Lara-Tejero M, Sutterwala FS, Ogura Y, Grant EP, Bertin J, Coyle AJ, Flavell RA, Galan JE (2006) Role of the caspase-1 inflammasome in Salmonella typhimurium pathogenesis. J Exp Med 203:1407–1412
38. Jesenberger V, Procyk KJ, Yuan J, Reipert S, Baccarini M (2000) Salmonella-induced caspase-2 activation in macrophages: a novel mechanism in pathogen-mediated apoptosis. J Exp Med 192:1035–1046
39. Cervantes J, Nagata T, Uchijima M, Shibata K, Koide Y (2008) Intracytosolic Listeria monocytogenes induces cell death through caspase-1 activation in murine macrophages. Cell Microbiol 10:41–52
40. Molofsky AB, Byrne BG, Whitfield NN, Madigan CA, Fuse ET, Tateda K, Swanson MS (2006) Cytosolic recognition of flagellin by mouse macrophages restricts Legionella pneumophila infection. J Exp Med 203:1093–1104
41. Zamboni DS, Kobayashi KS, Kohlsdorf T, Ogura Y, Long EM, Vance RE, Kuida K, Mariathasan S, Dixit VM, Flavell RA et al (2006) The Birc1e cytosolic pattern-recognition receptor contributes to the detection and control of Legionella pneumophila infection. Nat Immunol 7:318–325
42. Bergsbaken T, Cookson BT (2007) Macrophage activation redirects yersinia-infected host cell death from apoptosis to caspase-1-dependent pyroptosis. PLoS Pathog 3:e161
43. Sutterwala FS, Mijares LA, Li L, Ogura Y, Kazmierczak BI, Flavell RA (2007) Immune recognition of Pseudomonas aeruginosa mediated by the IPAF/NLRC4 inflammasome. J Exp Med 204:3235–3245

44. Sun GW, Lu J, Pervaiz S, Cao WP, Gan YH (2005) Caspase-1 dependent macrophage death induced by Burkholderia pseudomallei. Cell Microbiol 7:1447–1458
45. Mariathasan S, Weiss DS, Dixit VM, Monack DM (2005) Innate immunity against Francisella tularensis is dependent on the ASC/caspase-1 axis. J Exp Med 202:1043–1049
46. Dinarello CA (1992) Role of interleukin-1 in infectious diseases. Immunol Rev 127:119–146
47. Monack DM, Navarre WW, Falkow S (2001) Salmonella-induced macrophage death: the role of caspase-1 in death and inflammation. Microbes Infect 3:1201–1212
48. Fantuzzi G, Zheng H, Faggioni R, Benigni F, Ghezzi P, Sipe JD, Shaw AR, Dinarello CA (1996) Effect of endotoxin in IL-1 beta-deficient mice. J Immunol 157:291–296
49. Sarkar A, Hall MW, Exline M, Hart J, Knatz N, Gatson NT, Wewers MD (2006) Caspase-1 regulates Escherichia coli sepsis and splenic B cell apoptosis independently of interleukin-1beta and interleukin-18. Am J Respir Crit Care Med 174:1003–1010
50. Cookson BT, Brennan MA (2001) Pro-inflammatory programmed cell death. Trends Microbiol 9:113–114
51. Fernandes-Alnemri T, Wu J, Yu JW, Datta P, Miller B, Jankowski W, Rosenberg S, Zhang J, Alnemri ES (2007) The pyroptosome: a supramolecular assembly of ASC dimers mediating inflammatory cell death via caspase-1 activation. Cell Death Differ 14:1590–1604
52. Taylor RC, Cullen SP, Martin SJ (2008) Apoptosis: controlled demolition at the cellular level. Nat Rev Mol Cell Biol 9:231–241
53. Fink SL, Cookson BT (2006) Caspase-1-dependent pore formation during pyroptosis leads to osmotic lysis of infected host macrophages. Cell Microbiol 8:1812–1825
54. Ha HC, Snyder SH (1999) Poly(ADP-ribose) polymerase is a mediator of necrotic cell death by ATP depletion. Proc Natl Acad Sci U S A 96:13978–13982
55. Shao W, Yeretssian G, Doiron K, Hussain SN, Saleh M (2007) The Caspase-1 Digestome Identifies the Glycolysis Pathway as a Target during Infection and Septic Shock. J Biol Chem 282:36321–36329
56. Malireddi RK, Ippagunta S, Lamkanfi M, Kanneganti TD Cutting edge: proteolytic inactivation of poly(ADP-ribose) polymerase 1 by the Nlrp3 and Nlrc4 inflammasomes. J Immunol 185:3127–3130
57. Sakahira H, Enari M, Nagata S (1998) Cleavage of CAD inhibitor in CAD activation and DNA degradation during apoptosis. Nature 391:96–99
58. Lamkanfi M, Kanneganti TD, Van Damme P, Vanden Berghe T, Vanoverberghe I, Vandekerckhove J, Vandenabeele P, Gevaert K, Nunez G (2008) Targeted peptidecentric proteomics reveals caspase-7 as a substrate of the caspase-1 inflammasomes. Mol Cell Proteomics 7:2350–2363
59. Akhter A, Gavrilin MA, Frantz L, Washington S, Ditty C, Limoli D, Day C, Sarkar A, Newland C, Butchar J et al (2009) Caspase-7 activation by the Nlrc4/Ipaf inflammasome restricts Legionella pneumophila infection. PLoS Pathog 5: e1000361
60. Brodsky IE, Monack D (2009) NLR-mediated control of inflammasome assembly in the host response against bacterial pathogens. Semin Immunol 21:199–207
61. Wilkins C, Gale M, Jr. Recognition of viruses by cytoplasmic sensors. Curr Opin Immunol
62. Labbe K, Miu J, Yeretssian G, Serghides L, Tam M, Finney CA, Erdman LK, Goulet ML, Kain KC, Stevenson MM et al (2010) Caspase-12 Dampens the Immune Response to Malaria Independently of the Inflammasome by Targeting NF-{kappa}B Signaling. J Immunol
63. McIntire CR, Yeretssian G, Saleh M (2009) Inflammasomes in infection and inflammation. Apoptosis 14:522–535
64. Miao EA, Alpuche-Aranda CM, Dors M, Clark AE, Bader MW, Miller SI, Aderem A (2006) Cytoplasmic flagellin activates caspase-1 and secretion of interleukin 1beta via Ipaf. Nat Immunol 7:569–575
65. Miao EA, Mao DP, Yudkovsky N, Bonneau R, Lorang CG, Warren SE, Leaf IA, Aderem A (2010) Innate immune detection of the type III secretion apparatus through the NLRC4 inflammasome. Proc Natl Acad Sci U S A 107:3076–3080

66. Mariathasan S, Newton K, Monack DM, Vucic D, French DM, Lee WP, Roose-Girma M, Erickson S, Dixit VM (2004) Differential activation of the inflammasome by caspase-1 adaptors ASC and Ipaf. Nature 430:213–218
67. Franchi L, Stoolman J, Kanneganti TD, Verma A, Ramphal R, Nunez G (2007) Critical role for Ipaf in Pseudomonas aeruginosa-induced caspase-1 activation. Eur J Immunol 37:3030–3039
68. Miao EA, Ernst RK, Dors M, Mao DP, Aderem A (2008) Pseudomonas aeruginosa activates caspase 1 through Ipaf. Proc Natl Acad Sci U S A 105:2562–2567
69. Suzuki T, Franchi L, Toma C, Ashida H, Ogawa M, Yoshikawa Y, Mimuro H, Inohara N, Sasakawa C, Nunez G (2007) Differential regulation of caspase-1 activation, pyroptosis, and autophagy via Ipaf and ASC in Shigella-infected macrophages. PLoS Pathog 3: e111
70. Franchi L, Amer A, Body-Malapel M, Kanneganti TD, Ozoren N, Jagirdar R, Inohara N, Vandenabeele P, Bertin J, Coyle A et al (2006) Cytosolic flagellin requires Ipaf for activation of caspase-1 and interleukin 1beta in salmonella-infected macrophages. Nat Immunol 7:576–582
71. Mariathasan S, Weiss DS, Newton K, McBride J, O'Rourke K, Roose-Girma M, Lee WP, Weinrauch Y, Monack DM, Dixit VM (2006) Cryopyrin activates the inflammasome in response to toxins and ATP. Nature 440:228–232
72. Case CL, Shin S, Roy CR (2009) Asc and Ipaf Inflammasomes direct distinct pathways for caspase-1 activation in response to Legionella pneumophila. Infect Immun 77:1981–1991
73. Lightfield KL, Persson J, Brubaker SW, Witte CE, von Moltke J, Dunipace EA, Henry T, Sun YH, Cado D, Dietrich WF et al (2008) Critical function for Naip5 in inflammasome activation by a conserved carboxy-terminal domain of flagellin. Nat Immunol 9:1171–1178
74. Silveira TN, Zamboni DS. Pore formation triggered by Legionella is an Nlrc4 inflammasome-dependent host cell response that precedes pyroptosis. Infect Immun
75. Diez E, Lee SH, Gauthier S, Yaraghi Z, Tremblay M, Vidal S, Gros P (2003) Birc1e is the gene within the Lgn1 locus associated with resistance to Legionella pneumophila. Nat Genet 33:55–60
76. Young JA, Collier RJ (2007) Anthrax toxin: receptor binding, internalization, pore formation, and translocation. Annu Rev Biochem 76:243–265
77. Chopra AP, Boone SA, Liang X, Duesbery NS (2003) Anthrax lethal factor proteolysis and inactivation of MAPK kinase. J Biol Chem 278:9402–9406
78. Boyden ED, Dietrich WF (2006) Nalp1b controls mouse macrophage susceptibility to anthrax lethal toxin. Nat Genet 38:240–244
79. Kang TJ, Basu S, Zhang L, Thomas KE, Vogel SN, Baillie L, Cross AS (2008) Bacillus anthracis spores and lethal toxin induce IL-1beta via functionally distinct signaling pathways. Eur J Immunol 38:1574–1584
80. Muruve DA, Petrilli V, Zaiss AK, White LR, Clark SA, Ross PJ, Parks RJ, Tschopp J (2008) The inflammasome recognizes cytosolic microbial and host DNA and triggers an innate immune response. Nature 452:103–107
81. Hise AG, Tomalka J, Ganesan S, Patel K, Hall BA, Brown GD, Fitzgerald KA (2009) An essential role for the NLRP3 inflammasome in host defense against the human fungal pathogen Candida albicans. Cell Host Microbe 5:487–497
82. Gross O, Poeck H, Bscheider M, Dostert C, Hannesschlager N, Endres S, Hartmann G, Tardivel A, Schweighoffer E, Tybulewicz V et al (2009) Syk kinase signalling couples to the Nlrp3 inflammasome for anti-fungal host defence. Nature 459:433–436
83. Duncan JA, Gao X, Huang MT, O'Connor BP, Thomas CE, Willingham SB, Bergstralh DT, Jarvis GA, Sparling PF, Ting JP (2009) Neisseria gonorrhoeae activates the proteinase cathepsin B to mediate the signaling activities of the NLRP3 and ASC-containing inflammasome. J Immunol 182:6460–6469
84. Hornung V, Bauernfeind F, Halle A, Samstad EO, Kono H, Rock KL, Fitzgerald KA, Latz E (2008) Silica crystals and aluminum salts activate the NALP3 inflammasome through phagosomal destabilization. Nat Immunol 9:847–856

85. Dostert C, Petrilli V, Van Bruggen R, Steele C, Mossman BT, Tschopp J (2008) Innate immune activation through Nalp3 inflammasome sensing of asbestos and silica. Science 320:674–677
86. Li H, Willingham SB, Ting JP, Re F (2008) Cutting edge: inflammasome activation by alum and alum's adjuvant effect are mediated by NLRP3. J Immunol 181:17–21
87. Martinon F, Petrilli V, Mayor A, Tardivel A, Tschopp J (2006) Gout-associated uric acid crystals activate the NALP3 inflammasome. Nature 440:237–241
88. Yamasaki K, Muto J, Taylor KR, Cogen AL, Audish D, Bertin J, Grant EP, Coyle AJ, Misaghi A, Hoffman HM et al (2009) NLRP3/cryopyrin is necessary for interleukin-1beta (IL-1beta) release in response to hyaluronan, an endogenous trigger of inflammation in response to injury. J Biol Chem 284:12762–12771
89. Halle A, Hornung V, Petzold GC, Stewart CR, Monks BG, Reinheckel T, Fitzgerald KA, Latz E, Moore KJ, Golenbock DT (2008) The NALP3 inflammasome is involved in the innate immune response to amyloid-beta. Nat Immunol 9:857–865
90. Willingham SB, Bergstralh DT, O'Connor W, Morrison AC, Taxman DJ, Duncan JA, Barnoy S, Venkatesan MM, Flavell RA, Deshmukh M et al (2007) Microbial pathogen-induced necrotic cell death mediated by the inflammasome components CIAS1/cryopyrin/NLRP3 and ASC. Cell Host Microbe 2:147–159
91. Scaffidi P, Misteli T, Bianchi ME (2002) Release of chromatin protein HMGB1 by necrotic cells triggers inflammation. Nature 418:191–195
92. Willingham SB, Allen IC, Bergstralh DT, Brickey WJ, Huang MT, Taxman DJ, Duncan JA, Ting JP (2009) NLRP3 (NALP3, Cryopyrin) facilitates in vivo caspase-1 activation, necrosis, and HMGB1 release via inflammasome-dependent and -independent pathways. J Immunol 183:2008–2015
93. Burckstummer T, Baumann C, Bluml S, Dixit E, Durnberger G, Jahn H, Planyavsky M, Bilban M, Colinge J, Bennett KL et al (2009) An orthogonal proteomic-genomic screen identifies AIM2 as a cytoplasmic DNA sensor for the inflammasome. Nat Immunol 10:266–272
94. Fernandes-Alnemri T, Yu JW, Datta P, Wu J, Alnemri ES (2009) AIM2 activates the inflammasome and cell death in response to cytoplasmic DNA. Nature 458:509–513
95. Hornung V, Ablasser A, Charrel-Dennis M, Bauernfeind F, Horvath G, Caffrey DR, Latz E, Fitzgerald KA (2009) AIM2 recognizes cytosolic dsDNA and forms a caspase-1-activating inflammasome with ASC. Nature 458:514–518
96. Roberts TL, Idris A, Dunn JA, Kelly GM, Burnton CM, Hodgson S, Hardy LL, Garceau V, Sweet MJ, Ross IL et al (2009) HIN-200 proteins regulate caspase activation in response to foreign cytoplasmic DNA. Science 323:1057–1060
97. Fernandes-Alnemri T, Yu JW, Juliana C, Solorzano L, Kang S, Wu J, Datta P, McCormick M, Huang L, McDermott E et al (2010) The AIM2 inflammasome is critical for innate immunity to Francisella tularensis. Nat Immunol 11:385–393
98. Sauer JD, Witte CE, Zemansky J, Hanson B, Lauer P, Portnoy DA (2010) Listeria monocytogenes triggers AIM2-mediated pyroptosis upon infrequent bacteriolysis in the macrophage cytosol. Cell Host Microbe 7:412–419
99. Broz P, von Moltke J, Jones JW, Vance RE, Monack DM Differential requirement for Caspase-1 autoproteolysis in pathogen-induced cell death and cytokine processing. Cell Host Microbe 8:471–483
100. Yu M, Wang H, Ding A, Golenbock DT, Latz E, Czura CJ, Fenton MJ, Tracey KJ, Yang H (2006) HMGB1 signals through toll-like receptor (TLR) 4 and TLR2. Shock 26:174–179
101. Ohashi K, Burkart V, Flohe S, Kolb H (2000) Cutting edge: heat shock protein 60 is a putative endogenous ligand of the toll-like receptor-4 complex. J Immunol 164:558–561
102. Hefeneider SH, Cornell KA, Brown LE, Bakke AC, McCoy SL, Bennett RM (1992) Nucleosomes and DNA bind to specific cell-surface molecules on murine cells and induce cytokine production. Clin Immunol Immunopathol 63:245–251

103. Schotte P, Denecker G, Van Den Broeke A, Vandenabeele P, Cornelis GR, Beyaert R (2004) Targeting Rac1 by the Yersinia effector protein YopE inhibits caspase-1-mediated maturation and release of interleukin-1beta. J Biol Chem 279:25134–25142
104. Master SS, Rampini SK, Davis AS, Keller C, Ehlers S, Springer B, Timmins GS, Sander P, Deretic V (2008) Mycobacterium tuberculosis prevents inflammasome activation. Cell Host Microbe 3:224–232
105. Johnston JB, Barrett JW, Nazarian SH, Goodwin M, Ricciuto D, Wang G, McFadden G (2005) A poxvirus-encoded pyrin domain protein interacts with ASC-1 to inhibit host inflammatory and apoptotic responses to infection. Immunity 23:587–598
106. Stasakova J, Ferko B, Kittel C, Sereinig S, Romanova J, Katinger H, Egorov A (2005) Influenza A mutant viruses with altered NS1 protein function provoke caspase-1 activation in primary human macrophages, resulting in fast apoptosis and release of high levels of interleukins 1beta and 18. J Gen Virol 86:185–195
107. Weiss DS, Brotcke A, Henry T, Margolis JJ, Chan K, Monack DM (2007) In vivo negative selection screen identifies genes required for Francisella virulence. Proc Natl Acad Sci U S A 104:6037–6042
108. van der Velden AW, Velasquez M, Starnbach MN (2003) Salmonella rapidly kill dendritic cells via a caspase-1-dependent mechanism. J Immunol 171:6742–6749
109. Nogueira CV, Lindsten T, Jamieson AM, Case CL, Shin S, Thompson CB, Roy CR (2009) Rapid pathogen-induced apoptosis: a mechanism used by dendritic cells to limit intracellular replication of Legionella pneumophila. PLoS Pathog 5: e1000478
110. Lotze MT, Tracey KJ (2005) High-mobility group box 1 protein (HMGB1): nuclear weapon in the immune arsenal. Nat Rev Immunol 5:331–342
111. Yang H, Ochani M, Li J, Qiang X, Tanovic M, Harris HE, Susarla SM, Ulloa L, Wang H, DiRaimo R et al (2004) Reversing established sepsis with antagonists of endogenous high-mobility group box 1. Proc Natl Acad Sci U S A 101:296–301
112. Kazama H, Ricci JE, Herndon JM, Hoppe G, Green DR, Ferguson TA (2008) Induction of immunological tolerance by apoptotic cells requires caspase-dependent oxidation of high-mobility group box-1 protein. Immunity 29:21–32
113. Miao EA, Leaf IA, Treuting PM, Mao DP, Dors M, Sarkar A, Warren SE, Wewers MD, Aderem A (2010) Caspase-1-induced pyroptosis is an innate immune effector mechanism against intracellular bacteria. Nat Immunol 11:1136–1142

Activation of the Inflammasome by Bacterial Pathogens

Tyler K. Ulland and Fayyaz S. Sutterwala

Abstract The innate immune system is capable of integrating a number of diverse signals in order to respond to both pathogenic threats and sterile inflammatory injury. The inflammasome is a multiprotein complex that when assembled mediates the activation of caspase-1 with the subsequent secretion of the proinflammatory cytokines IL-1β and IL-18 as well as the induction of the pyroptotic cell death pathway. Members of the NLR (nucleotide-binding domain leucine-rich repeat containing) family, NLRP1, NLRP3, and NLRC4, as well as the PYHIN (pyrin and HIN-200 domain-containing) family member AIM2 have all been shown to form functional inflammasomes capable of activating caspase-1. In this chapter, we focus on bacterial activators of the inflammasome and examine the mechanisms by which various pathogens can activate specific inflammasome pathways.

T.K. Ulland
Graduate Program in Immunology, University of Iowa, Iowa City, IA 52242, USA
and
Inflammation Program, University of Iowa, Iowa City, IA 52242, USA

F.S. Sutterwala (✉)
Graduate Program in Immunology, University of Iowa, Iowa City, IA 52242, USA
and
Inflammation Program, University of Iowa, Iowa City, IA 52242, USA
and
Division of Infectious Diseases, Department of Internal Medicine, University of Iowa, Iowa City, IA 52242, USA
and
Veterans Affairs Medical Center, Iowa City, IA 52241, USA
and
University of Iowa, 2501 Crosspark Road, D156 MTF, Coralville, IA 52241, USA
e-mail: fayyaz-sutterwala@uiowa.edu

I. Couillin et al. (eds.), *The Inflammasomes*, Progress in Inflammation Research,
DOI 10.1007/978-3-0348-0148-5_3,

1 Introduction

Pattern recognition receptors (PRR) are important cellular sensors that coordinate initial responses to a wide array of pathogens including fungi, viruses, parasites, and bacteria. Pattern recognition receptors include Toll-like receptors (TLR), RIG-I-like RNA helicases (RLH), C-type lectin receptors (CLR), and nucleotide-binding domain leucine-rich repeat containing receptors (NLRs) [1]. These germline-encoded PRRs are capable of recognizing highly conserved pathogen-associated molecular patterns (PAMPs) as well as endogenous danger signals. These danger signals, or danger-associated molecular patterns (DAMPs), are released under conditions of cellular damage or stress induced by either pathogens or sterile inflammatory insults. The presence of microbes in an otherwise sterile compartment of the host is always cause for alarm; however, much of an organism is in direct contact with the surrounding environment and may be colonized by a multitude of different microbes, many of which are highly beneficial. The concurrent presence of PAMPs and DAMPs may provide the immune system the information needed to discriminate between invasive pathogenic organisms and favorable commensal organisms.

The NLR family of cytoplasmic proteins contains 23 members in humans and 34 members in the mouse [2]. Sequencing of the sea urchin *Strongylocentrotus purpuratus* genome revealed over 200 NLR family members [3], demonstrating the importance of these innate immune receptors in lower species that lack adaptive immune systems. The NLR family has a unique domain structure consisting of a central nucleotide-binding domain (NACHT domain) and N-terminal protein-binding domain that is typically a caspase recruitment domain (CARD), a pyrin domain (PYD), or a baculovirus IAP repeat domain (BIR domain), and a C-terminal leucine-rich repeat domain (LRR) [2].

A number of NLR family members have been shown to effect caspase 1 processing and IL-1β production including NOD1, NOD2, NLRP1, NLRP2, NLRP3, NLRP6, NLRP7, NLRP10, NLRP12, and NLRC4 [4]. However, to date only NLRP1 (also known as NALP1 or DEFCAP), NLRP3 (also known as Cryopyrin, CIAS1 or NALP3), and NLRC4 (also known as CARD12 or IPAF) have been shown to assemble inflammasomes with clear physiological roles in vivo.

The inflammasome is a multiprotein complex consisting minimally of an NLR, the adaptor molecule ASC (apoptosis-associated speck-like protein containing a CARD), and caspase-1 [2, 4]. When assembled the inflammasome facilitates the activation of caspsae-1 with the subsequent processing and secretion of pro-IL-1β and pro-IL-18. Studies in vitro have also shown caspase-1 to possess additional functions such as the regulation of glycolysis pathways [5] and unconventional protein secretion [6], suggesting that inflammasome action is not restricted to modulation of proinflammatory cytokine production.

Plant R proteins bear striking similarities in domain structure and function to NLRs and have been shown to recognize, directly or indirectly, pathogen-derived molecules [7]. The ubiquitin ligase-associated protein SGT1 and the chaperone protein heat-shock protein 90 (HSP90) can associate with plant R proteins. Associations between SGT1 and HSP90 with NLRP2, NLRP3, NLRP4, NLRP12, NOD1, NOD2, and NLRC4 have also been found, highlighting the similarities between the plant R proteins and the NLR family [8–10].

Recently, the PYHIN family member AIM2 has also been shown to form a functional inflammasome [11–14]. AIM2 possesses a PYRIN domain through which it can interact with ASC and recruit caspase-1. Similar to the NLRs, AIM2 is also found in the cytoplasm where it recognizes cytosolic dsDNA via its HIN-200 domain. Upon activation of the AIM2 inflammasome by dsDNA, caspase-1 is activated resulting in the processing and secretion of IL-1β and IL-18. The PYHIN family also contains three other members: IFIX, IFI16, and MNDA; however, it is unclear if they are also capable of forming inflammasomes [15].

2 The NLRP3 Inflammasome

The NLRP3 inflammasome is the best characterized of the inflammasomes and can be activated by a diverse array of stimuli. The importance of NLRP3 was first recognized when mutations in *NLRP3* were shown to be responsible for the autoinflammatory syndromes, Muckle–Wells syndrome, familial cold autoinflammatory syndrome, and neonatal-onset multisystem inflammatory disease known collectively as cryopyrin-associated periodic syndrome (CAPS) [16–18]. Mutations in *NLRP3* associated with CAPS, over 40 of which have been identified, result in a constitutively active form of NLRP3 causing increased activation of caspase-1 and the resultant increased secretion of IL-1β [19]. Variants in the regulatory region downstream of NLRP3 have also been associated with susceptibility to Crohn's disease [20]. It is tempting to speculate that mutations associated with NLRP3 may lead to dysregulation of inflammation in response to intestinal flora contributing to the pathogenesis of Crohn's disease.

Biochemical analysis using human monocytic cells has shown that upon activation NLRP3 can associate with ASC, Cardinal, and caspase-1 to form an inflammasome (Fig. 1) [21]. The role for Cardinal in the inflammasome remains unclear [22]; as there is no murine homologue for Cardinal, it is likely that Cardinal is dispensable for NLRP3 inflammasome function. It is, however, clear that NLRP3 is critical for caspase-1 activation in response to a wide variety of stimuli including both microbial and nonmicrobial agents. Sterile mediators of NLRP3 inflammasome activation include endogenous danger signals such as ATP, uric acid crystals, and amyloid-β [23–26]. Exogenous particles such as silica, asbestos, and alum have also been shown to robustly activate the NLRP3 inflammasome [27–31]. In addition, a number of microbes and microbial products can also activate the NLRP3 inflammasome. For example, influenza virus, Sendai virus, Vaccinia virus, and adenovirus have all been shown to activate the NLRP3 inflammasome [22, 32–34]. Although both cytosolic DNA and RNA associated with viral infection have been proposed to activate the

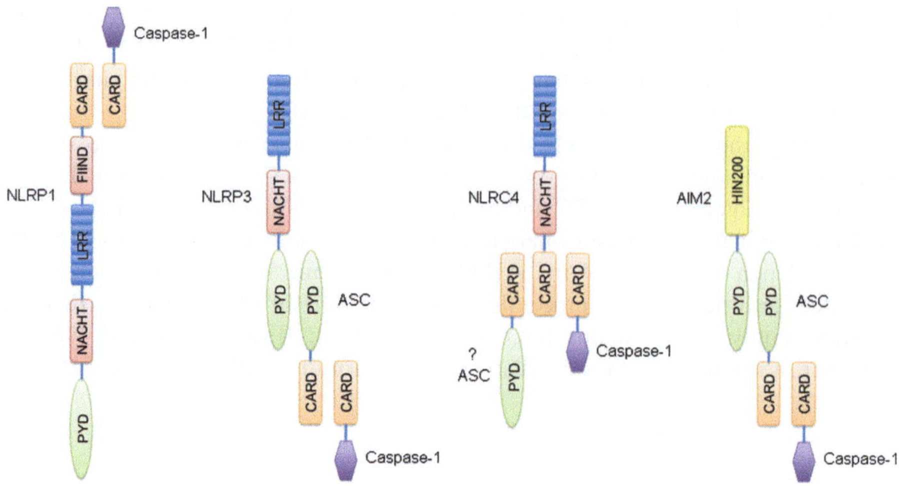

Fig. 1 Schematic representation of the NLRP1, NLRP3, NLRC4, and AIM2 inflammasomes. Unoligomerized murine inflammasome complexes are shown. Mouse NLRP1 does not possess a functional PYD; hence, caspase-1 is thought to directly interact with the C-terminal CARD. Human NLRP1 is thought to interact with ASC and caspase-1 via a functional N-terminal PYD and also recruit caspase-5 to the complex via the C-terminal CARD (not shown). How ASC interacts with the NLRC4 inflammasome complex remains unknown. Additionally, the role of NAIP5 in the NLRC4 inflammasome complex is still unclear. *CARD* caspase recruitment domain, *FIIND* domain with function to find, *NACHT* nucleotide-binding and oligomerization domain, *PYD* pyrin domain, *LRR* leucine-rich repeats, *HIN200* HIN-200 domain

NLRP3 inflammasome [35, 36], a recent study by Ichinohe and colleagues found that the ability of influenza virus to activate the NLRP3 inflammasome was independent of viral nucleic acid [37]. Instead, this study revealed that activation of the NLRP3 inflammasome was due to the action of the viral M2 protein, a proton-selective ion channel, on the Golgi apparatus [37]. The fungal pathogens *Candida albicans* and *Aspergillus fumigatus* can also activate the NLRP3 inflammasome [38–41]. In the case of *C. albicans*, the yeast to hyphae transition is key for its activation of the NLRP3 inflammasome, although the precise mechanism has not been elucidated [38]. The parasite plasmodium is also indirectly sensed by the NLRP3 inflammasome via the generation of the inorganic crystal hemozoin, which similar to uric acid crystals and silica can activate NLRP3 [42, 43].

Bacteria activate the NLRP3 inflammasome through a number of different mechanisms. Both Gram-positive and Gram-negative bacteria are capable of activating the NLRP3 inflammasome through the production of a pore-forming toxin. For example, the potassium ionophore nigericin derived from *Streptomyces hygroscopicus*, maitotoxin produced by the dinoflagellate *Gambierdiscus toxicus* [23], and bacterial pore-forming toxins such as listeriolysin O from *Listeria monocytogenes* [23, 44], aerolysin from *Aeromonas* species [45], streptolysin O from *Streptococcus pyogenes* [46], and hemolysin and multifunctional repeat-in-toxins from *Vibrio vulnificus* and *V. cholerae* can all activate caspase-1 in an NLRP3-dependent manner [47]. In addition, the bacterial pathogens *Porphyromonas gingivalis* [48], *Neisseria gonorrhoeae* [49], *Chlamydia pneumoniae* [50], and *C. trachomatis* [51] are also capable of activating the NLRP3 inflammasome, although the bacterial factor responsible for this remains to be identified for these pathogens.

Staphylococcus aureus is a Gram-positive organism that is responsible for causing skin and soft-tissue infections in humans. *S. aureus* can also cause severe invasive disease including pneumonia, endocarditis, and septic arthritis. Using an in vivo murine cutaneous model of *S. aureus* infection, Miller and colleagues demonstrated that mice deficient in IL-1β or ASC had increased bacterial burden and decreased neutrophil migration to the infection site [52]. Consistent with these findings, *S. aureus* has been shown to activate the NLRP3 inflammasome via its hemolysins in conjunction with *S. aureus*-derived lipoproteins [53, 54]. Shimada and colleagues have also shown that particulate peptidoglycan from *S. aureus* can activate the NLRP3 inflammasome; in addition the *S. aureus* enzyme PGN O-acetyltransferase A suppresses inflammasome activation [55], suggesting that degradation of the *S. aureus* cell wall may contribute to inflammasome activation.

Mycobacterium tuberculosis has also been shown to activate the NLRP3 inflammasome. ESAT-6, a protein secreted by *M. tuberculosis* through its ESX-1 secretion system, causes host cell membrane damage and has been shown to be involved in *M. tuberculosis*-induced inflammasome activation [56–59]. Master and colleagues have also demonstrated that the *M. tuberculosis* gene, *zmp1*, encoding a putative Zn^{2+} metalloprotease, can inhibit inflammasome activation [60]. Although the NLRP3 inflammasome is clearly activated by *M. tuberculosis* in vitro, its role in vivo is less clear. IL-1R-deficient and IL-1β-deficient mice displayed increased susceptibility to pulmonary infection with *M. tuberculosis*; however, ASC and caspase-1 appeared to be dispensable, as mice lacking these proteins have normal responses to infection with *M. tuberculosis* [61]. It is interesting that during an in vivo *M. tuberculosis* infection pro-IL-1β was processed to its mature form in a caspase-1-independent manner, suggesting that viable alternate pathways for IL-1β processing and release exist [61].

Given the wide range of seemingly unrelated stimuli that are capable of activating the NLRP3 inflammasome, it is unlikely that NLRP3 directly interacts with all of them; there is likely to be a single common endogenous ligand that mediates the response. A similarity that most bacterial pathogens that can activate the NLRP3 inflammasome share is their secretion of a membrane-damaging toxin. This membrane damage can induce two events that are required for NLRP3 inflammasome activation: a potassium efflux and the generation of reactive oxygen species, which are reviewed in detail elsewhere [2, 62]. Although significant gaps in our knowledge remain as to the mechanism of NLRP3 inflammasome activation, it is most likely that the cellular damage caused by invading pathogens is sensed by NLRP3 and not the microbial components themselves.

3 The NLRC4 Inflammasome

The NLRC4 inflammasome is also capable of activating caspase-1 with the subsequent processing and secretion of IL-1β and IL-18 and the initiation of pyroptotic cell death pathways. The microbial component flagellin, as well as Gram-negative bacteria possessing either a type III or type IV secretion system, such as *Salmonella*, *Shigella*, *Legionella*, and *Pseudomonas,* induces the activation of the NLRC4 inflammasome. NLRC4 contains an N-terminal CARD, a central NACHT domain, and C-terminal LRRs (Fig. 1) [2]. The presence of an N-terminal CARD potentially allows NLRC4 to interact directly with pro-caspase-1 through a CARD–CARD interaction. Despite this theoretical direct interaction, ASC plays an important role in NLRC4-mediated caspase-1 activation as ASC-deficient macrophages have defective caspase-1 activation and IL-1β secretion in response to pathogens such as *Salmonella*, *Shigella*, and *Pseudomonas* [63–65]. However, ASC-deficient macrophages infected with these pathogens still undergo cell death with similar kinetics as WT macrophages despite no detectable caspase-1 activation, suggesting that in the absence of ASC NLRC4 may trigger an as yet unidentified cell death pathway.

The first study to clearly demonstrate a role for NLRC4 in host defense showed that NLRC4 was required for *S. typhimurium*-induced activation of macrophage caspase-1 and secretion of IL-1β and IL-18 in vitro [63]. *S. typhimurium*-induced pyroptotic death of macrophages was also delayed in NLRC4-deficient macrophages [63]. However, in contrast to caspase-1-deficient mice, NLRC4-deficient mice infected orally with *S. typhimurium* were not more susceptible to infection [66]. The likely explanation for the in vivo difference seen between caspase-1-deficient mice and NLRC4-deficient mice is the presence of additional as yet undefined pathways that lead to *Salmonella*-induced caspase-1 activation in vivo, but are independent of NLRC4.

Legionella pneumophila is also capable of inducing the activation of caspase-1 in infected macrophages in an NLRC4-dependent manner [67, 68]. In addition to NLRC4 another NLR family member, Naip5 (also known as Birc1e) plays a role in susceptibility to infection with *L. pneumophila* [67, 69]. Naip5 and NLRC4 can interact directly with each other, suggesting that Naip5 may be part of the NLRC4 inflammasome complex activated in response to infection with *L. pneumophila* [67]. NLRC4 is also important for *L. pneumophila*-mediated phagosome maturation, and in NLRC4- and caspase-1-deficient macrophages, the *L. pneumophila*-containing phagosomes avoid fusion with the lysosome allowing increased bacterial replication [68]. Activation of caspase-1 in an NLRC4-dependent manner by *L. pneumophila* also results in caspase-7 activation [70]. Caspase-7-deficient macrophages displayed increased permissiveness to *L. pneumophila* through defective phagosomal maturation [70].

Unlike the NLRP3 inflammasome that appears to recognize the adverse effects of a pathogen on the host cell, it appears that NLRC4 is directly activated in response to microbial moieties. *S. typhimurium* and *L. pneumophila* strains deficient in flagellin were impaired in their ability to activate macrophage caspase-1, suggesting that flagellin played an important role in activation of the NLRC4 inflammasome [69, 71–73]. This was confirmed through the direct delivery of purified flagellin into the macrophage cytosol by transfection, which resulted in NLRC4-dependent caspase-1 activation [72–75]. Pathogens, such as *Salmonella*, *Legionella*, *Shigella*, and *Pseudomonas*, also require a functional bacterial type III (T3SS) or type IV (T4SS) secretion system to activate the NLRC4 inflammasome [63–65]. These findings led to a proposed model, whereby flagellin monomers passively enter the cytosol of infected cells through T3SS or T4SS leading to the activation of caspase-1 [72, 73]. Using retroviral transduction of *L. pneumophila* flagellin, in the absence of transfection agents and any other bacterial components, Lightfield and colleagues demonstrated that flagellin alone was sufficient to trigger NLRC4-dependent macrophage death [76]. However, the activation of the NLRC4 inflammasome in response to a C-terminal 35 amino acid fragment of *L. pneumophila* flagellin also required the presence of Naip5, suggesting again that Naip5 may be a required component of the NLRC4 inflammasome complex [76]. In contrast, the expression of full-length flagellin and infection with *S. typhimurium* and *P. aeruginosa* induced macrophage death in a manner that is dependent on NLRC4 but only partially Naip5 dependent [76].

The NLRC4 inflammasome can also be activated independently of flagellin. The nonflagellated bacterium *S. flexneri* as well as flagellin-deficient strains of *P. aeruginosa* can efficiently activate caspase-1 in an NLRC4-dependent manner [64, 65]. This discrepancy as to the requirement for flagellin in NLRC4 inflammasome activation was recently explained in an elegant study by Miao and colleagues [77]. They found the NLRC4 inflammasome was activated in response to the basal body rod component of the T3SS apparatus from *S. typhimurium* (PrgJ), *Burkholderia pseudomallei* (BsaK), *Escherichia coli* (EprJ and EscI), *S. flexneri* (MxiI), and *P. aeruginosa* (PscI) [77]. These rod proteins contain a sequence motif that resembles one found in flagellin; hence, NLRC4 is activated by both of these similar stimuli.

4 The NLRP1 Inflammasome

NLRP1 has a unique domain structure containing a C-terminal FIIND and CARD domain allowing it to interact directly with pro-caspase-1 (Fig. 1) [2]. The mouse has three orthologues of *NLRP1*, and unlike human NLRP1, the murine orthologues lack a functional PYD and hence fail to interact with ASC. Using a cell free system, Faustin and colleagues demonstrated that the minimum components required to activate caspase-1 were NLRP1 and pro-caspase-1 [78]. As predicted the adaptor molecule ASC was not required for NLRP1 inflammasome activation: however, its presence did enhance caspase-1 activation [78].

Bacillus anthracis lethal toxin (LeTx) can induce caspase-1-dependent death of macrophages from susceptible mouse strains. Boyden and Dietrich identified the

Nlrp1b gene as being responsible for macrophage susceptibility to LeTx [79]. Hsu et al. also described an important role for NLRP1 in LeTx-induced IL-1β secretion in response to the whole pathogen *B. anthracis* [80]. In this study, they also found a requirement for NOD2 in *B. anthracis*-induced IL-1β secretion, suggesting that NOD2 may be part of an NLRP1 inflammasome complex [80]. The NLRP1 inflammasome can also be directly activated by muramyldipeptide (MDP), although the physiologic role of this activation pathway remains to be examined [78]. Sequence variants in the *NLRP1* gene have been linked to autoimmune and autoinflammatory diseases associated with vitiligo in humans [81]. However, as with NLRP3, to date no infectious pathologies in humans have been associated with defects in the NLRP1 inflammasome pathway.

5 The AIM2 Inflammasome

AIM2 (absent in melanoma 2) belongs to a family of interferon-inducible HIN-200 domain possessing proteins and is the first non-NLR family member identified that can form a caspase-1 activating inflammasome [11–14]. AIM2 possesses a C-terminal HIN-200 domain that is responsible for the recognition of dsDNA; this facilitates the interaction of the AIM2 PYD with ASC through homotypic PYD–PYD interactions allowing ASC to recruit caspase-1 via its CARD (Fig. 1). AIM2 inflammasome activation is triggered by the presence of cytosolic DNA [11–14]. The source of this DNA appears to be less important than the cytosolic location as artificial delivery of DNA into the cytoplasm by transfection or infection with Vaccinia virus or mouse cytomegalovirus is capable of activating the AIM2 inflammasome [82].

Francisella tularensis, the causative agent of tularemia, is an intracellular pathogen that persists and replicates within phagocytes. Infection of activated macrophages with *F. tularensis* results in the activation of caspase-1 leading to pyroptosis of the infected cell along with the secretion of IL-1β and IL-18 [83]. Following phagocytosis, *F. tularensis* escapes the phagosome into the cytosol triggering the activation of the AIM2 inflammasome [82, 84]. AIM2 inflammasome activation plays an important role in the innate immune response against *F. tularensis* as mice deficient in AIM2, ASC, or caspase-1 have increased bacterial burden and mortality following infection with *F. tularensis* [83, 84]. As AIM2 is activated by cytosolic DNA, it has been postulated that the cell wall of some bacteria is damaged within the phagosome, and following escape from the phagosome these bacteria undergo further lysis releasing their DNA into the cytosol.

L. monocytogenes is capable of activating caspase-1 following infection of macrophages. However, this activation is rather complex and utilizes multiple inflammasomes; both NLRP3 and NLRC4 inflammasomes are activated in response to cytosolic *L. monocytogenes* [85, 86]. NLRP3 inflammasome activation is dependent on the presence of listeriolysin O, which is also required for the escape of the bacteria from the vacuole into the cytosol [23]. *L. monocytogenes* flagellin is also capable of activating the NLRC4 inflammasome following the entry of the bacterium into the cytosol [68, 69, 71]. In addition to NLRP3 and NLRC4, the AIM2 inflammasome is also capable of sensing infection with *L. monocytogenes* [82, 86, 87]. Lysis of bacteria within the cytosol may result in the release of bacterial DNA that subsequently triggers AIM2 inflammasome activation. Further studies are required to determine when and how bacterial DNA is released into the cytosol; in addition, it will be important to determine if activation of the AIM2 inflammasome pathway is specific to certain pathogens or whether all intracellular pathogens can activate this pathway to some extent.

6 Concluding Remarks

Understanding of inflammasome biology has advanced greatly over the past few years and four distinct inflammasome complexes, NLRP1, NLRP3, NLRC4, and AIM2, have been identified that regulate caspase-1 activation. The need for strict regulation of caspase-1 can be demonstrated by a number of disease states associated with dysregulated caspase-1 activity [19]. Although bacterial pathogens can clearly activate all four of these inflammasome complexes in a variety of animal models, the role of the inflammasome in combating pathogens in humans remains to be seen. A growing number of therapeutics under development are aimed at disruption of inflammasome pathways for the treatment of autoimmune and autoinflammatory disorders [88]. It seems likely that increased use of these agents will reveal a role for inflammasomes in controlling infectious disease, manifest as increased infectious complications in the patients undergoing these therapies.

A number of fascinating questions in inflammasome biology remain to be answered. Identification of the ligands that directly interact with NLRP1, NLRP3, and NLRC4 will shed light on the mechanism by which these inflammasomes are activated. In addition, the characterization and function of the majority of NLR family members remain unexplored. Further study will reveal if these family members are also capable of forming inflammasome complexes.

Acknowledgments We thank Suzanne Cassel for helpful discussion and critical review of this manuscript. A Merit Review Grant from the Veterans Administration, IBX000167A (F.S.S.) and NIH grant R01 AI087630 (F.S.S.) supported this work.

References

1. Medzhitov R (2009) Approaching the asymptote: 20 years later. Immunity 30:766–775
2. Martinon F, Mayor A, Tschopp J (2009) The inflammasomes: guardians of the body. Annu Rev Immunol 27:229–265

3. Rast JP, Smith LC, Loza-Coll M, Hibino T, Litman GW (2006) Genomic insights into the immune system of the sea urchin. Science 314:952–956
4. Mariathasan S, Monack DM (2007) Inflammasome adaptors and sensors: intracellular regulators of infection and inflammation. Nat Rev Immunol 7:31–40
5. Shao W, Yeretssian G, Doiron K, Hussain SN, Saleh M (2007) The caspase-1 digestome identifies the glycolysis pathway as a target during infection and septic shock. J Biol Chem 282:36321–36329
6. Keller M, Ruegg A, Werner S, Beer HD (2008) Active caspase-1 is a regulator of unconventional protein secretion. Cell 132:818–831
7. Altenbach D, Robatzek S (2007) Pattern recognition receptors: from the cell surface to intracellular dynamics. Mol Plant Microbe Interact 20:1031–1039
8. Mayor A, Martinon F, De Smedt T, Petrilli V, Tschopp J (2007) A crucial function of SGT1 and HSP90 in inflammasome activity links mammalian and plant innate immune responses. Nat Immunol 8:497–503
9. da Silva CJ, Miranda Y, Leonard N, Ulevitch R (2007) SGT1 is essential for Nod1 activation. Proc Natl Acad Sci USA 104:6764–6769
10. Arthur JC, Lich JD, Aziz RK, Kotb M, Ting JP (2007) Heat shock protein 90 associates with monarch-1 and regulates its ability to promote degradation of NF-kappaB-inducing kinase. J Immunol 179:6291–6296
11. Burckstummer T, Baumann C, Bluml S, Dixit E, Durnberger G, Jahn H, Planyavsky M, Bilban M, Colinge J, Bennett KL et al (2009) An orthogonal proteomic-genomic screen identifies AIM2 as a cytoplasmic DNA sensor for the inflammasome. Nat Immunol 10:266–272
12. Fernandes-Alnemri T, Yu JW, Datta P, Wu J, Alnemri ES (2009) AIM2 activates the inflammasome and cell death in response to cytoplasmic DNA. Nature 458:509–513
13. Hornung V, Ablasser A, Charrel-Dennis M, Bauernfeind F, Horvath G, Caffrey DR, Latz E, Fitzgerald KA (2009) AIM2 recognizes cytosolic dsDNA and forms a caspase-1-activating inflammasome with ASC. Nature 458:514–518
14. Roberts TL, Idris A, Dunn JA, Kelly GM, Burnton CM, Hodgson S, Hardy LL, Garceau V, Sweet MJ, Ross IL et al (2009) HIN-200 proteins regulate caspase activation in response to foreign cytoplasmic DNA. Science 323:1057–1060
15. Hornung V, Latz E (2010) Intracellular DNA recognition. Nat Rev Immunol 10:123–130
16. Hoffman HM, Mueller JL, Broide DH, Wanderer AA, Kolodner RD (2001) Mutation of a new gene encoding a putative pyrin-like protein causes familial cold autoinflammatory syndrome and Muckle-Wells syndrome. Nat Genet 29:301–305
17. Dode C, Le Du N, Cuisset L, Letourneur F, Berthelot JM, Vaudour G, Meyrier A, Watts RA, Scott DG, Nicholls A et al (2002) New mutations of CIAS1 that are responsible for Muckle-Wells syndrome and familial cold urticaria: a novel mutation underlies both syndromes. Am J Hum Genet 70:1498–1506
18. Feldmann J, Prieur AM, Quartier P, Berquin P, Certain S, Cortis E, Teillac-Hamel D, Fischer A, de Saint BG (2002) Chronic infantile neurological cutaneous and articular syndrome is caused by mutations in CIAS1, a gene highly expressed in polymorphonuclear cells and chondrocytes. Am J Hum Genet 71:198–203
19. Ting JP, Kastner DL, Hoffman HM (2006) CATERPILLERs, pyrin and hereditary immunological disorders. Nat Rev Immunol 6:183–195
20. Villani AC, Lemire M, Fortin G, Louis E, Silverberg MS, Collette C, Baba N, Libioulle C, Belaiche J, Bitton A et al (2009) Common variants in the NLRP3 region contribute to Crohn's disease susceptibility. Nat Genet 41:71–76
21. Martinon F, Burns K, Tschopp J (2002) The inflammasome: a molecular platform triggering activation of inflammatory caspases and processing of proIL-beta. Mol Cell 10:417–426
22. Allen IC, Scull MA, Moore CB, Holl EK, McElvania-TeKippe E, Taxman DJ, Guthrie EH, Pickles RJ, Ting JP (2009) The NLRP3 inflammasome mediates in vivo innate immunity to influenza A virus through recognition of viral RNA. Immunity 30:556–565

23. Mariathasan S, Weiss DS, Newton K, McBride J, O'Rourke K, Roose-Girma M, Lee WP, Weinrauch Y, Monack DM, Dixit VM (2006) Cryopyrin activates the inflammasome in response to toxins and ATP. Nature 440:228–232
24. Sutterwala FS, Ogura Y, Szczepanik M, Lara-Tejero M, Lichtenberger GS, Grant EP, Bertin J, Coyle AJ, Galan JE, Askenase PW et al (2006) Critical role for NALP3/CIAS1/Cryopyrin in innate and adaptive immunity through its regulation of caspase-1. Immunity 24:317–327
25. Martinon F, Petrilli V, Mayor A, Tardivel A, Tschopp J (2006) Gout-associated uric acid crystals activate the NALP3 inflammasome. Nature 440:237–241
26. Halle A, Hornung V, Petzold GC, Stewart CR, Monks BG, Reinheckel T, Fitzgerald KA, Latz E, Moore KJ, Golenbock DT (2008) The NALP3 inflammasome is involved in the innate immune response to amyloid-beta. Nat Immunol 9:857–865
27. Cassel SL, Eisenbarth SC, Iyer SS, Sadler JJ, Colegio OR, Tephly LA, Carter AB, Rothman PB, Flavell RA, Sutterwala FS (2008) The Nalp3 inflammasome is essential for the development of silicosis. Proc Natl Acad Sci USA 105:9035–9040
28. Eisenbarth SC, Colegio OR, O'Connor W, Sutterwala FS, Flavell RA (2008) Crucial role for the Nalp3 inflammasome in the immunostimulatory properties of aluminium adjuvants. Nature 453:1122–1126
29. Dostert C, Petrilli V, Van Bruggen R, Steele C, Mossman BT, Tschopp J (2008) Innate immune activation through Nalp3 inflammasome sensing of asbestos and silica. Science 320:674–677
30. Hornung V, Bauernfeind F, Halle A, Samstad EO, Kono H, Rock KL, Fitzgerald KA, Latz E (2008) Silica crystals and aluminum salts activate the NALP3 inflammasome through phagosomal destabilization. Nat Immunol 9:847–856
31. Li H, Willingham SB, Ting JP, Re F (2008) Cutting edge: inflammasome activation by alum and alum's adjuvant effect are mediated by NLRP3. J Immunol 181:17–21
32. Kanneganti TD, Body-Malapel M, Amer A, Park JH, Whitfield J, Franchi L, Taraporewala ZF, Miller D, Patton JT, Inohara N et al (2006) Critical role for Cryopyrin/Nalp3 in activation of caspase-1 in response to viral infection and double-stranded RNA. J Biol Chem 281:36560–36568
33. Thomas PG, Dash P, Aldridge JR Jr, Ellebedy AH, Reynolds C, Funk AJ, Martin WJ, Lamkanfi M, Webby RJ, Boyd KL et al (2009) The intracellular sensor NLRP3 mediates key innate and healing responses to influenza A virus via the regulation of caspase-1. Immunity 30:566–575
34. Ichinohe T, Lee HK, Ogura Y, Flavell R, Iwasaki A (2009) Inflammasome recognition of influenza virus is essential for adaptive immune responses. J Exp Med 206:79–87
35. Kanneganti TD, Ozoren N, Body-Malapel M, Amer A, Park JH, Franchi L, Whitfield J, Barchet W, Colonna M, Vandenabeele P et al (2006) Bacterial RNA and small antiviral compounds activate caspase-1 through cryopyrin/Nalp3. Nature 440:233–236
36. Muruve DA, Petrilli V, Zaiss AK, White LR, Clark SA, Ross PJ, Parks RJ, Tschopp J (2008) The inflammasome recognizes cytosolic microbial and host DNA and triggers an innate immune response. Nature 452:103–107
37. Ichinohe T, Pang IK, Iwasaki A (2010) Influenza virus activates inflammasomes via its intracellular M2 ion channel. Nat Immunol 11:404–410
38. Joly S, Ma N, Sadler JJ, Soll DR, Cassel SL, Sutterwala FS (2009) Cutting edge: Candida albicans hyphae formation triggers activation of the Nlrp3 inflammasome. J Immunol 183: 3578–3581
39. Gross O, Poeck H, Bscheider M, Dostert C, Hannesschlager N, Endres S, Hartmann G, Tardivel A, Schweighoffer E, Tybulewicz V et al (2009) Syk kinase signalling couples to the Nlrp3 inflammasome for anti-fungal host defence. Nature 459:433–436
40. Hise AG, Tomalka J, Ganesan S, Patel K, Hall BA, Brown GD, Fitzgerald KA (2009) An essential role for the NLRP3 inflammasome in host defense against the human fungal pathogen Candida albicans. Cell Host Microbe 5:487–497

41. Said-Sadier N, Padilla E, Langsley G (2010) Ojcius DM Aspergillus fumigatus stimulates the NLRP3 inflammasome through a pathway requiring ROS production and the Syk tyrosine kinase. PLoS One 5:e10008
42. Shio MT, Eisenbarth SC, Savaria M, Vinet AF, Bellemare MJ, Harder KW, Sutterwala FS, Bohle DS, Descoteaux A, Flavell RA et al (2009) Malarial hemozoin activates the NLRP3 inflammasome through Lyn and Syk kinases. PLoS Pathog 5:e1000559
43. Dostert C, Guarda G, Romero JF, Menu P, Gross O, Tardivel A, Suva ML, Stehle JC, Kopf M, Stamenkovic I et al (2009) Malarial hemozoin is a Nalp3 inflammasome activating danger signal. PLoS One 4:e6510
44. Meixenberger K, Pache F, Eitel J, Schmeck B, Hippenstiel S, Slevogt H, N'Guessan P, Witzenrath M, Netea MG, Chakraborty T et al (2010) Listeria monocytogenes-infected human peripheral blood mononuclear cells produce IL-1beta, depending on listeriolysin O and NLRP3. J Immunol 184:922–930
45. Gurcel L, Abrami L, Girardin S, Tschopp J, van der Goot FG (2006) Caspase-1 activation of lipid metabolic pathways in response to bacterial pore-forming toxins promotes cell survival. Cell 126:1135–1145
46. Harder J, Franchi L, Munoz-Planillo R, Park JH, Reimer T, Nunez G (2009) Activation of the Nlrp3 inflammasome by Streptococcus pyogenes requires streptolysin O and NF-kappa B activation but proceeds independently of TLR signaling and P2X7 receptor. J Immunol 183:5823–5829
47. Toma C, Higa N, Koizumi Y, Nakasone N, Ogura Y, McCoy AJ, Franchi L, Uematsu S, Sagara J, Taniguchi S et al (2010) Pathogenic Vibrio activate NLRP3 inflammasome via cytotoxins and TLR/nucleotide-binding oligomerization domain-mediated NF-kappaB signaling. J Immunol 184:5287–5297
48. Huang MT, Taxman DJ, Holley-Guthrie EA, Moore CB, Willingham SB, Madden V, Parsons RK, Featherstone GL, Arnold RR, O'Connor BP et al (2009) Critical role of apoptotic speck protein containing a caspase recruitment domain (ASC) and NLRP3 in causing necrosis and ASC speck formation induced by Porphyromonas gingivalis in human cells. J Immunol 182:2395–2404
49. Duncan JA, Gao X, Huang MT, O'Connor BP, Thomas CE, Willingham SB, Bergstralh DT, Jarvis GA, Sparling PF, Ting JP (2009) Neisseria gonorrhoeae activates the proteinase cathepsin B to mediate the signaling activities of the NLRP3 and ASC-containing inflammasome. J Immunol 182:6460–6469
50. He X, Mekasha S, Mavrogiorgos N, Fitzgerald KA, Lien E, Ingalls RR (2010) Inflammation and fibrosis during Chlamydia pneumoniae infection is regulated by IL-1 and the NLRP3/ASC inflammasome. J Immunol 184(10):5743–5754
51. Abdul-Sater AA, Koo E, Hacker G, Ojcius DM (2009) Inflammasome-dependent caspase-1 activation in cervical epithelial cells stimulates growth of the intracellular pathogen Chlamydia trachomatis. J Biol Chem 284:26789–26796
52. Miller LS, Pietras EM, Uricchio LH, Hirano K, Rao S, Lin H, O'Connell RM, Iwakura Y, Cheung AL, Cheng G et al (2007) Inflammasome-mediated production of IL-1beta is required for neutrophil recruitment against Staphylococcus aureus in vivo. J Immunol 179:6933–6942
53. Munoz-Planillo R, Franchi L, Miller LS, Nunez G (2009) A critical role for hemolysins and bacterial lipoproteins in Staphylococcus aureus-induced activation of the Nlrp3 inflammasome. J Immunol 183:3942–3948
54. Craven RR, Gao X, Allen IC, Gris D, Bubeck Wardenburg J, McElvania-Tekippe E, Ting JP, Duncan JA (2009) Staphylococcus aureus alpha-hemolysin activates the NLRP3-inflammasome in human and mouse monocytic cells. PLoS One 4:e7446
55. Shimada T, Park BG, Wolf AJ, Brikos C, Goodridge HS, Becker CA, Reyes CN, Miao EA, Aderem A, Gotz F et al (2010) Staphylococcus aureus evades lysozyme-based peptidoglycan digestion that links phagocytosis, inflammasome activation, and IL-1beta secretion. Cell Host Microbe 7:38–49
56. Derrick SC, Morris SL (2007) The ESAT6 protein of Mycobacterium tuberculosis induces apoptosis of macrophages by activating caspase expression. Cell Microbiol 9:1547–1555

57. Kurenuma T, Kawamura I, Hara H, Uchiyama R, Daim S, Dewamitta SR, Sakai S, Tsuchiya K, Nomura T, Mitsuyama M (2009) The RD1 locus in the Mycobacterium tuberculosis genome contributes to activation of caspase-1 via induction of potassium ion efflux in infected macrophages. Infect Immun 77:3992–4001
58. Mishra BB, Moura-Alves P, Sonawane A, Hacohen N, Griffiths G, Moita LF, Anes E (2010) Mycobacterium tuberculosis protein ESAT-6 is a potent activator of the NLRP3/ASC inflammasome. Cell Microbiol 12(8):1046–1063
59. Koo IC, Wang C, Raghavan S, Morisaki JH, Cox JS, Brown EJ (2008) ESX-1-dependent cytolysis in lysosome secretion and inflammasome activation during mycobacterial infection. Cell Microbiol 10:1866–1878
60. Master SS, Rampini SK, Davis AS, Keller C, Ehlers S, Springer B, Timmins GS, Sander P, Deretic V (2008) Mycobacterium tuberculosis prevents inflammasome activation. Cell Host Microbe 3:224–232
61. Mayer-Barber KD, Barber DL, Shenderov K, White SD, Wilson MS, Cheever A, Kugler D, Hieny S, Caspar P, Nunez G et al (2010) Caspase-1 independent IL-1beta production is critical for host resistance to mycobacterium tuberculosis and does not require TLR signaling in vivo. J Immunol 184:3326–3330
62. Martinon F (2010) Signaling by ROS drives inflammasome activation. Eur J Immunol 40:616–619
63. Mariathasan S, Newton K, Monack DM, Vucic D, French DM, Lee WP, Roose-Girma M, Erickson S, Dixit VM (2004) Differential activation of the inflammasome by caspase-1 adaptors ASC and Ipaf. Nature 430:213–218
64. Suzuki T, Franchi L, Toma C, Ashida H, Ogawa M, Yoshikawa Y, Mimuro H, Inohara N, Sasakawa C, Nunez G (2007) Differential regulation of caspase-1 activation, pyroptosis, and autophagy via Ipaf and ASC in Shigella-infected macrophages. PLoS Pathog 3:e111
65. Sutterwala FS, Mijares LA, Li L, Ogura Y, Kazmierczak BI, Flavell RA (2007) Immune recognition of Pseudomonas aeruginosa mediated by the IPAF/NLRC4 inflammasome. J Exp Med 204:3235–3245
66. Lara-Tejero M, Sutterwala FS, Ogura Y, Grant EP, Bertin J, Coyle AJ, Flavell RA, Galan JE (2006) Role of the caspase-1 inflammasome in Salmonella typhimurium pathogenesis. J Exp Med 203:1407–1412
67. Zamboni DS, Kobayashi KS, Kohlsdorf T, Ogura Y, Long EM, Vance RE, Kuida K, Mariathasan S, Dixit VM, Flavell RA et al (2006) The Birc1e cytosolic pattern-recognition receptor contributes to the detection and control of Legionella pneumophila infection. Nat Immunol 7:318–325
68. Amer A, Franchi L, Kanneganti TD, Body-Malapel M, Ozoren N, Brady G, Meshinchi S, Jagirdar R, Gewirtz A, Akira S et al (2006) Regulation of Legionella phagosome maturation and infection through flagellin and host Ipaf. J Biol Chem 281:35217–35223
69. Ren T, Zamboni DS, Roy CR, Dietrich WF, Vance RE (2006) Flagellin-deficient Legionella mutants evade caspase-1- and Naip5-mediated macrophage immunity. PLoS Pathog 2:e18
70. Akhter A, Gavrilin MA, Frantz L, Washington S, Ditty C, Limoli D, Day C, Sarkar A, Newland C, Butchar J et al (2009) Caspase-7 activation by the Nlrc4/Ipaf inflammasome restricts Legionella pneumophila infection. PLoS Pathog 5:e1000361
71. Molofsky AB, Byrne BG, Whitfield NN, Madigan CA, Fuse ET, Tateda K, Swanson MS (2006) Cytosolic recognition of flagellin by mouse macrophages restricts Legionella pneumophila infection. J Exp Med 203:1093–1104
72. Franchi L, Amer A, Body-Malapel M, Kanneganti TD, Ozoren N, Jagirdar R, Inohara N, Vandenabeele P, Bertin J, Coyle A et al (2006) Cytosolic flagellin requires Ipaf for activation of caspase-1 and interleukin 1beta in Salmonella-infected macrophages. Nat Immunol 7:576–582
73. Miao EA, Alpuche-Aranda CM, Dors M, Clark AE, Bader MW, Miller SI, Aderem A (2006) Cytoplasmic flagellin activates caspase-1 and secretion of interleukin 1beta via Ipaf. Nat Immunol 7:569–575

74. Miao EA, Ernst RK, Dors M, Mao DP, Aderem A (2008) Pseudomonas aeruginosa activates caspase 1 through Ipaf. Proc Natl Acad Sci USA 105:2562–2567
75. Franchi L, Stoolman J, Kanneganti TD, Verma A, Ramphal R, Nunez G (2007) Critical role for Ipaf in Pseudomonas aeruginosa-induced caspase-1 activation. Eur J Immunol 37:3030–3039
76. Lightfield KL, Persson J, Brubaker SW, Witte CE, von Moltke J, Dunipace EA, Henry T, Sun YH, Cado D, Dietrich WF et al (2008) Critical function for Naip5 in inflammasome activation by a conserved carboxy-terminal domain of flagellin. Nat Immunol 9:1171–1178
77. Miao EA, Mao DP, Yudkovsky N, Bonneau R, Lorang CG, Warren SE, Leaf IA (2010) Aderem A Innate immune detection of the type III secretion apparatus through the NLRC4 inflammasome. Proc Natl Acad Sci USA 107:3076–3080
78. Faustin B, Lartigue L, Bruey JM, Luciano F, Sergienko E, Bailly-Maitre B, Volkmann N, Hanein D, Rouiller I, Reed JC (2007) Reconstituted NALP1 inflammasome reveals two-step mechanism of caspase-1 activation. Mol Cell 25:713–724
79. Boyden ED, Dietrich WF (2006) Nalp1b controls mouse macrophage susceptibility to anthrax lethal toxin. Nat Genet 38:240–244
80. Hsu LC, Ali SR, McGillivray S, Tseng PH, Mariathasan S, Humke EW, Eckmann L, Powell JJ, Nizet V, Dixit VM et al (2008) A NOD2-NALP1 complex mediates caspase-1-dependent IL-1beta secretion in response to Bacillus anthracis infection and muramyl dipeptide. Proc Natl Acad Sci USA 105:7803–7808
81. Jin Y, Mailloux CM, Gowan K, Riccardi SL, LaBerge G, Bennett DC, Fain PR, Spritz RA (2007) NALP1 in vitiligo-associated multiple autoimmune disease. N Engl J Med 356: 1216–1225
82. Rathinam VA, Jiang Z, Waggoner SN, Sharma S, Cole LE, Waggoner L, Vanaja SK, Monks BG, Ganesan S, Latz E et al (2010) The AIM2 inflammasome is essential for host defense against cytosolic bacteria and DNA viruses. Nat Immunol 11:395–402
83. Mariathasan S, Weiss DS, Dixit VM, Monack DM (2005) Innate immunity against Francisella tularensis is dependent on the ASC/caspase-1 axis. J Exp Med 202:1043–1049
84. Fernandes-Alnemri T, Yu JW, Juliana C, Solorzano L, Kang S, Wu J, Datta P, McCormick M, Huang L, McDermott E et al (2010) The AIM2 inflammasome is critical for innate immunity to Francisella tularensis. Nat Immunol 11:385–393
85. Warren SE, Mao DP, Rodriguez AE, Miao EA, Aderem A (2008) Multiple Nod-like receptors activate caspase 1 during Listeria monocytogenes infection. J Immunol 180:7558–7564
86. Kim S, Bauernfeind F, Ablasser A, Hartmann G, Fitzgerald KA, Latz E, Hornung V (2010) Listeria monocytogenes is sensed by the NLRP3 and AIM2 inflammasome. Eur J Immunol 40(6):1545–1551
87. Sauer JD, Witte CE, Zemansky J, Hanson B, Lauer P, Portnoy DA (2010) Listeria monocytogenes triggers AIM2-mediated pyroptosis upon infrequent bacteriolysis in the macrophage cytosol. Cell Host Microbe 7(5):412–419
88. Hoffman HM (2009) Therapy of autoinflammatory syndromes. J Allergy Clin Immunol 124:1129–1138, quiz 1139–1140

The Role of Inflammasomes in Viral Infection

Christopher R. Lupfer and Thirumala-Devi Kanneganti

Abstract There are three known receptors capable of inducing inflammasome formation in response to viral infections: NLRP3, AIM2, and RIG-I. Here, we discuss the viruses and mechanisms of activation involved in inflammasome signaling. We also briefly discuss some of the viruses known to activate caspase-1 and induce secretion of IL-1β or IL-18, but for which the upstream inflammasome signaling pathway is not fully elucidated. Finally, the ability of several virus-encoded proteins to counteract inflammasome activation and their role in the pathogenesis of disease is reviewed.

1 Introduction

The role of the innate immune system is to recognize invading pathogens and initiate a broad spectrum defense until the adaptive immune system can be brought online. Three main classes of innate sensors have been described: Toll-like receptors (TLRs), RIG-I-like receptors (RLRs), and nucleotide oligomerization and binding domain (NOD)-like receptors (NLRs) [1, 2]. These classes of proteins are known as pattern recognition receptors (PRRs) as a result of their ability to recognize pathogen-associated molecular patterns (PAMPs). In the case of viruses, these patterns include double-stranded RNA, uncapped RNA, cytosolic DNA, hypomethylated CpG DNA, as well as several viral fusion glycoproteins. Whereas TLRs serve to sample the extracellular space and endosomes, RLRs and NLRs function as sensors of the cytosol [1–3].

Three cytosolic PRRs, namely NLRP3, RIG-I (Retinoic Acid Inducible Gene-I), and AIM2 (Absent in Melanoma 2), are known to recognize viruses and signal

C.R. Lupfer • T.-D. Kanneganti (✉)
Department of Immunology, St. Jude Children's Research Hospital, 262 Danny Thomas Place, MS 351, Memphis, TN 38105, USA
e-mail: Thirumala-Devi.Kanneganti@StJude.org

I. Couillin et al. (eds.), *The Inflammasomes*, Progress in Inflammation Research,
DOI 10.1007/978-3-0348-0148-5_4,

through a macromolecular complex called the inflammasome (Fig. 1). NLRP3 (also known as cryopyrin, CIAS1, PYPAF1, CATERPILLER 1.1, or NALP3) consists of (1) an amino-terminal pyrin domain (PYD) mediating protein–protein interactions; (2) an intermediary nucleotide binding and oligomerization domain (NOD) that is required for self-oligomerization; and (3) a carboxy-terminal Leucine-Rich Repeat (LRR) motif involved in sensing PAMPS or damage to the cell. AIM2 also possesses a PYD domain but relies on a HIN-200 DNA binding domain for recognition of pathogens. RIG-I contains a RNA helicase domain for binding to virus-specific RNA moieties as well as two caspase activation and recruitment domains (CARD) for protein–protein interaction. Recognition of viruses by any

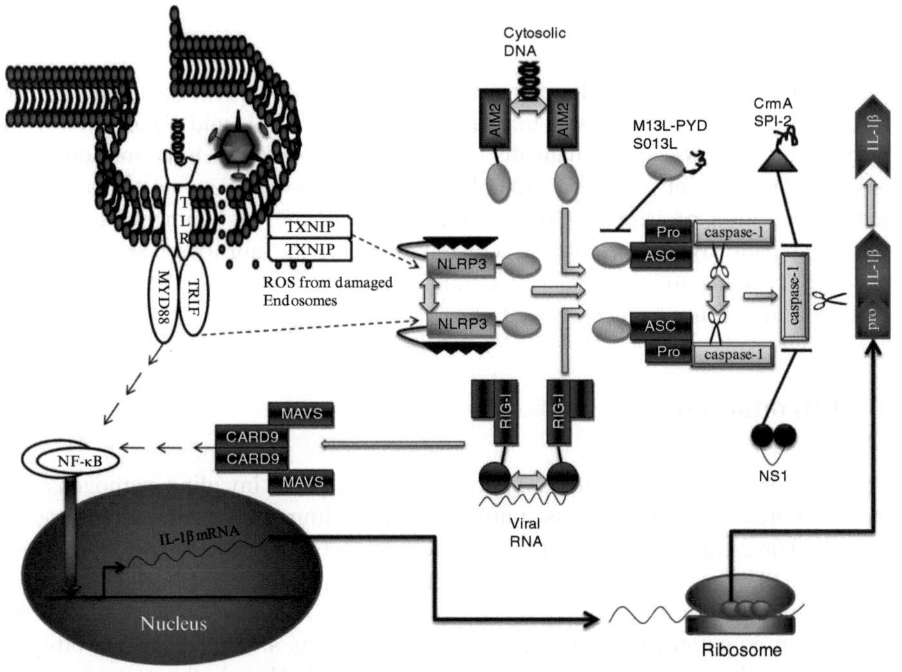

Fig. 1 Stimuli, inhibitors, and composition of distinct inflammasomes. NLRP3, AIM2, and RIG-I assemble a caspase-1 activating inflammasome complex in response to specific viral PAMPs or danger signals. The NLRP3 inflammasome recognizes multiple PAMPs in combination with ATP, ion flux, or ROS. Viral PAMPs may trigger assembly of the NLRP3 inflammasome through TLRs and/or a ROS-sensitive protein TXNIP. In contrast, AIM2 directly binds dsDNA and RIG-I directly binds viral RNA in the cytosol to induce caspase-1 activation. The CARD/pyrin-containing adaptor protein ASC is essential for all inflammasome complexes. Once activated, caspase-1 processes pro-IL-1β and pro-IL-18 into mature cytokines (for simplicity, only IL-1β is shown). CARD, caspase recruitment domain; NOD, nucleotide binding and oligomerization domain; LRR, leucine-rich repeat; PYD, pyrin domain; HIN-200, DNA binding domain; DExD/H-box, RNA helicase domain; dimerization; activation; inhibition; cleavage; potential activation; CrmA/SPI-2, serpine-like protease inhibitors of poxviruses; M13L-PYD/S013L, pyrin domain containing inhibitors of poxviruses; NS1, influenza A/PR/8/34 H1N1 NS1 protein

of these PRRs triggers self-oligomerization and interaction with the adaptor protein ASC (Apoptotic speck-like protein containing a CARD), which serves as a bridge between PRRs and caspase-1 by way of its PYD and CARD domains [4–9].

Inflammasome activation plays key roles in inflammation. First, the proinflammatory cytokines IL-1β and IL-18 are produced as inactive proforms that must be cleaved by caspase-1 before they are secreted from cells in their biologically active forms [10]. Secreted IL-1β and IL-18 then signal through cell-surface receptors and upregulate a myriad of additional cytokines and chemokines including KC, MIP-1α, IL-6, TNF-α, and IFN-γ [11–15]. Second, inflammasome activation can lead to a caspase-1-dependent proinflammatory form of cell death known as pyroptosis, which prevents further replication of invading pathogens [16, 17].

2 NLRP3 Inflammasome

Infection of primary macrophages by sendai virus and influenza A virus stimulates caspase-1, IL-1β, and IL-18 activation [11, 18]. This was later shown to occur in an NLRP3 inflammasome-dependent manner [19]. Influenza A virus is capable of activating NLRP3 in a wide variety of cells including mouse bone marrow-derived dendritic cells (BMDC) and macrophages (BMM), JME cells (human nasal airway epithelial cells), and THP-1cells (human monocytes) [13, 20]. In addition to influenza A, infections with modified vaccinia virus Ankara [21] or encephalomyocarditis virus [22] have also been shown to activate the NLRP3 inflammasome.

The requirement for NLRP3 inflammasome signaling in vivo during virus infection is best characterized for influenza A. Deletion of the NLRP3 inflammasome, or inhibition of IL-1β signaling, results in reduced infiltration of neutrophils and monocytes to the lung [13, 14, 23] as well as in reduced adaptive immune responses. This includes reduced isotype-specific antibodies and CD4 and CD8 T-cell responses [20, 23]. In part, this immunosuppression is explained by the diminished cytokine (IL-1β, IL-18, IL-6, and TNF-α) and chemokine (KC, Mip-1α, and Mip2) responses of NLRP3 inflammasome-deficient mice [13, 14, 20]. These same mice suffer from impaired healing processes during the resolution of infection, including increased necrosis and collagen deposition [14]. Finally, viral clearance late in infection is hindered in NLRP3 inflammasome or IL-1 receptor (IL1R1)-deficient mice [13, 20, 23] and they are more susceptible than wild-type counterparts to infection with influenza A [13, 14, 20, 23]. Even though cell culture experiments indicate that both epithelial cells and immune cells can sense influenza A, the generation of chimeric mice demonstrated that hematopoietic cells are the primary producers of IL-1β during influenza infection [20]. Although the reports listed above differ in some of their outcomes compared to others, the totality of data from NLRP3-, ASC-, caspase-1-, or IL1R1-deficient mice demonstrates that recognition of influenza A infection by the NLRP3 inflammasome and the concomitant production and signaling of IL-1β are critical for defense against, resolution of, and repair of damage as a result of influenza A infection.

Despite the fact that NLRP3 is capable of sensing a wide range of viruses, the exact mechanism by which it recognizes these invaders is poorly understood. Many lines of evidence suggest that viral RNA and DNA are capable of activating the NLRP3 inflammasome. Treatment of human and mouse cell lines with ssRNA or dsRNA analogues such as poly (I:C) is sufficient to activate NLRP3 [13]. Treatment of cells with purified rotavirus or brome mosaic virus dsRNA or influenza A ssRNA also activates NLRP3 [13, 19, 20]. This pathway has been further implicated in the detection of viral DNA from adenovirus in cell culture [24]. Adenovirus vector capsids alone did not activate NLRP3; however, replication-deficient adenovirus vectors, with their genome intact, were able to activate NLRP3. In vivo, the administration of poly (I:C) or purified influenza A viral RNA into mice also results in IL-1β secretion and inflammation through NLRP3 [13, 14, 19].

NLRP3 is not limited to the recognition of nucleic acids. In fact, it is unique among PRRs in its ability to initiate inflammasome formation by a myriad of stimuli ranging from bacterial PAMPs and uric acid crystals to viral nucleic acids and changes in intracellular ion or reactive oxygen species (ROS) levels [25–28]. The means by which NLRP3 is capable of detecting such a myriad of divergent PAMPs and danger signals from infectious and noninfectious agents has lead to the hypothesis that NLRP3 may actually sense a common upstream effect, rather than PAMPs themselves. It has been suggested that ROS [13, 28] or ion flux [29–31] may be the common signal [28, 32–35]. Alternatively, TLRs have been proposed to activate the NLRP3 inflammasome directly [36, 37]. However, most reports indicate that TLR activation or ROS alone is insufficient [35, 38], and a two-signal hypothesis has emerged [39, 40]. The addition of a TLR ligand and ATP/ions/ROS simultaneously provides sufficient signaling cues to activate the inflammasome and produce robust amounts of active IL-1β [30, 41–44]. In the case of viral infection, dual signaling cues are common. In influenza A infection, the viral RNA can trigger signaling through TLR 3 or 7 [45–48] and the process of virus entry, where endosomes are compromised resulting in ion and ROS release, provides the necessary signaling to activate the NLRP3 inflammasome [13]. Vaccinia virus was also shown to require the TLR 2-dependent production of pro-IL-1β and the endocytic internalization of the virus for robust IL-1β production by NLRP3 inflammasomes [21]. The fact that so many pathogens can activate the NLRP3 inflammasome has likely led to evolutionary pressures to keep it from being prematurely or continuously activated. A dual mode of activation allows for a wide range of stimuli to signal through NLRP3 and at the same time prevents unnecessary inflammation from this pathway by requiring two signals to be fully functional.

3 AIM2 Inflammasome

TLR9 is responsible for recognition of pathogen-associated CpG DNA in the endosomal compartment [2]. However, little is known about potential cytosolic DNA sensors which may activate the inflammasome. Adenovirus DNA, though partially recognized by NLRP3, also has an NLRP3-independent pathway for caspase-1 activation [24]. A specific cytosolic DNA sensor has been described previously (DAI, also known as DLM-1/ZBP1), but it induces IFN-α and IFN-β through IRF-3/7 transcription factors rather than activating caspase-1 [49, 50]. Using a combined approach of proteomics and gene expression profiling [51], or bioinformatics [52, 53], three separate groups recently identified AIM2 as a DNA binding protein which can activate the inflammasome [54]. AIM2 recognizes cytosolic DNA via a HIN-200 DNA binding domain and interacts with ASC through its PYD domain. These interactions are required for AIM2 inflammasome formation and release of active IL-1β. Pyroptosis was also increased by cytosolic DNA in an AIM2-dependent manner [52, 53]. Unlike NLRP3, AIM2 appears to directly recognize the presence of cytosolic DNA [51–53], making the AIM2 inflammasome the first to demonstrate a definitive interaction with a specific viral PAMP.

Due to the recentness of discovery, little is known about the range of pathogens that may be sensed by AIM2. The ability of AIM2 to detect cytosolic DNA does not guarantee its involvement in the host response to virus infection. However, Hornung et al. [53] showed that AIM2 is needed for inflammasome activation and pyroptosis in response to vaccinia virus in cell culture.

4 RIG-I Inflammasome

RIG-I is the prototypical member of the RLR family of cytosolic pathogen-associated RNA sensors and is known to recognize specific RNA viruses [55, 56]. Until now, its main function was proposed to be the activation of interferon genes through the adaptor MAVS and IRF-3/7 transcription factor pathway [3, 56]. Two recent reports, however, show that RIG-I plays a dual role in IL-1β production [22, 57]. RIG-I activation by transfected 5′-triphosphate RNA or vesicular stomatitis virus (VSV) infection resulted in the production of pro-IL-1β through a MAVS–CARD9–NF-κB signaling pathway. In response to these same stimuli, RIG-I also directly activates the inflammasome pathway by binding the adaptor ASC. This interaction is not universal for all RLRs as Mda5 and ASC do not interact. These results indicate the presence of an additional cytosolic nucleic acid sensing inflammasome for which a specific PAMP (5′-triphosphate RNA) is responsible for activation.

5 Inflammasome Orphaned Viruses

As inflammasome biology is an emerging field, it is likely that many other viruses activate this pathway. The inflammasomes are likely key to proper antiviral responses as caspase-1, IL-1β, and IL-18 are involved in many viral infections.

However, the upstream receptors such as NLRP3, RIG-I, or AIM2 have not yet been identified in most instances. In this section, we discuss some viruses known to activate caspase-1, IL-1β, and IL-18.

HIV-1-associated dementia (HAD) is the result of virus dissemination to the central nervous system where increased apoptosis of neuronal cells and resident immune cells is thought to mediate disease. In vivo experiments demonstrate that neurons, as well as microglia, produce IL-1β during HIV infection of the central nervous system [58]. Treatment of neuroblastoma cells with IL-1β alone results in increased cell death, which is inhibited by IL-1 receptor antagonist (IL1ra) or IL1r1 neutralizing antibodies, indicating that IL-1β signaling is at least partially responsible for neuron death. Inhibition of caspase-1 activation in cell culture with pharmacological inhibitors also reduces IL-1β release and cell death [59]. Interestingly, the HIV surface glycoprotein gp120 alone appears to be sufficient to activate caspase-1 and IL-1β in human neuroblastoma cells [59]. Intracerebral injection of purified gp120 increases the levels of active caspase-1 and IL-1β as well as apoptosis in the neocortex of rats [60]. This increase in apoptosis can be reduced by concomitant treatment with gp120 and inhibitors of caspase-1 or IL1ra [58, 61, 62]. Based on gp120's ability to stimulate calcium ion influx and production of ROS in neurons through excessive glutamate receptor (NMDA) activation and COX-2 expression [29, 63], it is possible that NLRP3 is involved in gp120 inflammasome activation. Additionally, gp120 binding to its coreceptors CXCR4 and CCR5 may induce IL-1β activation [64, 65]. Further research into the ability of HIV to activate one or more inflammasome pathways will therefore be essential to understanding the pathogenesis of HAD.

Several other virus-induced encephalitis are also linked to caspase-1and IL-1β. Reminiscent of the results seen for influenza A virus, IL-1β-deficient mice are more susceptible to herpes simplex virus-1 encephalitis due to increased viral load and reduced inflammation including fewer infiltrating immune cells and reduced cytokine/chemokine levels [66]. Ectromelia virus, an orthopoxvirus, activates caspase-1 in brain tissue of infected mice and this is presumed to play a role in inflammation-induced and pyroptosis-induced cell death [67]. Japanese encephalitis virus induces IL-1β and IL-18 secretion by microglial cells and astrocytes and neuronal cell death occurs by direct infection of neurons as well as IL-1β- and IL-18-mediated neuronal death [68]. It is evident that IL-1β and IL-18 play important roles in the development of many virus encephalitis viruses. Determining the role of inflammasomes in these infections is greatly needed to provide a complete picture of the mechanisms of encephalitic disease, as well as opening the door to potential treatment options.

A diverse array of viruses may activate the inflammasome pathways in addition to those discussed above. West Nile virus (WNV) infection causes migration of resident skin macrophages, known as Langerhans cells, to migrate to draining lymph nodes. The migration of Langerhans cells requires IL-1β, as treatment with an IL-1β neutralizing antibody reduces Langerhans cell migration as well as reduces the total number of cells in the lymph nodes of infected mice [69]. Hepatitis B virus (HBV) stimulates IL-18 production in human PBMCs by means of the histone-like domain of the virus' core antigen (HBVcAg) [70]. Interestingly, HBV deficient in the precore antigen (HBVeAg) stimulated higher levels of IL-18, possibly indicating a role for this protein in the inhibition of IL-18 signaling. Respiratory syncytial virus (RSV) induces the expression of caspase-1 and IL-1β as well as secretion of IL-1β from neonatal human monocytes, possibly implicating one or more inflammasomes in RSV infections [71]. The fact that virus-associated molecular patterns other than DNA or RNA (e.g., HBVcAg and HIV gp120) may activate inflammasomes speaks to the diversity of stimuli and possible recognition mechanisms and should lead to new insight into the activation of inflammasomes. The reports above are only a sampling of the viruses known to stimulate IL-1β and IL-18 production, or caspase-1 activation, and are by no means all inclusive. Indeed, there are many viruses that will likely be shown to activate inflammasomes as this field matures.

6 Virus-Encoded Inhibitors of Inflammasomes

In the evolutionary arms race between pathogens and their hosts, the host has selected for a variety of mechanisms to detect the presence of viruses and closely regulates the production of master cytokines such as IL-1β and IL-18. In response, viruses have undergone selective pressure to counter the inflammasome. Several viruses are known to encode proteins capable of interfering with inflammasome signaling. Not surprisingly, the point at which inflammasome signaling is interrupted is at the adaptor protein ASC or caspase-1 itself [72].

With one of the largest genomes in the virus world, poxviruses have evolved multiple inhibitors that interfere with innate and adaptive immunity [73]. Bioinformatics shows that many poxviruses contain a potential PYD domain protein hypothesized to interact with ASC; however, the myxoma virus M13L-PYD and Shope Fibroma virus S013L proteins are the only examples to have been studied in detail [74–76]. M13L-PYD is required for pathogenesis of myxoma virus and deletion results in severe attenuation in vivo. A decrease in viremia due to inefficient replication in lymphocytes and leukocytes and increased inflammation at the initial site of infection lead to more rapid resolution of disease. In cell culture, myxoma virus lacking M13L-PYD increases activation of caspase-1 and increases the levels of IL-1β and IL-18 [75]. M13L-PYD and S013L directly interact with ASC in cells. This interaction is mediated by PYD-PYD domain-specific interactions and expression of either M13L-PYD or S013L alone inhibits caspase-1 activation and IL-1β maturation from NLRP3 inflammasomes in cell culture [74, 75]. Poxvirus PYD proteins therefore inhibit inflammasomes at the level of the adaptor ASC and prevent all upstream PRRs from activating the inflammasomes. In addition, ASC may serve several inflammasome-independent functions such as MHC class-II containing exosome production, pyronecrosis, or release of additional proinflammatory molecules such as HMGB-1 [77–79]. ASC deficiency further results in diminished T-cell and lymph node cell proliferation independent of caspases-1 [80]. It is therefore possible that poxvirus PYD proteins block inflammation by inhibiting these processes as well.

Poxviruses also encode a variety of serpine-like protease inhibitors. The CrmA protein (also known as SPI-2) of cowpox virus specifically inhibits caspase-1 and subsequent cleavage of pro-IL-1β. Furthermore, deletion of CrmA reduces the number and severity of pox on the allantoic membrane when the virus is grown in chicken eggs [12, 81]. Infecting the respiratory tract of mice with CrmA mutants of cowpox or rabbit poxvirus results in diminished inflammation and viral loads compared to wild-type virus [85, 86]. During intradermal inoculation with cowpox, a more rapid viral clearance and more robust inflammatory response are observed with virus lacking CrmA [82]. It is interesting to note that no differences in IL-1β levels in the lungs of C57BL/6 mice infected with cowpox were observed [82], even though the studies in cell culture or in vitro indicate that CrmA is able to inhibit IL-1β maturation and secretion.

Additional pox viruses also encode SPI-2 homologues. Serp2 is the SPI-2 homologue found in myxoma virus [83]. Deletion of Serp2 results in severe attenuation of the virus in rabbits [84]. Some vaccinia viruses (VV) also encode a SPI-2 protein; however, deletion of VV SPI-2 did not reduce the IL-1β-induced fever seen in mice and VV SPI-2 mutants were not attenuated [85, 86]. Instead, fever reduction and weight loss were dependent on VV-encoded IL-1β scavenger receptor (vIL-1βR) [85]. In addition, several poxviruses possess an IL-18 binding protein that blocks IL-18 signaling [87]. It is apparent that poxviruses have a variety of inhibitors for inflammasomes and downstream IL-1β and IL-18 signaling. The differential requirement for SPI-2 family proteins may, therefore, be the result of differentially encoded additional inhibitors or the result of different natural host ranges of these viruses, as rodents are the natural host for cowpox but not VV [88].

Inflammasome inhibitors are not limited to poxviruses though. Influenza A/PR/8/34 H1N1 virus NS1 (PR/8 NS1) protein is also capable of blocking caspase-1 activation, apoptosis, and IL-1β maturation [89]. The N-terminus of PR/8 NS1 possesses an inflammasome inhibitory function and influenza A/PR/8/34 H1N1 virus lacking the N-terminus of NS1 is attenuated in cell culture and induces higher levels of IL-1β and apoptosis. Inhibitors of caspases reduce IL-1β maturation and are able to compensate for PR/8 NS1. Interestingly, inhibition of the dsRNA kinase, PKR, also reduces IL-1β maturation in PR/8 NS1 mutants. Though these results do not provide a clear inhibitory mechanism for PR/8 NS1, they do demonstrate the importance of this protein for virus replication. However, the ability of NS1 to block caspase-1 appears to be strain specific as NS1 from highly pathogenic H5N1 bird flu activates caspases and induces apoptosis [90]. One possible explanation for the ability of H5N1 virus to grow despite activating caspases is that it downregulates NLRP3 inflammasome components upon infection [91] and therefore may not require inhibition of caspase-1 to suppress inflammation. On the other hand, PR/8 influenza has been shown to upregulate inflammasome components [13] and may therefore require NS1 inhibition of caspase-1 to prevent inflammation and replicate successfully.

7 Conclusions

It is evident from the information presented here that inflammasome signaling is necessary for innate inflammatory control of virus infection as well as healing and adaptive immune responses. In addition, the presence of multiple viruses that encode inflammasome inhibitory proteins speaks to the effectiveness of this pathway at inhibiting virus replication. However, inflammasome biology is still in its infancy, as demonstrated by the fact that in recent years the field of inflammasomes has grown to include RIG-I (RLR family) and AIM2 (PYHIN family), as well as members of the NLR family. As research continues into the role of inflammasomes in antiviral immunity, it is likely that the list of viruses that activate and suppress inflammasome signaling will grow.

Acknowledgments The authors' own work is supported by NIH (AR056296), a Cancer Center Support Grant (CCSG 2 P30 CA 21765), Centers of Excellence for Influenza Research and Surveillance (CEIRS) project, and the American Lebanese Syrian Associated Charities (to T-D. K).

References

1. Creagh EM, O'Neill LA (2006) TLRs, NLRs and RLRs: a trinity of pathogen sensors that co-operate in innate immunity. Trends Immunol 27:352–357
2. Takeuchi O, Akira S (2009) Innate immunity to virus infection. Immunol Rev 227:75–86
3. Wilkins C, Gale M Jr (2010) Recognition of viruses by cytoplasmic sensors. Curr Opin Immunol 22(1):41–47
4. Yang X, Chang HY, Baltimore D (1998) Autoproteolytic activation of pro-caspases by oligomerization. Mol Cell 1:319–325
5. Srinivasula SM, Poyet JL, Razmara M, Datta P, Zhang Z, Alnemri ES (2002) The PYRIN-CARD protein ASC is an activating adaptor for caspase-1. J Biol Chem 277:21119–21122
6. Kanneganti TD, Lamkanfi M, Nunez G (2007) Intracellular NOD-like receptors in host defense and disease. Immunity 27:549–559
7. Lamkanfi M, Dixit VM (2009) Inflammasomes: guardians of cytosolic sanctity. Immunol Rev 227:95–105
8. Lamkanfi M, Kanneganti TD, Franchi L, Nunez G (2007) Caspase-1 inflammasomes in infection and inflammation. J Leukoc Biol 82:220–225
9. Petrilli V, Dostert C, Muruve DA, Tschopp J (2007) The inflammasome: a danger sensing complex triggering innate immunity. Curr Opin Immunol 19:615–622
10. Black RA, Kronheim SR, Merriam JE, March CJ, Hopp TP (1989) A pre-aspartate-specific protease from human leukocytes that cleaves pro-interleukin-1 beta. J Biol Chem 264: 5323–5326
11. Pirhonen J, Sareneva T, Kurimoto M, Julkunen I, Matikainen S (1999) Virus infection activates IL-1 beta and IL-18 production in human macrophages by a caspase-1-dependent pathway. J Immunol 162:7322–7329
12. Ray CA, Black RA, Kronheim SR, Greenstreet TA, Sleath PR, Salvesen GS, Pickup DJ (1992) Viral inhibition of inflammation: cowpox virus encodes an inhibitor of the interleukin-1 beta converting enzyme. Cell 69:597–604
13. Allen IC, Scull MA, Moore CB, Holl EK, McElvania-TeKippe E, Taxman DJ, Guthrie EH, Pickles RJ, Ting JP (2009) The NLRP3 inflammasome mediates in vivo innate immunity to influenza A virus through recognition of viral RNA. Immunity 30:556–565

14. Thomas PG, Dash P, Aldridge JR Jr, Ellebedy AH, Reynolds C, Funk AJ, Martin WJ, Lamkanfi M, Webby RJ, Boyd KL et al (2009) The intracellular sensor NLRP3 mediates key innate and healing responses to influenza A virus via the regulation of caspase-1. Immunity 30:566–575
15. Dinarello CA, Novick D, Puren AJ, Fantuzzi G, Shapiro L, Muhl H, Yoon DY, Reznikov LL, Kim SH, Rubinstein M (1998) Overview of interleukin-18: more than an interferon-gamma inducing factor. J Leukoc Biol 63:658–664
16. Fernandes-Alnemri T, Wu J, Yu JW, Datta P, Miller B, Jankowski W, Rosenberg S, Zhang J, Alnemri ES (2007) The pyroptosome: a supramolecular assembly of ASC dimers mediating inflammatory cell death via caspase-1 activation. Cell Death Differ 14:1590–1604
17. Miura M, Zhu H, Rotello R, Hartwieg EA, Yuan J (1993) Induction of apoptosis in fibroblasts by IL-1 beta-converting enzyme, a mammalian homolog of the C. elegans cell death gene ced-3. Cell 75:653–660
18. Pirhonen J, Sareneva T, Julkunen I, Matikainen S (2001) Virus infection induces proteolytic processing of IL-18 in human macrophages via caspase-1 and caspase-3 activation. Eur J Immunol 31:726–733
19. Kanneganti TD, Body-Malapel M, Amer A, Park JH, Whitfield J, Franchi L, Taraporewala ZF, Miller D, Patton JT, Inohara N et al (2006) Critical role for Cryopyrin/Nalp3 in activation of caspase-1 in response to viral infection and double-stranded RNA. J Biol Chem 281:36560–36568
20. Ichinohe T, Lee HK, Ogura Y, Flavell R, Iwasaki A (2009) Inflammasome recognition of influenza virus is essential for adaptive immune responses. J Exp Med 206:79–87
21. Delaloye J, Roger T, Steiner-Tardivel QG, Le Roy D, Knaup Reymond M, Akira S, Petrilli V, Gomez CE, Perdiguero B, Tschopp J et al (2009) Innate immune sensing of modified vaccinia virus Ankara (MVA) is mediated by TLR2-TLR6, MDA-5 and the NALP3 inflammasome. PLoS Pathog 5:e1000480
22. Poeck H, Bscheider M, Gross O, Finger K, Roth S, Rebsamen M, Hannesschlager N, Schlee M, Rothenfusser S, Barchet W et al (2009) Recognition of RNA virus by RIG-I results in activation of CARD9 and inflammasome signaling for interleukin 1 beta production. Nat Immunol 11:63–69
23. Schmitz N, Kurrer M, Bachmann MF, Kopf M (2005) Interleukin-1 is responsible for acute lung immunopathology but increases survival of respiratory influenza virus infection. J Virol 79:6441–6448
24. Muruve DA, Petrilli V, Zaiss AK, White LR, Clark SA, Ross PJ, Parks RJ, Tschopp J (2008) The inflammasome recognizes cytosolic microbial and host DNA and triggers an innate immune response. Nature 452:103–107
25. Benko S, Philpott DJ, Girardin SE (2008) The microbial and danger signals that activate Nod-like receptors. Cytokine 43:368–373
26. Cassel SL, Joly S, Sutterwala FS (2009) The NLRP3 inflammasome: a sensor of immune danger signals. Semin Immunol 21:194–198
27. Bryant C, Fitzgerald KA (2009) Molecular mechanisms involved in inflammasome activation. Trends Cell Biol 19:455–464
28. Dostert C, Petrilli V, Van Bruggen R, Steele C, Mossman BT, Tschopp J (2008) Innate immune activation through Nalp3 inflammasome sensing of asbestos and silica. Science 320:674–677
29. Corasaniti MT, Strongoli MC, Piccirilli S, Nistico R, Costa A, Bilotta A, Turano P, Finazzi-Agro A, Bagetta G (2000) Apoptosis induced by gp120 in the neocortex of rat involves enhanced expression of cyclooxygenase type 2 and is prevented by NMDA receptor antagonists and by the 21-aminosteroid U-74389G. Biochem Biophys Res Commun 274: 664–669
30. Franchi L, Kanneganti TD, Dubyak GR, Nunez G (2007) Differential requirement of P2X7 receptor and intracellular K+ for caspase-1 activation induced by intracellular and extracellular bacteria. J Biol Chem 282:18810–18818

31. Petrilli V, Papin S, Dostert C, Mayor A, Martinon F, Tschopp J (2007) Activation of the NALP3 inflammasome is triggered by low intracellular potassium concentration. Cell Death Differ 14:1583–1589
32. Halle A, Hornung V, Petzold GC, Stewart CR, Monks BG, Reinheckel T, Fitzgerald KA, Latz E, Moore KJ, Golenbock DT (2008) The NALP3 inflammasome is involved in the innate immune response to amyloid-beta. Nat Immunol 9:857–865
33. Hornung V, Bauernfeind F, Halle A, Samstad EO, Kono H, Rock KL, Fitzgerald KA, Latz E (2008) Silica crystals and aluminum salts activate the NALP3 inflammasome through phagosomal destabilization. Nat Immunol 9:847–856
34. Martinon F, Petrilli V, Mayor A, Tardivel A, Tschopp J (2006) Gout-associated uric acid crystals activate the NALP3 inflammasome. Nature 440:237–241
35. Zhou R, Tardivel A, Thorens B, Choi I, Tschopp J (2010) Thioredoxin-interacting protein links oxidative stress to inflammasome activation. Nat Immunol 11:136–140
36. Imamura M, Tsutsui H, Yasuda K, Uchiyama R, Yumikura-Futatsugi S, Mitani K, Hayashi S, Akira S, Taniguchi S, Van Rooijen N et al (2009) Contribution of TIR domain-containing adapter inducing IFN-beta-mediated IL-18 release to LPS-induced liver injury in mice. J Hepatol 51:333–341
37. Kanneganti TD, Ozoren N, Body-Malapel M, Amer A, Park JH, Franchi L, Whitfield J, Barchet W, Colonna M, Vandenabeele P et al (2006) Bacterial RNA and small antiviral compounds activate caspase-1 through cryopyrin/Nalp3. Nature 440:233–236
38. Maitra R, Clement CC, Scharf B, Crisi GM, Chitta S, Paget D, Purdue PE, Cobelli N, Santambrogio L (2009) Endosomal damage and TLR2 mediated inflammasome activation by alkane particles in the generation of aseptic osteolysis. Mol Immunol 47:175–184
39. Meylan E, Tschopp J, Karin M (2006) Intracellular pattern recognition receptors in the host response. Nature 442:39–44
40. Ogura Y, Sutterwala FS, Flavell RA (2006) The inflammasome: first line of the immune response to cell stress. Cell 126:659–662
41. Kanneganti TD, Lamkanfi M, Kim YG, Chen G, Park JH, Franchi L, Vandenabeele P, Nunez G (2007) Pannexin-1-mediated recognition of bacterial molecules activates the cryopyrin inflammasome independent of Toll-like receptor signaling. Immunity 26:433–443
42. Pelegrin P, Surprenant A (2006) Pannexin-1 mediates large pore formation and interleukin-1beta release by the ATP-gated P2X7 receptor. EMBO J 25:5071–5082
43. Piccini A, Carta S, Tassi S, Lasiglie D, Fossati G, Rubartelli A (2008) ATP is released by monocytes stimulated with pathogen-sensing receptor ligands and induces IL-1beta and IL-18 secretion in an autocrine way. Proc Natl Acad Sci USA 105:8067–8072
44. Schroder K, Zhou R, Tschopp J (2010) The NLRP3 inflammasome: a sensor for metabolic danger? Science 327:296–300
45. Guillot L, Le Goffic R, Bloch S, Escriou N, Akira S, Chignard M, Si-Tahar M (2005) Involvement of toll-like receptor 3 in the immune response of lung epithelial cells to double-stranded RNA and influenza A virus. J Biol Chem 280:5571–5580
46. Koyama S, Ishii KJ, Kumar H, Tanimoto T, Coban C, Uematsu S, Kawai T, Akira S (2007) Differential role of TLR- and RLR-signaling in the immune responses to influenza A virus infection and vaccination. J Immunol 179:4711–4720
47. Heer AK, Shamshiev A, Donda A, Uematsu S, Akira S, Kopf M, Marsland BJ (2007) TLR signaling fine-tunes anti-influenza B cell responses without regulating effector T cell responses. J Immunol 178:2182–2191
48. Le Goffic R, Balloy V, Lagranderie M, Alexopoulou L, Escriou N, Flavell R, Chignard M, Si-Tahar M (2006) Detrimental contribution of the Toll-like receptor (TLR)3 to influenza A virus-induced acute pneumonia. PLoS Pathog 2:e53
49. Takaoka A, Wang Z, Choi MK, Yanai H, Negishi H, Ban T, Lu Y, Miyagishi M, Kodama T, Honda K et al (2007) DAI (DLM-1/ZBP1) is a cytosolic DNA sensor and an activator of innate immune response. Nature 448:501–505

50. Wang Z, Choi MK, Ban T, Yanai H, Negishi H, Lu Y, Tamura T, Takaoka A, Nishikura K, Taniguchi T (2008) Regulation of innate immune responses by DAI (DLM-1/ZBP1) and other DNA-sensing molecules. Proc Natl Acad Sci USA 105:5477–5482
51. Burckstummer T, Baumann C, Bluml S, Dixit E, Durnberger G, Jahn H, Planyavsky M, Bilban M, Colinge J, Bennett KL et al (2009) An orthogonal proteomic-genomic screen identifies AIM2 as a cytoplasmic DNA sensor for the inflammasome. Nat Immunol 10:266–272
52. Fernandes-Alnemri T, Yu JW, Datta P, Wu J, Alnemri ES (2009) AIM2 activates the inflammasome and cell death in response to cytoplasmic DNA. Nature 458:509–513
53. Hornung V, Ablasser A, Charrel-Dennis M, Bauernfeind F, Horvath G, Caffrey DR, Latz E, Fitzgerald KA (2009) AIM2 recognizes cytosolic dsDNA and forms a caspase-1-activating inflammasome with ASC. Nature 458:514–518
54. Schroder K, Muruve DA, Tschopp J (2009) Innate immunity: cytoplasmic DNA sensing by the AIM2 inflammasome. Curr Biol 19:R262–R265
55. Schlee M, Hartmann E, Coch C, Wimmenauer V, Janke M, Barchet W, Hartmann G (2009) Approaching the RNA ligand for RIG-I? Immunol Rev 227:66–74
56. Yoneyama M, Fujita T (2009) RNA recognition and signal transduction by RIG-I-like receptors. Immunol Rev 227:54–65
57. Rintahaka J, Wiik D, Kovanen PE, Alenius H, Matikainen S (2008) Cytosolic antiviral RNA recognition pathway activates caspases 1 and 3. J Immunol 180:1749–1757
58. Bagetta G, Corasaniti MT, Berliocchi L, Nistico R, Giammarioli AM, Malorni W, Aloe L, Finazzi-Agro A (1999) Involvement of interleukin-1beta in the mechanism of human immunodeficiency virus type 1 (HIV-1) recombinant protein gp120-induced apoptosis in the neocortex of rat. Neuroscience 89:1051–1066
59. Corasaniti MT, Bilotta A, Strongoli MC, Navarra M, Bagetta G, Di Renzo G (2001) HIV-1 coat protein gp120 stimulates interleukin-1beta secretion from human neuroblastoma cells: evidence for a role in the mechanism of cell death. Br J Pharmacol 134:1344–1350
60. Corasaniti MT, Turano P, Bilotta A, Malorni W, Stringaro AR, Nistico R, Finazzi-Agro A, Bagetta G (2001) Evidence that increases of mitochondrial immunoreactive IL-1beta by HIV-1 gp120 implicate in situ cleavage of pro-IL-1beta in the neocortex of rat. J Neurochem 78:611–618
61. Corasaniti MT, Russo R, Amantea D, Gliozzi M, Siviglia E, Stringaro AR, Malorni W, Melino G, Bagetta G (2005) Neuroprotection by the caspase-1 inhibitor Ac-YVAD-(acyloxy)mk in experimental neuroAIDS is independent from IL-1beta generation. Cell Death Differ 12(Suppl 1):999–1001
62. Corasaniti MT, Bellizzi C, Russo R, Colica C, Amantea D, Di Renzo G (2003) Caspase-1 inhibitors abolish deleterious enhancement of COX-2 expression induced by HIV-1 gp120 in human neuroblastoma cells. Toxicol Lett 139:213–219
63. Corasaniti MT, Melino G, Navarra M, Garaci E, Finazzi-Agro A, Nistico G (1995) Death of cultured human neuroblastoma cells induced by HIV-1 gp120 is prevented by NMDA receptor antagonists and inhibitors of nitric oxide and cyclooxygenase. Neurodegeneration 4:315–321
64. Catani MV, Corasaniti MT, Navarra M, Nistico G, Finazzi-Agro A, Melino G (2000) gp120 induces cell death in human neuroblastoma cells through the CXCR4 and CCR5 chemokine receptors. J Neurochem 74:2373–2379
65. Corasaniti MT, Piccirilli S, Paoletti A, Nistico R, Stringaro A, Malorni W, Finazzi-Agro A, Bagetta G (2001) Evidence that the HIV-1 coat protein gp120 causes neuronal apoptosis in the neocortex of rat via a mechanism involving CXCR4 chemokine receptor. Neurosci Lett 312:67–70
66. Sergerie Y, Rivest S, Boivin G (2007) Tumor necrosis factor-alpha and interleukin-1 beta play a critical role in the resistance against lethal herpes simplex virus encephalitis. J Infect Dis 196:853–860

67. Krzyzowska M, Cymerys J, Winnicka A, Niemialtowski M (2006) Involvement of Fas and FasL in Ectromelia virus-induced apoptosis in mouse brain. Virus Res 115:141–149
68. Das S, Mishra MK, Ghosh J, Basu A (2008) Japanese Encephalitis Virus infection induces IL-18 and IL-1beta in microglia and astrocytes: correlation with in vitro cytokine responsiveness of glial cells and subsequent neuronal death. J Neuroimmunol 195:60–72
69. Byrne SN, Halliday GM, Johnston LJ, King NJ (2001) Interleukin-1beta but not tumor necrosis factor is involved in West Nile virus-induced Langerhans cell migration from the skin in C57BL/6 mice. J Invest Dermatol 117:702–709
70. Manigold T, Bocker U, Chen J, Gundt J, Traber P, Singer MV, Rossol S (2003) Hepatitis B core antigen is a potent inductor of interleukin-18 in peripheral blood mononuclear cells of healthy controls and patients with hepatitis B infection. J Med Virol 71:31–40
71. Takeuchi R, Tsutsumi H, Osaki M, Sone S, Imai S, Chiba S (1998) Respiratory syncytial virus infection of neonatal monocytes stimulates synthesis of interferon regulatory factor 1 and interleukin-1beta (IL-1beta)-converting enzyme and secretion of IL-1beta. J Virol 72:837–840
72. Johnston JB, Rahman MM, McFadden G (2007) Strategies that modulate inflammasomes: insights from host-pathogen interactions. Semin Immunopathol 29:261–274
73. Seet BT, Johnston JB, Brunetti CR, Barrett JW, Everett H, Cameron C, Sypula J, Nazarian SH, Lucas A, McFadden G (2003) Poxviruses and immune evasion. Annu Rev Immunol 21: 377–423
74. Dorfleutner A, Talbott SJ, Bryan NB, Funya KN, Rellick SL, Reed JC, Shi X, Rojanasakul Y, Flynn DC, Stehlik C (2007) A Shope Fibroma virus PYRIN-only protein modulates the host immune response. Virus Genes 35:685–694
75. Johnston JB, Barrett JW, Nazarian SH, Goodwin M, Ricciuto D, Wang G, McFadden G (2005) A poxvirus-encoded pyrin domain protein interacts with ASC-1 to inhibit host inflammatory and apoptotic responses to infection. Immunity 23:587–598
76. Benedict CA, Ware CF (2005) Poxviruses aren't stuPYD. Immunity 23:553–555
77. Qu Y, Ramachandra L, Mohr S, Franchi L, Harding CV, Nunez G, Dubyak GR (2009) P2X7 receptor-stimulated secretion of MHC class II-containing exosomes requires the ASC/NLRP3 inflammasome but is independent of caspase-1. J Immunol 182:5052–5062
78. Willingham SB, Allen IC, Bergstralh DT, Brickey WJ, Huang MT, Taxman DJ, Duncan JA, Ting JP (2009) NLRP3 (NALP3, Cryopyrin) facilitates in vivo caspase-1 activation, necrosis, and HMGB1 release via inflammasome-dependent and -independent pathways. J Immunol 183:2008–2015
79. Willingham SB, Bergstralh DT, O'Connor W, Morrison AC, Taxman DJ, Duncan JA, Barnoy S, Venkatesan MM, Flavell RA, Deshmukh M et al (2007) Microbial pathogen-induced necrotic cell death mediated by the inflammasome components CIAS1/cryopyrin/NLRP3 and ASC. Cell Host Microbe 2:147–159
80. Kolly L, Karababa M, Joosten LA, Narayan S, Salvi R, Petrilli V, Tschopp J, van den Berg WB, So AK, Busso N (2009) Inflammatory role of ASC in antigen-induced arthritis is independent of caspase-1, NALP-3, and IPAF. J Immunol 183:4003–4012
81. Komiyama T, Ray CA, Pickup DJ, Howard AD, Thornberry NA, Peterson EP, Salvesen G (1994) Inhibition of interleukin-1 beta converting enzyme by the cowpox virus serpin CrmA. An example of cross-class inhibition. J Biol Chem 269:19331–19337
82. MacNeill AL, Moldawer LL, Moyer RW (2009) The role of the cowpox virus crmA gene during intratracheal and intradermal infection of C57BL/6 mice. Virology 384:151–160
83. Petit F, Bertagnoli S, Gelfi J, Fassy F, Boucraut-Baralon C, Milon A (1996) Characterization of a myxoma virus-encoded serpin-like protein with activity against interleukin-1 beta-converting enzyme. J Virol 70:5860–5866
84. Messud-Petit F, Gelfi J, Delverdier M, Amardeilh MF, Py R, Sutter G, Bertagnoli S (1998) Serp2, an inhibitor of the interleukin-1beta-converting enzyme, is critical in the pathobiology of myxoma virus. J Virol 72:7830–7839

85. Kettle S, Alcami A, Khanna A, Ehret R, Jassoy C, Smith GL (1997) Vaccinia virus serpin B13R (SPI-2) inhibits interleukin-1beta-converting enzyme and protects virus-infected cells from TNF- and Fas-mediated apoptosis, but does not prevent IL-1beta-induced fever. J Gen Virol 78(Pt 3):677–685
86. Kettle S, Blake NW, Law KM, Smith GL (1995) Vaccinia virus serpins B13R (SPI-2) and B22R (SPI-1) encode M(r) 38.5 and 40K, intracellular polypeptides that do not affect virus virulence in a murine intranasal model. Virology 206:136–147
87. Smith VP, Bryant NA, Alcami A (2000) Ectromelia, vaccinia and cowpox viruses encode secreted interleukin-18-binding proteins. J Gen Virol 81:1223–1230
88. Chantrey J, Meyer H, Baxby D, Begon M, Bown KJ, Hazel SM, Jones T, Montgomery WI, Bennett M (1999) Cowpox: reservoir hosts and geographic range. Epidemiol Infect 122: 455–460
89. Stasakova J, Ferko B, Kittel C, Sereinig S, Romanova J, Katinger H, Egorov A (2005) Influenza A mutant viruses with altered NS1 protein function provoke caspase-1 activation in primary human macrophages, resulting in fast apoptosis and release of high levels of interleukins 1beta and 18. J Gen Virol 86:185–195
90. Lam WY, Tang JW, Yeung AC, Chiu LC, Sung JJ, Chan PK (2008) Avian influenza virus A/HK/483/97(H5N1) NS1 protein induces apoptosis in human airway epithelial cells. J Virol 82:2741–2751
91. Cilloniz C, Shinya K, Peng X, Korth MJ, Proll SC, Aicher LD, Carter VS, Chang JH, Kobasa D, Feldmann F et al (2009) Lethal influenza virus infection in macaques is associated with early dysregulation of inflammatory related genes. PLoS Pathog 5:e1000604

Inflammasome Activation in Response to Eukaryotic Pathogens

Olaf Gross, Christina J. Thomas, and Laura E. Layland

Abstract Inflammasomes are multi-protein complexes that serve as platforms for caspase-1 activation and subsequent proteolytic maturation of interleukin 1β (IL-1β) within innate immune cells. The Nlrp3 inflammasome is the most fully characterised. It is activated by various endogenous and exogenous danger signals such as environmental irritants, signals of tissue damage and pathogens. The broad spectrum of activators is reflected at the physiological level in its implication in normal and dysregulated immune responses, including various autoinflammatory diseases and the defence against numerous pathogens. Here, we summarise the present data on the activation of the Nlrp3 inflammasome by eukaryotic pathogens. Recent genetic studies using mice deficient in inflammasome components demonstrate the involvement of the inflammasome in the outcome of infection with the fungus *Candida albicans*, the helminth *Schistosoma mansoni*, as well as the malarial parasite *Plasmodium berghei*. Altered immune responses were respectively linked to the ability of live fungi, schistosomal egg antigen (SEA) or malarial hemozoin to activate the inflammasome and induce secretion of mature IL-1β. These initial findings suggest that inflammasome activation may serve as a common and potentially druggable pathway in the defence against eukaryotic pathogens.

O. Gross (✉) • C.J. Thomas
Département de Biochimie, Université de Lausanne, Chemin des Boveresses 155, 1066 Epalinges, Switzerland
e-mail: ogross4926@lrz.tum.de

L.E. Layland
Institut für Medizinische Mikrobiologie, Immunologie und Hygiene, Technische Universität München, 81675 Munich, Germany
and
Institut für Medizinische Mikrobiologie, Immunologie und Parasitologie, Universität Bonn, 53127 Bonn, Germany

I. Couillin et al. (eds.), *The Inflammasomes*, Progress in Inflammation Research,
DOI 10.1007/978-3-0348-0148-5_5, © Springer Basel AG 2011

1 Introduction

Since their first description in 2002 [1], many studies have demonstrated the involvement of inflammasomes in normal and pathological immune function. Because dominant gain-of-function mutations in Nlrp3 were associated with hereditary periodic fever syndromes, early studies characterised the role of the inflammasome in autoinflammatory diseases [2–4]. Particle-induced sterile inflammation has also been linked to inflammasome action. Host-derived particles such as monosodium urate (MSU) crystals (which are involved in the pathogenesis of gout [5]), as well as environmental and industrial particles such as asbestos and silica [6] (which induce lung inflammation upon inhalation) are examples of particulate agents that, at least in part, perform their pathogenic action by triggering the inflammasome. New evidence is accumulating that implicates inflammasome activity in other chronic inflammatory processes such as the development of cancer [7] and metabolic (and inflammatory) diseases like type 2 diabetes [8].

In a second wave of research, inflammasomes were identified as important players in the immune activation against bacterial pathogens, primarily by inducing local inflammation and fever. *Salmonella*, *Shigella* [9], *Legionella* [10] and *Francisella,* and a growing list of other bacterial species have all been shown to activate the inflammasome [7]. Since these topics are focal points within other chapters of this issue we shall not discuss them here in further detail. Recently, viruses and eukaryotic pathogens have been added to the list of inflammasome activators. While the previous chapter focuses on viruses, we here will discuss the available data on eukaryotes including fungi, protozoa and helminths.

Inflammasomes are cytoplasmic multi-protein complexes that contain and activate caspase-1. They are present in innate immune cells of the myeloid linage and are best characterised in dendritic cells and macrophages. Minimally, inflammasomes contain an oligomerising sensor/receptor protein connected to caspase-1 through homotypic protein interactions [11]. In most cases, this connection is mediated by the adaptor protein Asc (Pycard) [12]. Among the described cytoplasmic sensors are proteins of the NOD-like receptors (NLR) superfamily including Nlrp1, Nlrp3, Nlrc4, Nod1 or Nod2 [13]; receptors for viral RNA (Rig-I and Mda5 [14]); and the cytoplasmic DNA receptor Aim2 [15–18]. The Nlrp3 (Cryopyrin/Nalp3/Cias1/Pypaf1) inflammasome is currently the most fully characterised.

Via upstream pathways that are not fully understood, the Nlrp3 inflammasome senses signals of cellular stress and drives IL-1β-mediated inflammation in response to certain pathogens, environmental irritants, and signals or tissue damage or metabolic stress. Though the precise mechanism of Nlrp3 inflammasome activation remains a matter of debate, the plethora of activating stimuli has given insight to some common pathways that may be involved. The cumulative literature suggests that potassium efflux, reactive oxygen species (ROS) generation and lysosomal damage serve as proximal signals for inflammasome activation, either independently or in concert with one another. Other chapers of this issue give a detailed description of the mechanisms implicated here. These signals are believed to be integrated by the C-terminal LRR-domain of Nlrp3, which normally maintains the protein in an autorepressed state [1, 19, 20]. Upon signal detection, LRR-mediated autorepression is relieved, and the PYD and NACHT domains are exposed. This allows homotypic interactions between the NACHT domains leading to oligomerisation of Nlrp3 and recruitment of the PYD- and CARD-domain containing adaptor protein Asc and caspase-1 (Fig. 2). Clustering of caspase-1 leads to its autocatalytic activation, and active caspase-1 proteolytically cleaves a variety of cellular targets, including the IL-1 family members IL-1β, IL-18, IL-1F7 and potentially IL-33. These cytokines are produced in an inactive pro-form and require cleavage for activation [21]. Bioactive IL-1β is a highly potent pro-inflammatory cytokine and pyrogen, which orchestrates local and systemic effects of inflammasome activation [11].

While caspase-1 and Asc are constitutively expressed in dendritic cells and macrophages, Nlrp3 and pro-IL-1β are expressed at a very low level or not at all. The latter two must be upregulated in a process called priming, in which activation of transcription factors such as NF-κB downstream of various pattern recognition receptors (PRRs) leads to augmentation of transcript levels [22]. Whereas some authors define priming as a part of inflammasome activation, others see it as a prerequisite. Within this review, we define inflammasome activation (or signal 2) as the process by which the inflammasome components are assembled leading to the cleavage and activation of caspase-1, while processes modulating gene expression of Nlrp3, pro-IL-1β, and potentially others are referred to as priming (or signal 1). Additionally, we see the subsequent export of both cleaved caspase-1 and IL-1β as another, principally independent process downstream of inflammasome activation.

2 Fungi

While aggressive fungal pathogens that infect healthy individuals are rare, there are many opportunistic fungi such as *Candida albicans*, *Aspergillus fumigatus* or *Pneumocystis jirovecii* that colonise immunocompromised patients. The most widely studied fungal pathogen is *C. albicans*. In its milder form, infections with *Candida* species mainly affect mucocutaneous tissues causing thrush and onychomycosis in susceptible hosts such as patients with diabetes mellitus, Job's hyper-IgE syndrome or APECED (autoimmune polyendocrinopathy–candidiasis –ectodermal dystrophy) syndrome [23, 24]. However, in severely immunocompromised individuals, such as persons with AIDS or those undergoing immunosuppressive regimes for organ transplantation or cancer treatment, the organism can disseminate from its mucocutaneous niche, leading to sepsis and life-threatening systemic organ infections [23, 25, 26]. Several studies have indicated that IL-1β and IL-18 play a key role in the immune defence against *C. albicans* [27–33]. Until recently however, it was unclear what role, if any, the inflammasome played in anti-fungal immunity. Three groups have concurrently and independently reported that live *C. albicans* triggers Nlrp3 inflammasome activation [34–36]. They found a requirement for Nlrp3, Asc and caspase-1 for the cleavage and release of IL-1β and IL-18 in vitro. Furthermore, Nlrp3-deficient mice showed increased susceptibility to *C. albicans* infection, leading to higher fungal loads and increased mortality.

2.1 Receptors and Pathways Involved in Inflammasome Priming by Fungi

The classical PRR for fungal recognition is the C-type lectin-like receptor (CLR) Dectin-1 [37]. It recognises β-glucans, a prominent fungal cell wall component, and is involved in the phagocytosis of fungal cells, the production of ROS for fungal killing and the secretion of pro-inflammatory cytokines such as IL-6 and tumour necrosis factor (TNF). Dectin-1 utilises an immune-receptor tyrosine-based activation motif (ITAM) to couple to the kinase Syk for downstream signalling to NF-κB via Card9 [38, 39] (see Fig. 1). Other C-type lectin receptors like Dectin-2 and Mincle, which utilise the same downstream pathway as Dectin-1, have also been implicated in fungal recognition [40, 41].

Mice deficient in Dectin-1 [42] or Card9 [39] display higher susceptibility to fungal infection, leading to increased mortality. Genetic studies on families that suffer from chronic mucocutaneous candidiasis revealed an association with mutations in Dectin-1 (Y238X) [43] and in Card9 (Q295X) [44], demonstrating the applicability of these findings in mice to the human situation. Recently, it was found that Dectin-1 is required for inflammasome priming (pro-IL-1β production) upon stimulation with *C. albicans* [35], and this process was also found to be Syk- and Card9-dependent [21, 34].

Other studies have shown that TLR2 is also important for transmitting the priming signal downstream of fungal recognition. In addition to Dectin-1, TLR2 is also required for pro-IL-1β production by bone marrow-derived macrophages (BMDMs) upon stimulation with *C. albicans* [35]. In bone marrow dendritic cells (BMDCs) however, MyD88, a central downstream adaptor protein for TLR signalling, is largely dispensable for this process [34]. This discrepancy is likely explained by the use of different cell types, as fungal-initiated TLR2 signalling is required for NF-κB activation in macrophages, but not in dendritic cells [45]. In support of this explanation, the full requirement of TLR2 is only observed in in vitro experiments using BMDMs, while in an in vivo *Candida* infection model, IL-1β production is only partly dependent on TLR2, but fully dependent on Dectin-1 [35]. Experiments using blocking antibodies on human peripheral blood mononuclear cells (PBMCs) stimulated with fungi show that inhibition of Dectin-1 or TLR2, but not TLR4 or -9, leads to reduced IL-1β production, again demonstrating the relevance to the human system [46].

Additional TLRs aside from TLR2 may also be involved in providing a priming signal in response to fungi. Studies investigating the requirement of TLRs for inflammasome priming in response to the fungal cell wall components highlight

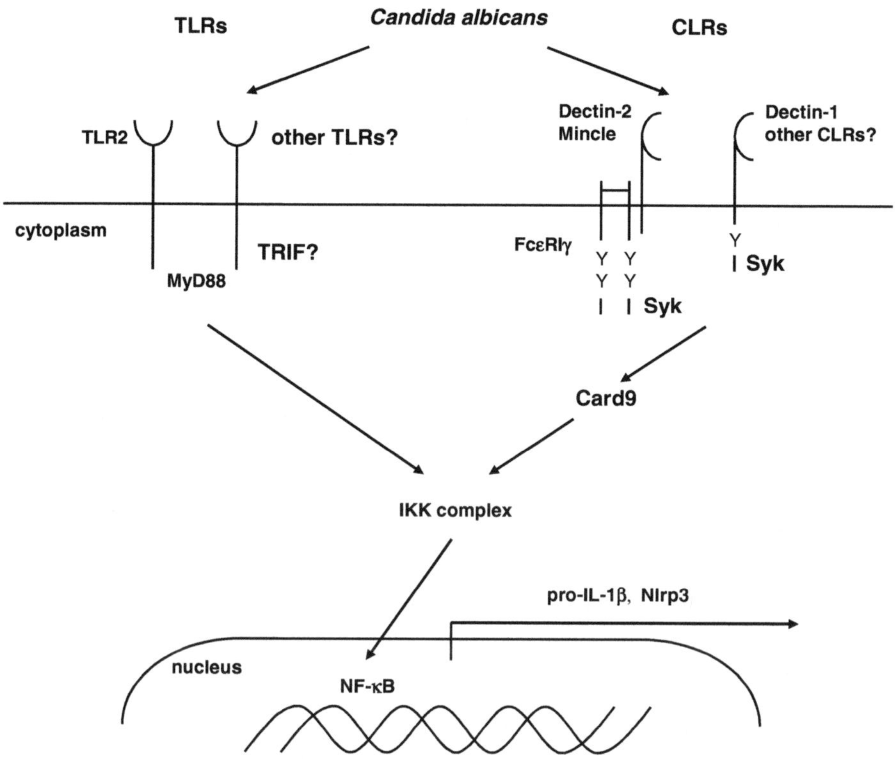

Fig. 1 Inflammasome priming by fungi. Pattern recognition receptors (PRRs) of the Toll-like receptor (TLR) and C-type lectin like receptor (CLR) families activate gene transcription of IL-1β and inflammasome components including Nlrp3 by activation of canonical NF-κB signalling using independent upstream pathways. *IKK* inhibitor of NF-κB kinase; *Y* immune receptor tyrosine-based activation motif, ITAM; for details, see text

some discrepancy in this field. Zymosan (a crude fungal cell wall preparation) and mannan are both capable of providing a strong priming signal that, in BMDMs, is fully MyD88- and partially TRIF-dependent. However, it is unclear which TLR(s) are upstream, as in this study, TLR2 was only partially required [47]. The lack of dependence on TLR2 might be explained by the use of fungal cell wall preparations like zymosan instead of inactivated but intact fungal cells. During the preparation of zymosan, some components that are recognised by certain TLRs may be excluded, while other TLR-stimulating compounds may become more accessible. For example, fungal DNA contamination could be recognised by endosomal TLR9 after uptake of zymosan. These findings underline the general limitations of using purified or fractionated stimuli, as compared to full pathogens. Though these reagents have been useful for determining the receptors and molecular pathways involved, one must exercise caution in extending these observations to infections with live pathogens. This notion becomes even more apparent when investigating inflammasome activation (signal 2).

2.2 Fungus-Induced Inflammasome Activation

The hallmark of inflammasome activation is the cleavage and subsequent secretion of both cleaved caspase-1 and cleaved IL-1β. Three studies have independently shown that stimulation of mouse BMDCs or BMDMs with *C. albicans* leads to the secretion of both factors within 2–6 h, as determined by the presence of the p10 and p20 cleavage fragments of caspase-1, and the mature IL-1β p17 fragment in the cell-free supernatant by Western blotting [34–36]. Genetic studies showed that proteolytic maturation of IL-1β in response to *Candida* is caspase-1-dependent, thereby suggesting the involvement of an inflammasome [34]. This was confirmed in all three studies by the observation that Nlrp3- and Asc-deficient cells fail to cleave and secrete caspase-1 and IL-1β after stimulation with *Candida*. This process was independent of all other tested sensors known to engage caspase-1. Furthermore, in an in vivo model of fungal sepsis, Nlrp3-deficient mice show ablation of serum IL-1β, increased fungal burden and increased mortality when compared to wild-type mice [34, 36], highlighting the relevance of the in vitro findings. Importantly, these reports were the first to demonstrate that eukaryotes can activate the Nlrp3 inflammasome.

In order to mimic the early stages of infection in humans and a more natural mode of entry, one of the aforementioned studies induced oral candidiasis in mice [35]. Using this model, infected Nlrp3-, Asc- and caspase-1-deficient mice showed slight increases in fungal burden of the tongues and increased clinical scores over a 3-week infection period. This result is in contrast to an older study in which the caspase-1 knockout mice, in a *C. albicans*-induced sepsis model, showed no difference in survival, but only a role for caspase-1-dependent IL-18 cleavage in the activation of T_H1 responses [28]. The differences may be due to the use of another *C. albicans* strain or a differential requirement for caspase-1 in Nlrp3-dependent IL-1β activation depending on the route of infection. More strikingly, in the oral infection system, all three knockout mice show dramatically increased dissemination, as measured by the fungal burden in the kidney, and also mortality. Interestingly, infected IL-1 receptor knockout mice, which were tested in parallel, show close resemblance to this phenotype, suggesting that the Nlrp3 inflammasome is the major factor for IL-1 production in this system. It remains unclear however, whether the increased fungal burdens in the kidneys occurred due to increased amounts of fungal cells reaching the bloodstream or simply attributed to impaired immune defences following the onset of sepsis. Nevertheless, these data clearly demonstrate that the Nlrp3 inflammasome is crucial in controlling local mucosal fungal infections, potentially preventing dissemination, as well as for the responses elicited during sepsis and the control of systemic infection. Indications for the applicability of these findings in humans come from a report associating a polymorphism in Nlrp3 with recurrent vulvovaginal candidiasis [48]. The authors correlate this phenotype with Nlrp3 expression in mucosal keratinocytes, which was previously demonstrated in another study [49]. Epithelial cells play an important role in orchestrating anti-fungal immunity to mucosal infections. Thus, keratinocyte inflammasome activation may be important for early onset anti-fungal immunity and especially for the prevention of chronic and disseminating infection. However, further study is required to conclusively demonstrate this role.

2.3 Molecular Mechanisms of Nlrp3-Inflammasome Activation in Response to Fungi

Potassium efflux, production of oxygen radicals (ROS), phagocytosis of particles and lysosomal damage have all been shown to be involved in Nlrp3 inflammasome activation, with different activators having differential requirements for each. With regards to fungi, inhibition of ROS, phagocytosis or potassium efflux prevents Nlrp3 inflammasome activation, while inhibition of lysosomal acidification or cathepsin-B deficiency has no effect [34]. However, the cathepsin-B inhibitor CA-074-Me has a mild reducing effect of *C. albicans*-induced inflammasome activation [36]. It remains to be established, whether this discrepancy is due to redundancy among cathepsins for inflammasome activation and/or off-target effects of the inhibitor.

As mentioned, receptor-proximal action of Syk and Card9 is required for priming after fungal recognition. To determine if these also play a role in inflammasome activation, one study used cells deficient for the factors. In addition for its role in inflammasome priming, Syk was found to be also essential for fungus-induced inflammasome activation, placing it at a unique crux in the regulation of the inflammasome [34]. Syk and its upstream receptor Dectin-1 are both required for ROS production after stimulation with fungal cells or with zymosan [50]. Therefore, the requirement of Syk for inflammasome activation seems, at least in part, to be a result of its activation of pathways leading to ROS production. In contrast, Card9 is only required for priming but dispensable for both ROS production and inflammasome activation (see Fig. 2).

Perhaps the most surprising finding with regards to inflammasome activation by *Candida* was that only live fungi can activate the inflammasome. Neither heat-killed *C. albicans*, nor preparations of the fungal cell wall components (zymosan or mannan) nor medium conditioned by live *C. albicans* was sufficient for inflammasome activation, even when cells were treated with LPS to provide a independent and strong priming signal [34, 36, 47]. *C. albicans* is a dimorphic fungus that can grow in both yeast and hyphal forms. As is the case for most pathogenic fungi, the ability to transition between these forms is crucial for virulence (in terms of its ability to establish chronic infection and to invade tissues) [51]. Sutterwala and colleagues hypothesised that the transition into the hyphal state was critical for inflammasome activation and proved this theory by using targeted mutations of *C. albicans* genes required for the development of hyphae. This leads to a potent reduction in their ability to activate the inflammasome, which correlates with data obtained using *Candida* species that naturally occur only in the yeast form

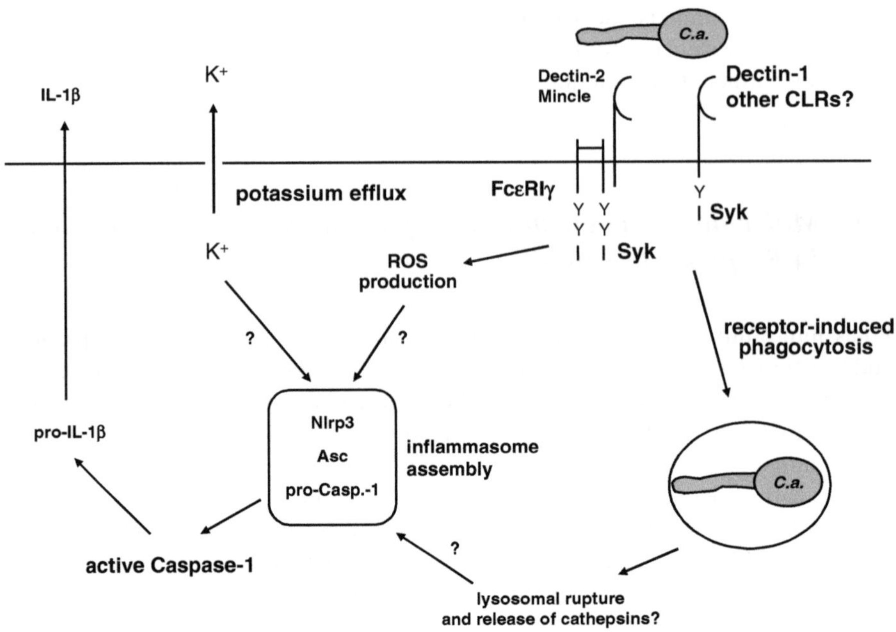

Fig. 2 Inflammasome activation by fungi. Several mechanisms have been implicated in fungal-induced inflammasome activation. Among them are receptor-mediated phagocytosis, production of ROS, and potassium (K+) efflux. If lysosomal damage and release of cathepsins into the cytoplasm are involved, the mechanisms by which this would occur remain unclear. These upstream mechanisms are envisioned to lead to the multimerisation of the inflammasome components Nlrp3, Asc and pro-caspase-1 and the activation of caspase-1 (Casp-1). Caspase-1 now can cleave pro-IL-1β where after IL-1β is secreted from the cells. *Question marks* indicated that, although an involvement was shown, the mechanistic details are unclear. *C.a.*, Candida albicans; for details, see text

[36]. Similarly, viable cells of the ahyphal yeast *Saccharomyces cerevisiae* only induce a mild inflammasome activation [34]. Under cell culture conditions, live *C. albicans* yeast cells begin transition into the hyphal stage after approximately 2–3 h [36], a timeframe exactly corresponding with the detectable inflammasome activation [34], again suggesting a connection between the two processes. Furthermore, by inactivating fungal cells at different hyphal germination stages using formalin instead of heat, Hise et al. could partially overcome the requirement of fungal viability for inflammasome activation [35], potentially due to increased lysosomal stress after uptake of the unwieldy hyphae, or even failure to take them up. To conclude, these data indicate that either heat inactivation destroys an inflammasome-activating component of the fungus or that only hyphal stages activate the inflammasome. Both could be true, as heat inactivated in contrast to life hyphae also lose their ability for inflammasome activation [34], while UV-irradiated yeast stages lose much but not all of their inflammasome activating potential [36]. To verify and further dissect these current findings, more experiments are required

which compare all stages of *C. albicans* and test viable as well as formalin, UV and heat inactivated as well as fungistatics-treated fungal cells and germination-deficient mutants within the same test parameters.

Regardless of the mechanism by which the immune system discriminates between a live and a dead fungus, it seems to be intuitive that a mechanism exists that enables the immune system to evaluate the level of danger posed by the organism. Dead fungal components or those derived from food or environmental non-pathogenic sources can enter the body via lesions but usually do not pose a threat and full immune activation upon recognition would be counterproductive. Only the entry of a viable pathogenic fungus that can infiltrate host tissues requires strong immune responses. Live and dead fungal cells or zymosan (derived from *S. cerevisiae*) induce comparable amounts of TNF, but only live pathogenic fungal cells producing hyphal stages after entering the host induce an inflammasome response. It therefore seems that inflammasome activation could provide the immune system with "information" about an elevated level of danger, which may elicit a full-blown local or systemic immune response. Future work elucidating the molecular basis of the mechanism(s) involved in inflammasome activation by live pathogenic fungi might also be applicable for other inflammasome stimuli and would help to understand inflammasome activation in general.

3 Malaria

Malaria is a tropical and subtropical infectious disease that affects several hundred million people and causes more than one million deaths each year [52]. It is caused by intracellular *Plasmodium* parasites that reside in erythrocytes and feeds on haemoglobin. The typical life history of each *Plasmodium* organism involves over two dozen morphological forms, but only six are commonly encountered during human infection. Several *Plasmodium* species cause malaria in humans and induce the classical symptoms such as periodic fever, chills, headache and anaemia. However, only the most severe form, caused by *P. falciparum,* can lead to organ damage and cerebral malaria. The murine equivalent to *P. falciparum* is *P. berghei* that, like the human form, can cause cerebral malaria involving micro-vascular damage, leukocyte infiltration and chronic inflammation in the brain and consequently death, presumably due to an excessive immune response. As a means to detoxify the free haeme liberated as it metabolises the erythrocyte's haemoglobin, *Plasmodium* produces insoluble hemozoin crystals. Hemozoin is released when the parasites break free of the erythrocytes and has been classified as a major causative agent of disease symptoms [52].

Several components of *Plasmodium* have been linked to the induction of pro-inflammatory cytokines after recognition by TLRs and CLRs [52]. Members of the TLR family, as well as their downstream signalling adapter MyD88, are involved in the production of pro-inflammatory cytokines such as TNF or IL-6 after stimulation of macrophages with *Plasmodium*-infected erythrocytes [53]. Both natural and synthetic preparations of hemozoin have been shown to activate TLR9 and TLR2 [54] and the latter also recognises *Plasmodium*-derived glycosylphosphatidylinositol [55]. In some but not all studies, a deficiency in certain TLRs or MyD88 has altered the outcome of malarial infection in mice in terms of survival and severity of cerebral malaria [53, 56]. TLR activity also plays a role in vaccination strategies. Vaccination with a whole-parasite preparation was TLR9 dependent. However, unlike MyD88-deficient mice, TLR9 knockouts did not display a defect in the activation of adaptive immune responses following immunisation using synthetic hemozoin as an adjuvant [54]. Though it has not been formally tested, it is conjectured that the aforementioned *Plasmodium* components or products that induce pro-inflammatory cytokines also induce pro-IL-1β production and thereby contribute to priming of the inflammasome.

While a role of IL-1β in anti-malarial immunity and malaria pathogenesis has been implicated, it is still controversial [57–61]. Nonetheless, the strong resemblance between hemozoin and MSU crystals [62] prompted researchers to investigate its potential to activate the inflammasome. Two studies have independently shown that synthetic hemozoin induces the release of cleaved caspase-1 and IL-1β in an Nlrp3- and Asc-dependent manner. In line with studies on other inflammasome-activating particles, inhibition of potassium efflux, ROS production or phagocytosis, but not cathepsin-B deficiency blocks inflammasome activation by hemozoin crystals [63]. In an effort to elucidate upstream pathways, Shio and colleagues showed that Syk and Src kinases (acting upstream of Syk) are also involved in hemozoin-induced inflammasome activation [64]. As fungi and recently MSU crystals have also been shown to activate Syk [34, 65, 66], this is the third report associating Syk signalling with inflammasome activation, implying that Syk might be a general player in particle-induced inflammasome activation. It remains unknown which, if any, Syk-coupled receptors recognise hemozoin, or whether like MSU, hemozoin triggers receptor-independent Syk activation by inducing the rearrangement of lipid rafts [65]. Pathways downstream of Syk, in addition to Card9/NF-κB, also involve PI3K and MAPK activation. Following hemozoin stimulation the latter two might contribute to inflammasome activation, as chemical inhibition of PI3K and the MAPK Erk results in reduction thereof. The authors of this study suggest a role for PI3K/Erk-dependent cathepsin-mediated activation of the inflammasome in the absence of lysosomal rupture, since hemozoin, unlike silica, does not induce lysosomal damage. It remains to be clarified how cathepsins contribute to inflammasome activation without being released to the cytoplasm or how they are released without lysosomal rupture [64, 67].

In an in vivo hemozoin-induced peritonitis mouse model, Nlrp3-deficiency results in reduced neutrophil influx into the peritoneal cavity after hemozoin injection [63, 68]. Moreover, *Plasmodium*-infected Nlrp3-deficient mice show significantly reduced IL-1β in the serum, and also less profound fever when compared to wild type [64]. Accordingly, two studies observed enhanced survival in malaria-infected mice deficient for components of the Nlrp3 inflammasome, suggesting a pathological role for the inflammasome during disease progression.

Interestingly, while one study using *Plasmodium chabaudi*, an organism that does not cause cerebral malaria, links this enhanced survival to reduced parasitemia in Nlrp3- or IL-1β-deficient animals [64], the other study, using *Plasmodium berghei,* associates the effect on survival with decreased symptoms of cerebral malaria, but reported unaltered parasitemia [63]. Using a similar model of *Plasmodium* infection, a third study did not see any differences in the survival rates of infected Nlrp3-deficent mice [69]. Although these differences could be explained by varying protocols, mouse strains or *Plasmodium* species, further research is required to firmly establish and specify the role of the inflammasome during malarial infections.

Since the in vitro-based data do agree that *Plasmodium*-derived stimuli can activate the inflammasome, this immune response may possibly be involved in inducing some of the symptoms experienced by infected patients. Given that it is well established that the host's dysregulated immune response contributes to the symptoms of severe malaria, treatment of patients with drugs interfering with the biological effects of IL-1, such as the anti-IL-1β monoclonal antibody *Canakinumab* or the IL-1 receptor antagonist (IL-1RA) *Anakinra* might be beneficial in reducing fever and even preventing the onset of cerebral malaria. In addition, these findings shed new light on the above-mentioned finding that MyD88, but not TLR9 or Nlrp3, is required for vaccination with hemozoin [54]. Since MyD88 is also mediating IL-1R signalling [21], hypothetically, both TLR9/MyD88- and IL-1R/MyD88-dependent signalling may be involved, and somewhat redundant, in the activation of adaptive immunity. This redundancy would be revealed only in the MyD88-deficient situation. Further research is required to clarify the relative contributions of TLRs, CLRs and the inflammasome in hemozoin-activated immune responses. It should also be tested if *Plasmodium* merozoites have an inflammasome-activation potential in addition to and independent of hemozoin.

4 Helminth Infection

Geohelminths infect one quarter of the world's population and the trematode genus *Schistosoma* is the fourth most prevalent afflicting over 200 million people worldwide. The ensuing immunopathological outcome of *S. mansoni* infection in man is mirrored in mouse. The life cycle of this parasitic worm is extremely complex. In brief, cercariae penetrate the mammalian host and migrate to the lung undergoing a transformation into schistosomula. After a further 5 weeks they develop into male and female juvenile worms that migrate to the portal venous system of liver, where they pair up and produce eggs. Eggs then penetrate into the intestinal lumen and are excreted with the stool. To complete the cycle, miracidia hatch from the eggs and enter the intermediate host snails. Astoundingly, with the exception of eggs, all other life stages are almost invisible to the host's immune system and antigen-specific immune responses are only detectable once eggs are shed from fecund females [70]. The major cause of morbidity actually results from $CD4^+$ T cell-mediated granulomas that develop in the liver and intestine due to eggs becoming trapped in the liver sinusoids and intestinal lamina propria. Interestingly, whereas acutely infected patients show strong T_H1 (IFN-γ) immune responses, those who progress into the chronic phase of infection develop stronger T_H2 responses (IL-5, IL-13, IL-4). In addition, although the onset of T_H1 responses coincides with the start of egg laying there is still no conclusive explanation as to why T_H2 responses become more dominant.

To investigate innate activation, as the first step towards T_H responses, the majority of in vitro studies have been performed with the soluble fraction of *Schistosoma* egg antigen (SEA), which is obtained after mechanically disrupting freshly isolated eggs from infected mouse or hamster livers. It is well established that SEA can inhibit the production of pro-inflammatory cytokines like TNF and IL-6 by dendritic cells or macrophages in response to TLR stimulation in vitro, but does not induce any cytokine production per se [71]. In contrast, in vivo studies using *S. mansoni*-infected mice deficient in TLRs or MyD88 show that these mice have altered immunopathology and antigen-specific immune responses upon T cell re-stimulation. For example, whereas infected MyD88-deficient mice fail to produce T_H1 responses and develop smaller granulomas, TLR2 knockouts have uncontrolled immunopathology and T_H responses [72, 73]. The latter phenotype was linked to the failure of helminth-specific regulatory T cells that expand during infection [74].

Despite the accumulation of data regarding the role of TLR triggering, the function of the inflammasome during this helminth infection has not been previously studied, even though it was shown that the endogenous IL-1 receptor antagonist (IL-1RA) could regulate granuloma development and influence regional cytokine production [75–77]. Very recently, a study has verified that SEA can indeed activate the Nlrp3 inflammasome [78]. Although SEA maintains its suppressive effect on TLR-triggered cytokine production from BMDCs, it simultaneously induces the production of IL-1β. This effect was only observed if BMDCs were primed with TLR ligands for several hours prior to SEA administration. Such requirements for priming have been shown for many other pure inflammasome stimuli like ATP that fail to activate a priming signal by themselves. Under such conditions, SEA induces substantial IL-1β release, which was also accompanied by the release of cleaved caspase-1, strongly suggesting inflammasome activation. This hypothesis was further verified by the dependency of this response on the presence of Nlrp3 and Asc. Using various inhibitors, it was further elucidated that this process required ROS production and potassium efflux but not phagocytosis. Using $P2X_7$-deficient cells, it could also be ruled out that SEA preparations contain ATP or that they indirectly activate the Nlrp3 inflammasome by inducing the release of intracellular ATP. Finally, it was also found that SEA-mediated IL-1β production involved Syk signalling, as pharmacological inhibition of Syk prevented cytokine secretion.

SEA is a complex mixture of proteins, glycoproteins and glycolipids and this diversity allows both the activation of antigen-specific $CD4^+$ T cells and NKT cells in a CD1-depedent manner. With regards to inflammasome activation, further experiments revealed that SEA was no longer functional following heat- or proteinase-inactivation. Interestingly however, such treatment did not affect the suppressive capacity of SEA on TLR-triggered cytokine production. This suggests that these effects are mediated through separate factors, and that a (glyco-)protein component might be involved in inflammasome activation. Along these lines of conjecture, a previous study has demonstrated that C-type lectin receptors (CLRs), which classically bind to sugar chains, can also bind to SEA and in turn play a role in its internalisation [71]. Since SEA-mediated inflammasome activation could be prevented by Syk inhibition, it is conceivable that a Syk-activating CLR might be involved in inflammasome activation by binding to a thus far unidentified SEA-derived glycoprotein.

As mentioned earlier, the morbidity in *S. mansoni*-infected patients arises from $CD4^+$ T cell mediated granulomas which result from finely balanced adaptive immune responses [79]. In line with the pro-inflammatory role of IL-1β, the size of granulomas in the liver and gut wall was smaller in infected Nlrp3- or Asc-deficient mice when compared to wild-type controls. In addition, upon antigen-specific re-stimulation, T cells isolated from both Nlrp3- and Asc-deficient infected mice show reduced Th1 and Th17 responses. With the exception of a somewhat increased IL-13 production, Th2/Treg responses remained unaffected. IL-13 promotes fibrosis, and further analysis of the livers from infected Asc knockout mice revealed a greater deposition of collagen fibres in the granulomas. These results strongly reflected those findings obtained with IL-1 signalling-deficient MyD88 knockout mice and emphasise the role of the Nlpr3 inflammasome during infection. This phenotype (reduced granuloma development and Th1 responses) is probably related to the ability of IL-1β to induce the production of other pro-inflammatory and T cell stimulating cytokines, but also potentially involves the direct effects of IL-1β and IL-18 on T cells. These findings represent the first report on the ability of a multi-cellular eukaryotic parasite to activate an inflammasome response. In addition, it gives fundamental progress to the field of *Schistosoma* research and provides an explanation for the long-standing discrepancy between an immune activation in vivo that could not be recapitulated with in vitro assays, which only demonstrated an inhibition of pro-inflammatory cytokine production. Current research focuses on the identification of the inflammasome-activating SEA component and the nature of receptors and signalling pathways involved in its action. These future findings may hold valuable information for the general understanding of inflammasome activation.

5 Concluding Remarks

The last few years have seen enormous progress in inflammasome research, reflected in a broadening spectrum of known activating agents and an ever-deepening understanding of the mechanisms involved. It is now apparent that the ability of a pathogen to elicit cytokine production at the level of transcription or at the level of inflammasome activation can often rely on different components of the pathogen and on its viability. This prompts the re-evaluation of the current one antigen – one adjuvant vaccination strategies. In order to ensure that an appropriate adaptive response is triggered, it would likely be advantageous if adjuvant compositions reflected the *pattern* of innate immune stimulation triggered by a pathogen. If a clearance of and memory against a given pathogen is inflammasome dependent, it could be beneficial if this was recapitulated in any vaccine strategy against this pathogen. Effective vaccination against such pathogen may therefore require more than one adjuvant component to ensure both upregulation of cytokine production and activation of an inflammasome. Moreover, the use of live attenuated pathogen vaccines should be reconsidered, as has recently been discussed in the literature for the case of vaccination against malaria [80]. Provided that patient safety is not compromised, increased side effects during a vaccination may be a price worth paying if it ensures protective immunity.

Herein we summarised the recent pilot studies on inflammasome activation in response to eukaryotic pathogens. Investigations with the model pathogens *C. albicans* (for fungi), *Plasmodium* spp. (for single-cell parasites) and *S. mansoni* (for helminths) have all demonstrated an important role of the Nlrp3 inflammasome. A role for IL-1β in mediating immune responses against other eukaryotes such as the fungus *Aspergillus* spp. [81], the single-cell parasite *Leishmania major* [82] and the tape worm *Echinococcus multilocularis* has also been suggested [82–84]. Future studies should aim to elucidate if these and other eukaryotic pathogens activate an inflammasome (Nlrp3- or other), and to what extent this is involved in the activation of protective or pathological immune responses.

Acknowledgements We thank Maximilian Kordes, Kai Matuschewski and Clarissa Prazeres da Costa for critical discussions and review of the manuscript. OGs work is funded by a Marie-Curie RTN Apoptrain fellowship and an EMBO long-term fellowship, CJvNTs by an NSERC postgraduate scholarship, and LELs by the DFG and SFB Tr22.

The authors declare no competing financial interests.

References

1. Martinon F, Burns K, Tschopp J (2002) The inflammasome: a molecular platform triggering activation of inflammatory caspases and processing of proIL-beta. Mol Cell 10:417–426
2. Agostini L, Martinon F, Burns K, McDermott MF, Hawkins PN, Tschopp J (2004) NALP3 forms an IL-1beta-processing inflammasome with increased activity in Muckle-Wells autoinflammatory disorder. Immunity 20:319–325
3. Meng G, Zhang F, Fuss I, Kitani A, Strober W (2009) A mutation in the Nlrp3 gene causing inflammasome hyperactivation potentiates Th17 cell-dominant immune responses. Immunity 30:860–874
4. Brydges SD, Mueller JL, McGeough MD, Pena CA, Misaghi A, Gandhi C, Putnam CD, Boyle DL, Firestein GS, Horner AA et al (2009) Inflammasome-mediated disease animal models reveal roles for innate but not adaptive immunity. Immunity 30:875–887

5. Martinon F, Petrilli V, Mayor A, Tardivel A, Tschopp J (2006) Gout-associated uric acid crystals activate the NALP3 inflammasome. Nature 440:237–241
6. Dostert C, Petrilli V, Van Bruggen R, Steele C, Mossman BT, Tschopp J (2008) Innate immune activation through Nalp3 inflammasome sensing of asbestos and silica. Science 320:674–677
7. Ghiringhelli F, Apetoh L, Tesniere A, Aymeric L, Ma Y, Ortiz C, Vermaelen K, Panaretakis T, Mignot G, Ullrich E et al (2009) Activation of the NLRP3 inflammasome in dendritic cells induces IL-1beta-dependent adaptive immunity against tumors. Nat Med 15:1170–1178
8. Zhou R, Tardivel A, Thorens B, Choi I, Tschopp J (2010) Thioredoxin-interacting protein links oxidative stress to inflammasome activation. Nat Immunol 11:136–140
9. Miao EA, Mao DP, Yudkovsky N, Bonneau R, Lorang CG, Warren SE, Leaf IA, Aderem A (2010) Innate immune detection of the type III secretion apparatus through the NLRC4 inflammasome. Proc Natl Acad Sci USA 107:3076–3080
10. Case CL, Shin S, Roy CR (2009) Asc and Ipaf Inflammasomes direct distinct pathways for caspase-1 activation in response to Legionella pneumophila. Infect Immun 77:1981–1991
11. Martinon F, Mayor A, Tschopp J (2009) The inflammasomes: guardians of the body. Annu Rev Immunol 27:229–265
12. Mariathasan S, Weiss DS, Newton K, McBride J, O'Rourke K, Roose-Girma M, Lee WP, Weinrauch Y, Monack DM, Dixit VM (2006) Cryopyrin activates the inflammasome in response to toxins and ATP. Nature 440:228–232
13. Yu HB, Finlay BB (2008) The caspase-1 inflammasome: a pilot of innate immune responses. Cell Host Microbe 4:198–208
14. Poeck H, Bscheider M, Gross O, Finger K, Roth S, Rebsamen M, Hannesschlager N, Schlee M, Rothenfusser S, Barchet W et al (2010) Recognition of RNA virus by RIG-I results in activation of CARD9 and inflammasome signaling for interleukin 1 beta production. Nat Immunol 11:63–69
15. Burckstummer T, Baumann C, Bluml S, Dixit E, Durnberger G, Jahn H, Planyavsky M, Bilban M, Colinge J, Bennett KL et al (2009) An orthogonal proteomic-genomic screen identifies AIM2 as a cytoplasmic DNA sensor for the inflammasome. Nat Immunol 10:266–272
16. Fernandes-Alnemri T, Yu JW, Datta P, Wu J, Alnemri ES (2009) AIM2 activates the inflammasome and cell death in response to cytoplasmic DNA. Nature 458:509–513
17. Hornung V, Ablasser A, Charrel-Dennis M, Bauernfeind F, Horvath G, Caffrey DR, Latz E, Fitzgerald KA (2009) AIM2 recognizes cytosolic dsDNA and forms a caspase-1-activating inflammasome with ASC. Nature 458:514–518
18. Roberts TL, Idris A, Dunn JA, Kelly GM, Burnton CM, Hodgson S, Hardy LL, Garceau V, Sweet MJ, Ross IL et al (2009) HIN-200 proteins regulate caspase activation in response to foreign cytoplasmic DNA. Science 323:1057–1060
19. Mayor A, Martinon F, De Smedt T, Petrilli V, Tschopp J (2007) A crucial function of SGT1 and HSP90 in inflammasome activity links mammalian and plant innate immune responses. Nat Immunol 8:497–503
20. Poyet JL, Srinivasula SM, Tnani M, Razmara M, Fernandes-Alnemri T, Alnemri ES (2001) Identification of Ipaf, a human caspase-1-activating protein related to Apaf-1. J Biol Chem 276:28309–28313
21. Dinarello CA (2009) Immunological and inflammatory functions of the interleukin-1 family. Annu Rev Immunol 27:519–550
22. Bauernfeind FG, Horvath G, Stutz A, Alnemri ES, MacDonald K, Speert D, Fernandes-Alnemri T, Wu J, Monks BG, Fitzgerald KA et al (2009) Cutting edge: NF-kappaB activating pattern recognition and cytokine receptors license NLRP3 inflammasome activation by regulating NLRP3 expression. J Immunol 183:787–791
23. Holland SM, Vinh DC (2009) Yeast infections–human genetics on the rise. N Engl J Med 361:1798–1801
24. Wisplinghoff H, Bischoff T, Tallent SM, Seifert H, Wenzel RP, Edmond MB (2004) Nosocomial bloodstream infections in US hospitals: analysis of 24,179 cases from a prospective nationwide surveillance study. Clin Infect Dis 39:309–317

25. Bodey GP, Mardani M, Hanna HA, Boktour M, Abbas J, Girgawy E, Hachem RY, Kontoyiannis DP, Raad II (2002) The epidemiology of Candida glabrata and Candida albicans fungemia in immunocompromised patients with cancer. Am J Med 112:380–385
26. Klein RS, Harris CA, Small CB, Moll B, Lesser M, Friedland GH (1984) Oral candidiasis in high-risk patients as the initial manifestation of the acquired immunodeficiency syndrome. N Engl J Med 311:354–358
27. Bellocchio S, Montagnoli C, Bozza S, Gaziano R, Rossi G, Mambula SS, Vecchi A, Mantovani A, Levitz SM, Romani L (2004) The contribution of the Toll-like/IL-1 receptor superfamily to innate and adaptive immunity to fungal pathogens in vivo. J Immunol 172:3059–3069
28. Mencacci A, Bacci A, Cenci E, Montagnoli C, Fiorucci S, Casagrande A, Flavell RA, Bistoni F, Romani L (2000) Interleukin 18 restores defective Th1 immunity to Candida albicans in caspase 1-deficient mice. Infect Immun 68:5126–5131
29. Stuyt RJ, Netea MG, Verschueren I, Fantuzzi G, Dinarello CA, Van Der Meer JW, Kullberg BJ (2002) Role of interleukin-18 in host defense against disseminated Candida albicans infection. Infect Immun 70:3284–3286
30. Stuyt RJ, Netea MG, van Krieken JH, van der Meer JW, Kullberg BJ (2004) Recombinant interleukin-18 protects against disseminated Candida albicans infection in mice. J Infect Dis 189:1524–1527
31. Steele C, Fidel PL Jr (2002) Cytokine and chemokine production by human oral and vaginal epithelial cells in response to Candida albicans. Infect Immun 70:577–583
32. Netea MG, Stuyt RJ, Kim SH, Van der Meer JW, Kullberg BJ, Dinarello CA (2002) The role of endogenous interleukin (IL)-18, IL-12, IL-1beta, and tumor necrosis factor-alpha in the production of interferon-gamma induced by Candida albicans in human whole-blood cultures. J Infect Dis 185:963–970
33. Vonk AG, Netea MG, van Krieken JH, Iwakura Y, van der Meer JW, Kullberg BJ (2006) Endogenous interleukin (IL)-1 alpha and IL-1 beta are crucial for host defense against disseminated candidiasis. J Infect Dis 193:1419–1426
34. Gross O, Poeck H, Bscheider M, Dostert C, Hannesschlager N, Endres S, Hartmann G, Tardivel A, Schweighoffer E, Tybulewicz V et al (2009) Syk kinase signalling couples to the Nlrp3 inflammasome for anti-fungal host defence. Nature 459:433–436
35. Hise AG, Tomalka J, Ganesan S, Patel K, Hall BA, Brown GD, Fitzgerald KA (2009) An essential role for the NLRP3 inflammasome in host defense against the human fungal pathogen Candida albicans. Cell Host Microbe 5:487–497
36. Joly S, Ma N, Sadler JJ, Soll DR, Cassel SL, Sutterwala FS (2009) Cutting edge: Candida albicans hyphae formation triggers activation of the Nlrp3 inflammasome. J Immunol 183: 3578–3581
37. Brown GD (2006) Dectin-1: a signalling non-TLR pattern-recognition receptor. Nat Rev Immunol 6:33–43
38. Rogers NC, Slack EC, Edwards AD, Nolte MA, Schulz O, Schweighoffer E, Williams DL, Gordon S, Tybulewicz VL, Brown GD et al (2005) Syk-dependent cytokine induction by Dectin-1 reveals a novel pattern recognition pathway for C type lectins. Immunity 22:507–517
39. Gross O, Gewies A, Finger K, Schafer M, Sparwasser T, Peschel C, Forster I, Ruland J (2006) Card9 controls a non-TLR signalling pathway for innate anti-fungal immunity. Nature 442:651–656
40. Robinson MJ, Osorio F, Rosas M, Freitas RP, Schweighoffer E, Gross O, Verbeek JS, Ruland J, Tybulewicz V, Brown GD et al (2009) Dectin-2 is a Syk-coupled pattern recognition receptor crucial for Th17 responses to fungal infection. J Exp Med 206:2037–2051
41. Wells CA, Salvage-Jones JA, Li X, Hitchens K, Butcher S, Murray RZ, Beckhouse AG, Lo YL, Manzanero S, Cobbold C et al (2008) The macrophage-inducible C-type lectin, mincle, is an essential component of the innate immune response to Candida albicans. J Immunol 180:7404–7413
42. Taylor PR, Tsoni SV, Willment JA, Dennehy KM, Rosas M, Findon H, Haynes K, Steele C, Botto M, Gordon S et al (2007) Dectin-1 is required for beta-glucan recognition and control of fungal infection. Nat Immunol 8:31–38

43. Ferwerda B, Ferwerda G, Plantinga TS, Willment JA, van Spriel AB, Venselaar H, Elbers CC, Johnson MD, Cambi A, Huysamen C et al (2009) Human dectin-1 deficiency and mucocutaneous fungal infections. N Engl J Med 361:1760–1767
44. Glocker EO, Hennigs A, Nabavi M, Schaffer AA, Woellner C, Salzer U, Pfeifer D, Veelken H, Warnatz K, Tahami F et al (2009) A homozygous CARD9 mutation in a family with susceptibility to fungal infections. N Engl J Med 361:1727–1735
45. Goodridge HS, Underhill DM (2008) Fungal recognition by TLR2 and Dectin-1. Handb Exp Pharmacol 183:87–109
46. van de Veerdonk FL, Joosten LA, Devesa I, Mora-Montes HM, Kanneganti TD, Dinarello CA, van der Meer JW, Gow NA, Kullberg BJ, Netea MG (2009) Bypassing pathogen-induced inflammasome activation for the regulation of interleukin-1beta production by the fungal pathogen Candida albicans. J Infect Dis 199:1087–1096
47. Lamkanfi M, Malireddi RK, Kanneganti TD (2009) Fungal zymosan and mannan activate the cryopyrin inflammasome. J Biol Chem 284:20574–20581
48. Lev-Sagie A, Prus D, Linhares IM, Lavy Y, Ledger WJ, Witkin SS (2009) Polymorphism in a gene coding for the inflammasome component NALP3 and recurrent vulvovaginal candidiasis in women with vulvar vestibulitis syndrome. Am J Obstet Gynecol 200(303):e301–e306
49. Kummer JA, Broekhuizen R, Everett H, Agostini L, Kuijk L, Martinon F, van Bruggen R, Tschopp J (2007) Inflammasome components NALP 1 and 3 show distinct but separate expression profiles in human tissues suggesting a site-specific role in the inflammatory response. J Histochem Cytochem 55:443–452
50. Underhill DM, Rossnagle E, Lowell CA, Simmons RM (2005) Dectin-1 activates Syk tyrosine kinase in a dynamic subset of macrophages for reactive oxygen production. Blood 106:2543–2550
51. Lo HJ, Kohler JR, DiDomenico B, Loebenberg D, Cacciapuoti A, Fink GR (1997) Nonfilamentous C. albicans mutants are avirulent. Cell 90:939–949
52. Schofield L, Grau GE (2005) Immunological processes in malaria pathogenesis. Nat Rev Immunol 5:722–735
53. Togbe D, Schofield L, Grau GE, Schnyder B, Boissay V, Charron S, Rose S, Beutler B, Quesniaux VF, Ryffel B (2007) Murine cerebral malaria development is independent of toll-like receptor signaling. Am J Pathol 170:1640–1648
54. Coban C, Igari Y, Yagi M, Reimer T, Koyama S, Aoshi T, Ohata K, Tsukui T, Takeshita F, Sakurai K et al (2010) Immunogenicity of whole-parasite vaccines against Plasmodium falciparum involves malarial hemozoin and host TLR9. Cell Host Microbe 7:50–61
55. Coban C, Ishii KJ, Horii T, Akira S (2007) Manipulation of host innate immune responses by the malaria parasite. Trends Microbiol 15:271–278
56. Coban C, Ishii KJ, Uematsu S, Arisue N, Sato S, Yamamoto M, Kawai T, Takeuchi O, Hisaeda H, Horii T et al (2007) Pathological role of Toll-like receptor signaling in cerebral malaria. Int Immunol 19:67–79
57. Ouma C, Davenport GC, Awandare GA, Keller CC, Were T, Otieno MF, Vulule JM, Martinson J, Ong'echa JM, Ferrell RE et al (2008) Polymorphic variability in the interleukin (IL)-1beta promoter conditions susceptibility to severe malarial anemia and functional changes in IL-1beta production. J Infect Dis 198:1219–1226
58. Brown H, Turner G, Rogerson S, Tembo M, Mwenechanya J, Molyneux M, Taylor T (1999) Cytokine expression in the brain in human cerebral malaria. J Infect Dis 180:1742–1746
59. Vogetseder A, Ospelt C, Reindl M, Schober M, Schmutzhard E (2004) Time course of coagulation parameters, cytokines and adhesion molecules in Plasmodium falciparum malaria. Trop Med Int Health 9:767–773
60. Armah HB, Wilson NO, Sarfo BY, Powell MD, Bond VC, Anderson W, Adjei AA, Gyasi RK, Tettey Y, Wiredu EK et al (2007) Cerebrospinal fluid and serum biomarkers of cerebral malaria mortality in Ghanaian children. Malar J 6:147
61. Lyke KE, Burges R, Cissoko Y, Sangare L, Dao M, Diarra I, Kone A, Harley R, Plowe CV, Doumbo OK et al (2004) Serum levels of the proinflammatory cytokines interleukin-1 beta

(IL-1beta), IL-6, IL-8, IL-10, tumor necrosis factor alpha, and IL-12(p70) in Malian children with severe Plasmodium falciparum malaria and matched uncomplicated malaria or healthy controls. Infect Immun 72:5630–5637
62. Jaramillo M, Bellemare MJ, Martel C, Shio MT, Contreras AP, Godbout M, Roger M, Gaudreault E, Gosselin J, Bohle DS et al (2009) Synthetic Plasmodium-like hemozoin activates the immune response: a morphology–function study. PLoS One 4:e6957
63. Dostert C, Guarda G, Romero JF, Menu P, Gross O, Tardivel A, Suva ML, Stehle JC, Kopf M, Stamenkovic I et al (2009) Malarial hemozoin is a Nalp3 inflammasome activating danger signal. PLoS One 4:e6510
64. Shio MT, Eisenbarth SC, Savaria M, Vinet AF, Bellemare MJ, Harder KW, Sutterwala FS, Bohle DS, Descoteaux A, Flavell RA et al (2009) Malarial hemozoin activates the NLRP3 inflammasome through Lyn and Syk kinases. PLoS Pathog 5:e1000559
65. Ng G, Sharma K, Ward SM, Desrosiers MD, Stephens LA, Schoel WM, Li T, Lowell CA, Ling CC, Amrein MW et al (2008) Receptor-independent, direct membrane binding leads to cell-surface lipid sorting and Syk kinase activation in dendritic cells. Immunity 29:807–818
66. Popa-Nita O, Rollet-Labelle E, Thibault N, Gilbert C, Bourgoin SG, Naccache PH (2007) Crystal-induced neutrophil activation. IX. Syk-dependent activation of class Ia phosphatidylinositol 3-kinase. J Leukocyt Biol 82:763–773
67. Hornung V, Bauernfeind F, Halle A, Samstad EO, Kono H, Rock KL, Fitzgerald KA, Latz E (2008) Silica crystals and aluminum salts activate the NALP3 inflammasome through phagosomal destabilization. Nat Immunol 9:847–856
68. Griffith JW, Sun T, McIntosh MT, Bucala R (2009) Pure Hemozoin is inflammatory in vivo and activates the NALP3 inflammasome via release of uric acid. J Immunol 183:5208–5220
69. Reimer T, Shaw MH, Franchi L, Coban C, Ishii KJ, Akira S, Horii T, Rodriguez A, Nunez G (2009) Experimental cerebral malaria progresses independently of the Nlrp3 inflammasome. Eur J Immunol 40(3):764–769
70. Pearce EJ, MacDonald AS (2002) The immunobiology of schistosomiasis. Nat Rev Immunol 2:499–511
71. van Liempt E, van Vliet SJ, Engering A, Garcia Vallejo JJ, Bank CM, Sanchez-Hernandez M, van Kooyk Y, van Die I (2007) Schistosoma mansoni soluble egg antigens are internalized by human dendritic cells through multiple C-type lectins and suppress TLR-induced dendritic cell activation. Mol Immunol 44:2605–2615
72. Layland LE, Rad R, Wagner H, da Costa CU (2007) Immunopathology in schistosomiasis is controlled by antigen-specific regulatory T cells primed in the presence of TLR2. Eur J Immunol 37:2174–2184
73. Layland LE, Wagner H, da Costa CU (2005) Lack of antigen-specific Th1 response alters granuloma formation and composition in Schistosoma mansoni-infected MyD88-/- mice. Eur J Immunol 35:3248–3257
74. Layland LE, Mages J, Loddenkemper C, Hoerauf A, Wagner H, Lang R, da Costa CU (2010) Pronounced phenotype in activated regulatory T cells during a chronic helminth infection. J Immunol 184:713–724
75. Shainheit MG, Smith PM, Bazzone LE, Wang AC, Rutitzky LI, Stadecker MJ (2008) Dendritic cell IL-23 and IL-1 production in response to schistosome eggs induces Th17 cells in a mouse strain prone to severe immunopathology. J Immunol 181:8559–8567
76. Lohning M, Grogan JL, Coyle AJ, Yazdanbakhsh M, Meisel C, Gutierrez-Ramos JC, Radbruch A, Kamradt T (1999) T1/ST2 expression is enhanced on CD4+ T cells from schistosome egg-induced granulomas: analysis of Th cell cytokine coexpression ex vivo. J Immunol 162:3882–3889
77. Chensue SW, Bienkowski M, Eessalu TE, Warmington KS, Hershey SD, Lukacs NW, Kunkel SL (1993) Endogenous IL-1 receptor antagonist protein (IRAP) regulates schistosome egg granuloma formation and the regional lymphoid response. J Immunol 151:3654–3662
78. Ritter M, Gross O, Kays S, Ruland J, Nimmerjahn F, Saijo S, Tschopp J, Layland LE, Prazeres da Costa C (2010) Schistosoma mansoni triggers Dectin-2, which activates the Nlrp3

inflammasome and alters adaptive immune responses. Proc Natl Acad Sci USA 107: 20459–20464
79. Pearce EJ, Caspar P, Grzych JM, Lewis FA, Sher A (1991) Downregulation of Th1 cytokine production accompanies induction of Th2 responses by a parasitic helminth, Schistosoma mansoni. J Exp Med 173:159–166
80. Palm NW, Medzhitov R (2009) Pattern recognition receptors and control of adaptive immunity. Immunol Rev 227:221–233
81. Sainz J, Perez E, Gomez-Lopera S, Jurado M (2008) IL1 gene cluster polymorphisms and its haplotypes may predict the risk to develop invasive pulmonary aspergillosis and modulate C-reactive protein level. J Clin Immunol 28:473–485
82. Xin L, Li Y, Soong L (2007) Role of interleukin-1beta in activating the CD11c(high) CD45RB- dendritic cell subset and priming Leishmania amazonensis-specific CD4+ T cells in vitro and in vivo. Infect Immun 75:5018–5026
83. Humphreys NE, Grencis RK (2009) IL-1-dependent, IL-1R1-independent resistance to gastrointestinal nematodes. Eur J Immunol 39:1036–1045
84. Eger A, Kirch A, Manfras B, Kern P, Schulz-Key H, Soboslay PT (2003) Pro-inflammatory (IL-1beta, IL-18) cytokines and IL-8 chemokine release by PBMC in response to Echinococcus multilocularis metacestode vesicles. Parasite Immunol 25:103–105

Inflammasome Activation by Pathogenic Crystals and Particles

Fabio Re

Abstract The NLRP3 inflammasome is a multiprotein complex that controls processing and secretion of bioactive IL-1β and IL-18, powerful proinflammatory cytokines involved in several human diseases. The NLRP3 inflammasome is activated by pathogens, danger signals released by necrotic cells, and crystals and particles of varied size and structure. These inflammasome-activating particles have a causative role in human diseases such as gout, asbestosis, silicosis, and Alzheimer's, or are useful as vaccine adjuvants, like in the case of alum. The molecular aspects involved in activation of the NLRP3 inflammasome by particles are being rapidly elucidated and emphasize the importance of phagocytosis, K^+ efflux, and generation of reactive oxygen species in this process. Here, the recent literature is reviewed and important questions to be answered in the future are identified.

1 Introduction

Interleukin-1β is one of the most powerful proinflammatory cytokines; it affects virtually all tissue types, and it is involved in several human diseases. Production of bioactive IL-1β is regulated at many levels, including the transcription and stability of the mRNA, the translation of the precursor pro-IL-1β, its maturation into bioactive IL-1β, and secretion (reviewed in ref. [1]). Maturation of pro-IL-1β, and the related cytokine pro-IL-18, requires a proteolytic cleavage that is carried out by caspase-1.

During the past few years, our understanding of the regulation of production of bioactive IL-1β and IL-18 has increased dramatically with the identification and characterization of the inflammasome as the multiprotein complex where the

F. Re (✉)
Department of Molecular Sciences, University of Tennessee Health Science Center, 858 Madison Avenue, MSB 501B, Memphis, TN 38163, USA
e-mail: fre@uthsc.edu

I. Couillin et al. (eds.), *The Inflammasomes*, Progress in Inflammation Research,
DOI 10.1007/978-3-0348-0148-5_6, © Springer Basel AG 2011

maturation of pro-IL-1β and pro-IL-18 occurs. Key components of the inflammasome are pattern recognition receptors that belong to the NLR family [2]. The inflammasome composed by the NLR molecule NLRP3 is presently the best-characterized inflammasome complex and mediates innate immune responses to various infectious agents including bacteria, viruses, fungi, and protozoa as well as endogenous danger signals released by stressed or necrotic cells. It is becoming clear that particles of varied size and chemical structure are also powerful activators of the NLRP3 inflammasome.

In this chapter, I review the recent literature concerning activation of the NLRP3 inflammasome by particles and crystals, discuss the different mechanisms proposed to explain the phenomenon, and highlight the major challenges for the future.

2 Particles and Crystals as Activators of the NLRP3 Inflammasome

2.1 Uric Acid Crystals

The first evidence that the NLRP3 inflammasome could be activated by crystals came from investigation into the mechanism of action of monosodium urate monohydrate (MSU) crystals. Since the eighteenth century, MSU crystals have been recognized as the etiologic agent of gout, a chronic inflammatory disease that affects the articular joints. Several studies have documented the proinflammatory potentials of MSU crystals and various models have been proposed to explain their immunostimulatory activities [3]. Neutrophils and monocytes exposed to MSU were known to secrete various cytokines and IL-1 was shown to play a key role in gout [4, 5]. MSU was also identified as an endogenous danger signal with adjuvant activity released by necrotic cells [6]. However, the molecular mechanism responsible for cytokine induction by MSU was unclear until the seminal work by Fabio Martinon and colleagues reported that the release of IL-1β by macrophages stimulated with MSU was dependent on the NLRP3 inflammasome [7]. In that paper, the authors also showed that calcium pyrophosphate dihydrate (CPPD) crystals, another type of pathogenic crystals that cause an analogous disease called pseudogout, induced caspase-1 activation and release of IL-1β. Surprisingly, other particles of similar size or composition, such as diamond crystals, aluminum powder, and crystals of allopurinol did not elicit inflammasome activation providing a specific connection to the pathogenic crystals. Using a mouse model of crystal-induced peritonitis, the authors also showed that neutrophil recruitment to the peritoneum following injection of MSU was impaired in mice deficient in NLRP3 or ASC, an inflammasome adaptor molecule. This work provided the first plausible explanation for the proinflammatory activities of MSU and suggested that inhibition of IL-1β activity could represent an effective therapeutic intervention for gout. Importantly, it also indicated a mechanism of action for other particles with proinflammatory properties.

2.2 Alum Crystals

The next crystalline structure found to activate the inflammasome was alum. Alum, in the form of aluminum hydroxide or aluminum phosphate crystals, is the only vaccine adjuvant approved for human vaccination in the USA. Prompted by the observation of Martinon et al. [7], we tested whether alum crystals could activate the inflammasome. Our results showed that in fact alum crystals were potent activators of caspase-1 and induced secretion of IL-1β and IL-18 in human dendritic cells and PBMC [8]. Subsequently, we and other groups demonstrated that inflammasome activation by alum and alum's adjuvant effect are mediated by the NLRP3 inflammasome [9–12]. Remarkably, our results showed that other particles with adjuvant properties such as chitosan, a polysaccharide particle derived from chitin, and QuilA, a saponin that forms particle emulsions, were also strong NLRP3 inflammasome stimulators. The observation that particulated adjuvants can activate the NLRP3 inflammasome was later extended to poly(lactide-co-glycolide) (PLG) and polystyrene microparticles [13], suggesting that inflammasome activation is a common mechanism of action of particulate adjuvants and indicating the important role of the inflammasome in adaptive immunity.

2.3 Asbestos and Silica

The notion that the NLRP3 inflammasome mediates innate immune response to particles was further reinforced with the discoveries that asbestos fibers and silica particles activated this pathway. Asbestosis and silicosis are chronic inflammatory conditions of the lung caused by inhalation of these mineral particles [14]. Asbestos is also a carcinogen responsible for development of mesothelioma, a fatal form of lung cancer. Several groups [15–17] demonstrated that asbestos and silica activate the NLRP3 infammasome in cultured cells and that in vivo pulmonary inflammation induced by inhalation of these particles was drastically reduced in NLRP3$^{-/-}$ mice. These results indicate a central role of the inflammasome in development of silicosis and asbestosis and suggest its possible involvement in development of mesothelioma.

2.4 Misfolded or Aggregated Peptides

A distinct type of particles that has been reported to activate the NLRP3 inflammasome is represented by amyloid-β aggregates. The fibrils composed of misfolded amyloid-β are responsible for the inflammation and tissue damage in the CNS that characterizes Alzheimer's disease. Amyloid-β fibrils were shown to stimulate secretion of IL-1β by microglial cells in a NLRP3-dependent way [18].

Moreover, migration of microglia cells into the brain and in vivo production of chemokines and nitric oxide radicals in response to amyloid-β were shown to require the inflammasome and the IL-1 pathway. Several human diseases are caused by deposition of different types of amyloid material and it is likely that in the near future other insoluble aggregates of misfolded peptides will be discovered as capable of inflammasome activation.

2.5 Malarial Hemozoin

Recently, a potentially important role of the NLRP3 inflammasome in the pathogenesis of malaria was unveiled with the discovery that hemozoin crystals activate the NLRP3 inflammasome [19, 20]. Hemozoin is a crystalline metabolite waste produced by the malaria parasite *Plasmodium* as it detoxifies red blood cells' heme. Hemozoin's proinflammatory activities and its association with the pathogenesis of malaria were documented by several studies, though the molecular basis for its immune stimulation was unknown. Interestingly, hemozoin was also shown to act as TLR agonist [21].

2.6 More Particles

The list of particles known to activate the inflammasome and its involvement in human diseases is rapidly growing. A major complication observed in 10% of patients undergoing articular joint replacement is the aseptic loosening of the implant. Microscopic debris composed of metallic alloy or polyethylene are constantly generated by attrition of the implant and have been shown to be able to activate macrophages inciting a sterile inflammatory response that is responsible for bone resorption and implant loosening. Recent work showed NLRP3-dependent caspase-1 activation and IL-1β secretion by these implant-derived debris [22].

Taken together these results indicate the potential involvement of the NLRP3 inflammasome in several human diseases. Precipitation and/or crystallization of proteins and other molecules in tissues are common occurrences of numerous diseases. Future studies should examine the role of the NLRP3 inflammasome in their pathogenesis. For instance, parenchymal and ectopic deposition of calcium oxalate crystals is known to cause several forms of renal diseases. Deposition of these crystals triggers production of chemokines, reactive oxygen species (ROS), and lead to chronic inflammation, fibrosis, loss of nephrons, and eventually chronic renal failure [23]. Still unpublished work provides evidence that cholesterol crystals found in atherosclerotic plaques are capable of activating the NLRP3 inflammasome (Eicke Latz, personal communication). The discovery of inflammasome involvement in human diseases has important therapeutic implications since inhibition of IL-1R signaling can be effectively achieved with administration of the

IL-1Ra (IL-1 receptor antagonist) Anakinra. Recent clinical trials have in fact demonstrated improvement of gout and pseudogout symptoms with administration of Anakinra [24, 25].

3 Mechanism of Action of Particles

The mechanistic aspect of inflammasome activation by particles is an intense area of research and is being rapidly unraveled. Because of the great variability in structure and composition of the particles reported to activate the inflammasome, their direct interaction with NLRP3 seems unlikely. Rather, a more probable scenario is that interaction of the particle/pathogen with the cell initiates events that result in activation of NLRP3. One favored idea is that NLRP3 is activated through interaction with an endogenous molecule that is generated/modified downstream of signaling pathways activated by interaction of the particle with cells. It is possible that more than one type of NLRP3 endogenous activator exists and that different NLRP3 activators may use distinct signaling pathways to generate the same endogenous activator. In addition, it is possible that activation of the NLRP3 inflammasome may require posttranslational modifications of NLRP3 or other molecules that compose the inflammasome.

A few events triggered by all particulated activators of NLRP3 have been identified (see Table 1) and models have been proposed for NLRP3 inflammasome activation. The different models are not necessarily mutually exclusive and likely coexist.

3.1 Phagocytosis

An essential requirement for NLRP3 inflammasome activation is the uptake of the particle through phagocytosis, as demonstrated by the fact that cytochalasins or other drugs that inhibit this process also prevent pro-IL-1β maturation in response

Table 1 Requirement of the indicated events for NLRP3 inflammasome activation by various activators

	Phagocytosis	K+ efflux	ROS	CatB	P2X7	Lysosome acidification	Pannexin	Syk
Nigericin, maitotoxin	No	Yes	Yes		Yes	No	Yes	No
ATP	No	Yes	Yes		Yes	No	Yes	
MSU	Yes	Yes	Yes	Yes	No	No	No	
Alum	Yes	Yes	Yes	Yes	No	Yes	No	
Silica, asbestos	Yes	Yes	Yes	Yes	No	Yes	No	
Hemozoin	Yes	Yes	Yes	Yes	No	No	No	Yes
Amyloid-β	Yes			Yes				
Candida	Yes	Yes	Yes	No	No	No		Yes

to NLRP3-activating particles but not ATP or bacterial toxins [7, 8, 15–20]. What characterizes the phagocytosis of the NLRP3-activating particles is that due to their size, shape, and composition, the process of phagocytosis cannot progress properly resulting in the so-called frustrated phenotype in which the particle cannot be completely engulfed or the resulting phagosome cannot mature correctly. The most glaring example of frustrated phagocytosis occurs when the particle size is greater than the cell trying to engulf it, as in the case of asbestos fibers. Frustrated phagocytosis is coupled to a powerful generation of ROS (see below). One aspect that is still unclear and should be examined is whether the phagocytosed particles need to be opsonized and therefore whether phagocytic receptors are directly involved in the process of inflammasome activation. Interestingly, both alum and MSU crystals were shown to fix complement and adsorb other serum proteins, suggesting possible involvement of phagocytic receptors [26, 27]. However, a recent study showed that interaction of MSU crystals with dendritic cells is receptor-independent and involves reorganization of membrane cholesterol [28]. Therefore, it would seem that phagocytosis of some particulated NLRP3 activators may involve membrane receptors, though the role of these receptors may be limited to the uptake process.

3.2 ROS Generation

One common property of the particles and agents that activate NLRP3 is their ability to induce a respiratory burst upon phagocytosis by myeloid cells. ROS are known to regulate various signaling pathways raising the possibility of their involvement in NLRP3 activation. In agreement with this notion are the observations that pharmacological inhibitors of NADPH oxidases severely impair inflammasome activation by most activators including MSU, asbestos, silica, alum, hemozoin, Nigericin, and ATP [7, 15–20] and that all known NLRP3 activators induce ROS generation [15, 16]. One important yet unanswered question is what is the source of the ROS during inflammasome activation. Silencing expression of the phox p22 subunit of the phagocytic NADPH oxidase significantly diminished NLRP3 activation by asbestos and MSU in THP1 macrophage cell line [15]. However, subsequent work has shown that inflammasome activation in response to MSU, silica, and hemozoin is not affected in macrophages derived from mice deficient in the gp91phox NADPH subunit [17, 19]. We also failed to observe difference in inflammasome activation by alum in wt or gp91phox$^{-/-}$ mice, although the oxidase inhibitor DPI significantly impaired inflammasome activation by alum [unpublished observation]. Inhibition of the mitochondrial complex I or complex II was reported not to be sufficient to block NLRP3 activation by MSU [15]. Thus, the oxidase(s) responsible for ROS generation during NLRP3 activation remains elusive. Another important question to be answered is how generation of ROS leads to NLRP3 activation. Recent work from the Tschopp group provided some important insight into this matter [29]. In this work, the authors showed that generation of ROS by MSU resulted in dissociation of thioredoxin interacting protein (TXNIP) from thioredoxin, a cellular protein with antioxidant functions. Freed TXNIP was then shown to interact directly with NLRP3. The LRR and NACHT domains of NLRP3 mediate this interaction, as predicted by models that postulated the existence of an endogenous activator. Importantly, NLRP3 inflammasome activation by MSU or ATP was defective in TXNIP-deficient mice or cultured cells. In agreement with a role of TXNIP in inflammasome activation, downregulation of thioredoxin expression by shRNA, which is predicted to increase the level of free TXNIP, or overexpression of TXNIP was reported to increase IL-1β processing in response to asbestos, MSU, and silica. Thus, TXNIP may represent the endogenous activator that couples ROS generation to activation of the NLRP3 inflammasome. Interestingly, glucose and insulin were shown to upregulate TXNIP expression and glucose was shown to trigger weak inflammasome activation in pancreas β-cells in a NLRP3/TXNIP-dependent way, suggesting inflammasome involvement into diabetes and a role for NLRP3 as sensor for metabolic danger [30]. Accordingly, anti-IL-1β therapy with Anakinra ameliorates diabetes in mouse and human clinical trials.

An important point that should not be overlooked is that although generation of ROS is clearly necessary for NLRP3 activation, by itself it is not sufficient. This is best illustrated by the fact that yeast zymosan, which is a powerful TLR agonist and inducer of ROS and phagocytosis, does not trigger activation of the inflammasome [31, 32]. The other event that is required is K^+ efflux.

3.3 *Potassium Efflux*

One of the earliest identified events triggered by all NLRP3 activator is K^+ efflux. Early in vitro experiments demonstrated that disruption of macrophages in buffer that contained subphysiologic concentration of K^+ led to a spontaneous activation of the inflammasome [33]. Prevention of K^+ efflux by supplementing the cell growth medium with 130 mM KCl is reported to block NLRP3 inflammasome activation in response to all known NLRP3 activators including nonparticulated ones such as extracellular ATP and the microbial toxins nigericin, maitotoxin, and gramicidin [34, 35]. These toxins are K^+ channels and cause a marked depletion of intracellular K^+. More recently, pharmacological inhibitors of K^+ channel glyburide (glibenclamide) were also shown to inhibit NLRP3 inflammasome, though this effect did not appear to depend on alteration of K^+ efflux [36]. Our unpublished observations indicate that K^+ efflux is also required for NLRP3 inflammasome activation induced in necrotic cells treated with indirubin derivative 7BIO [37].

It is interesting to note that low intracellular K^+ concentration triggers NLRP1 inflammasome activation [38] and it is required for assembly and activation of the APAF apoptosome by cytochrome *c* [39], suggesting that the regulation of intracellular K^+ concentration may be a mechanism to prevent the spontaneous activation of caspase-activating platforms. Potassium efflux was shown not to be required for activation of the NLRC4 inflammasome [34], though recent work showed the opposite [40]. Thus, one function of the inflammasomes may be to detect low intracellular K^+ concentration.

At present it is unclear how particles may trigger K^+ efflux. It is possible that phagocytosis of particles and crystals that lead to the frustrated phenotype may cause damage of cellular membranes and opening of channels and pores through which K^+ can efflux. Another possible mechanism is that K^+ efflux is in fact triggered by the LPS priming step that is commonly required for in vitro experimental inflammasome activation (see below). It has in fact been demonstrated that TLR4 associate and stimulate the large conductance K^+ channel MaxiK [41].

An important yet unanswered question is how K^+ concentration regulates NLRP3 assembly into a functional inflammasome. Interestingly, it has been demonstrated that agonist of Ca^{2+}-activated K^+ channels of small conductance (SK) and intermediate conductance (IK) causes production of superoxide and hydrogen peroxide in a NADPH-independent way in neutrophils [42], suggesting that K^+ efflux may be upstream and control ROS generation.

3.4 Pannexin

Extracellular ATP is one of the earliest identified activators of the NLRP3 inflammasome [43]. Extracellular ATP interacts with the purinergic receptor P2X7 and leads to the formation of a large hemichannel consisting of the protein pannexin. It has been proposed that this channel may allow microbial products to access the cytoplasm where they could subsequently activate the NLRP3 inflammasome [44, 45]. Of note, pannexin and P2X7 are not required for NLRP3 inflammasome activation by particles, ruling out extracellular ATP involvement for this type of NLRP3 activators [19, 31].

3.5 Lysosome Destabilization

One attractive model has been proposed where lysosome destabilization would play a central role in NLRP3 activation [17]. According to this model, phagocytosis of crystals or particles that the cell is unprepared to handle properly would damage the lysosome and lead to leakage of the protease cathepsin B into the cytosol leading to NLRP3 activation. In fact, lysosome rupture, release of lysosomal content, and cell death were observed during ingestion of MSU crystals by neutrophils [46]. Lysosome destabilization as inflammasome-triggering mechanism is attractive because it is consistent with the pathogenic mechanism of bacteria known to activate the NLRP3 inflammasome. NLRP3 activation by the intracellular bacterium *Listeria monocytogenes* depends on the action of the toxin listeriolysin O [43], which causes lysis of the phagosome membrane and allows the bacterium to colonize the cytoplasm. Similarly, NLRP3 inflammasome activation by *Streptococcus pyogenes* depends on expression of the pore-forming toxin streptolysin O [47]. Other microbial toxins such as nigericin or maitotoxin open pores in the membrane bilayer and are NLRP3 activators. In this sense, the NLRP3 inflammasome may be viewed as a guardian of membrane integrity, which in fact is breached by several infectious agents or in pathological situation such as necrosis. In agreement with this model is the early observation by the Tschopp group that mechanical disruption of cells leads to activation of the inflammasome [33]. We have also shown that specific forms of necrosis, during which the cell structure is drastically altered and lysosomes may be damaged, activate the NLRP3 inflammasome [37].

It is interesting to note that oxidative stress is known to damage lysosomal membranes and causes release of cathepsins and other hydrolases. Lysosomal destabilization was also shown to amplify ROS generation through a cathepsin B-dependent mechanism [48, 49]. Thus, ROS generation may be upstream and the cause of lysosomal damage during inflammasome activation by crystals and particles. Interestingly, cytosolic leakage of cathepsin B has a causative role in certain types of cell death and the cathepsin B inhibitor CA0744 was shown to inhibit pyroptosis, a form of necrosis mediated by NLRP3 [50]. However, the involvement of cathepsin B in NLRP3 activation has yet to be convincingly demonstrated. In fact, IL-1β processing in response to silica is only partially blocked by the cathepsin B inhibitor CA0744 [17] and inflammasome activation by amyloid-β was still observed, albeit to a lower degree, in cathepsin B-deficient macrophages [18]. However, more recent results and our own unpublished observation indicate that inflammasome activation in response to hemozoin, MSU, alum, silica, or R837 is not significantly affected in cathepsin B-deficient mice [19]. It has been proposed that cathepsin B and cathepsin L have overlapping functions and that cathepsin-deficient mice upregulate expression of other cathepsins as a compensatory mechanism [51]. It is therefore possible that more than one type of cathepsin is involved in inflammasome activation. The inhibitor CA0744 is specific for cathepsin B at low concentration, but displays a broader inhibitory spectrum with increasing concentration. The role of lysosomal acidification in inflammasome activation is also somewhat controversial since chloroquine or bafilomycinA1 blocks IL-1β maturation in response to silica [17] but not MSU, hemozoin, nigericin, or *Candida* [19, 52].

3.6 Syk Kinase

Syk is a nonreceptor tyrosine kinase that plays a critical role in signaling pathways activated by immunoreceptors that signal through the ITAM motif, such as Fc receptors, Complement receptor 3, and Dectin-1. Recent evidence demonstrated that activation of the NLRP3 inflammasome by *Candida albicans* [52] and the malaria hemozoin [20] requires the tyrosine kinase Syk. Surprisingly, in one of these studies Syk activation was not observed in response to MSU [20], an observation that contrasts to previous work that showed activation of Src and Syk kinases during interaction of MSU with neutrophils [53]. Moreover, MSU interaction with DC also causes Syk kinase activation through membrane cholesterol reorganization [28]. Our unpublished observations indicate that alum activates the Src > Syk > PI3K > Akt axis and that the activities of these kinases are required for inflammasome activation by alum. Interestingly, Syk-deficient neutrophils display very poor activation of the NADPH oxidase in response to bacteria [54] and Syk kinase was shown to be required for generation of ROS in macrophages stimulated with minimally oxidized low-density lipoproteins [55]. Several studies have demonstrated the involvement of Syk in phagocytosis mediated by different receptors. Taken together, these results suggest that the kinase Syk may participate in the early phases of inflammasome activation by particles by regulating phagocytosis and ROS generation.

3.7 Size (Shape and Composition) Matters

It is interesting to note that not all microscopic particles and crystals are able to activate the inflammasome. For example, diesel exhaust particles, which are known to possess proinflammatory activities, are not inflammasome activators [15]. As already mentioned, crystals of allopurinol and aluminum particles are inert [7]. Our work on alum has shown that aluminum hydroxide powder resuspended in PBS at the same concentration found in commercial Alum adjuvant does not stimulate IL-1β release [8]. These observations suggest that the size, shape, and composition of the phagocytosed particle are important determinants for inflammasome activation. This is best illustrated by a study that showed that titanium dioxide nanospheres and nanobelts of less than 5 μm are biologically inert, while nanobelts of length greater than 15 μm can activate the inflammasome in a cathepsin B-dependent way, suggesting the involvement of lysosomal destabilization and NLRP3 [56]. Recent works have shown that the form of *Candida albicans* that most strongly activates the NLRP3 inflammasome is the germ tube, the transition from the yeast to the hyphal form of the fungus [31, 32]. While the diameter of the yeast particle is around 5 μm, the average germ tube length is 12 μm. An effect of size and shape has also been demonstrated for MSU crystals' ability to induce joint inflammation [57]. Thus, the spherical form of MSU was less flogistic than the acicular crystals. It is also interesting to note that the pathologic MSU crystals retrieved from joint aspirations of gout patients are characterized morphologically by uniformity of size and constraint of crystal shape [58]. Interestingly, experimental evidence indicates that serum components regulate crystallization of uric acid in vivo [58, 59]. Studies on asbestos fibers have demonstrated that the chemical composition, and in particular the amount of Fe^{2+} on the surface of the fiber, correlates with its toxicity [60]. This likely reflects the ability of this element to efficiently trigger generation of ROS. Interestingly, treatment of asbestos fibers with the iron chelator deferoxamine abolishes their ability to activate the NLRP3 inflammasome [15].

4 More Open Questions

One major limitation of the in vitro studies that brought us to the present understanding of NLRP3 inflammasome activation by particles is that a priming step is required in order to observe experimental inflammasome activation in cultured cells. This priming step is most commonly represented by LPS stimulation, though we and others demonstrated that other proinflammatory stimuli such as TNF-α can enable inflammasome activation by alum or MSU [8, 61]. The function of the priming step is thought to be the induction of transcription and translation of pro-IL-1β. Recent work [62] has demonstrated that the priming step is also required for increasing transcription of NLRP3, which may be a limiting step for inflammasome activation, despite the fact that NLRP3 is expressed at detectable level in resting cells. Our work shows that in exceptional cases (i.e., cell lines that constitutively express IL-18 or IL-1β) inflammasome activation can proceed in the absence of proinflammatory stimulation [37]. It is important to note that inflammation associated with gout, silicosis, asbestosis, or Alzheimer's occurs in a sterile environment where stimulation of TLR or other inflammatory pathways should be minimal. It is therefore unclear whether this priming step is required in vivo and what it could be. We and others have shown that NLRP3-activating particles by themselves are incapable of promoting pro-IL-1β transcription and secretion of the mature cytokine. Although it has been proposed that CD14, TLR2, and TLR4 mediate the MSU-induced inflammation, suggesting MSU's potential to initiate NF-κB-mediated gene transcription [63], another study has clearly shown the inability of MSU to trigger TLR signaling [5]. We have proposed that IL-18, which is constitutively expressed by a variety of cells even in the absence of infection and that is capable of inducing IL-1β, could be processed in response to NLRP3-activating particles in a sterile environment and then, by acting on several cell types, induce pro-IL-1β transcription leading to the amplification of inflammation [8, 9]. It is also likely that danger signals released by necrotic cells are involved in this priming step. More work is required to clarify this aspect of inflammasome activation.

5 Toward a Unified Model

A plethora of experimental data indicates that particle uptake, ROS generation, and K^+ efflux are essential requirements for NLRP3 inflammasome activation by crystals and particles. On the basis of these observations, the following scenario can be envisioned (Fig. 1). Interaction of the particle with the cell membrane would activate Src and Syk kinases and trigger actin remodeling and phagocytosis (possibly through regulation of Rac GTPases activity). As a result, oxidase systems are activated leading to production of ROS. Concomitantly, K^+ efflux has to occur. The resulting oxidative stress may destabilize the lysosome and cause leakage of cathepsin B,

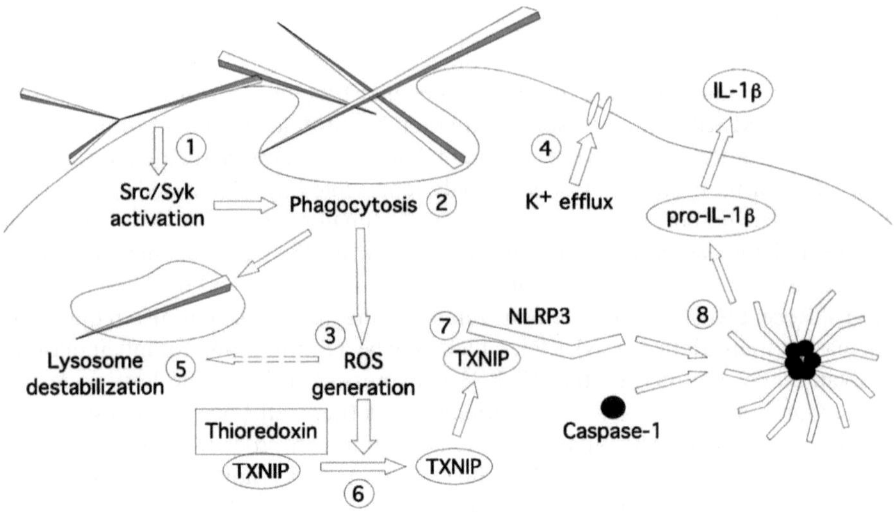

Fig. 1 A plausible model for activation of the NLRP3 inflammasome by particles and crystals. Interaction of the particle with the cell membrane activates Src and Syk kinases (1) and triggers actin remodeling and phagocytosis (2). As a result, oxidase systems are activated leading to the production of ROS (3). Concomitantly, K^+ efflux has to occur (4). The resulting oxidative stress may destabilize the lysosome (5) and cause leakage of cathepsin B, whose role in inflammasome activation remains to be elucidated. ROS generation causes dissociation of thioredoxin from TXNIP (6) and interaction of the latter with NLRP3 (7). This allows assembly of the inflammasome platform, caspase-1 activation, and processing and secretion of IL-1β (8)

whose role in inflammasome activation remains to be elucidated. ROS generation would cause dissociation of thioredoxin from TXNIP and interaction of the latter with NLRP3. This interaction would allow assembly of the inflammasome platform, caspase-1 activation, and processing and secretion of IL-1β.

References

1. Dinarello CA (2009) Immunological and inflammatory functions of the interleukin-1 family. Annu Rev Immunol 27:519–550
2. Martinon F, Mayor A, Tschopp J (2009) The inflammasomes: guardians of the body. Annu Rev Immunol 27:229–265
3. Popa-Nita O, Naccache PH (2010) Crystal-induced neutrophil activation. Immunol Cell Biol 88:32–40
4. Di Giovine FS, Malawista SE, Nuki G, Duff GW (1987) Interleukin 1 (IL 1) as a mediator of crystal arthritis. Stimulation of T cell and synovial fibroblast mitogenesis by urate crystal-induced IL 1. J Immunol 138:3213–3218
5. Chen CJ, Shi Y, Hearn A, Fitzgerald K, Golenbock D, Reed G, Akira S, Rock KL (2006) MyD88-dependent IL-1 receptor signaling is essential for gouty inflammation stimulated by monosodium urate crystals. J Clin Invest 116:2262–2271
6. Shi Y, Evans JE, Rock KL (2003) Molecular identification of a danger signal that alerts the immune system to dying cells. Nature 425:516–521

7. Martinon F, Petrilli V, Mayor A, Tardivel A, Tschopp J (2006) Gout-associated uric acid crystals activate the NALP3 inflammasome. Nature 440:237–241
8. Li H, Nookala S, Re F (2007) Aluminum hydroxide adjuvants activate caspase-1 and induce IL-1beta and IL-18 release. J Immunol 178:5271–5276
9. Li H, Willingham SB, Ting JP, Re F (2008) Cutting edge: inflammasome activation by alum and alum's adjuvant effect are mediated by NLRP3. J Immunol 181:17–21
10. Eisenbarth SC, Colegio OR, O'Connor W, Sutterwala FS, Flavell RA (2008) Crucial role for the Nalp3 inflammasome in the immunostimulatory properties of aluminium adjuvants. Nature 453:1122–1126
11. Kool M, Petrilli V, De Smedt T, Rolaz A, Hammad H, van Nimwegen M, Bergen IM, Castillo R, Lambrecht BN, Tschopp J (2008) Cutting edge: alum adjuvant stimulates inflammatory dendritic cells through activation of the NALP3 inflammasome. J Immunol 181:3755–3759
12. Franchi L, Nunez G (2008) The Nlrp3 inflammasome is critical for aluminium hydroxide-mediated IL-1beta secretion but dispensable for adjuvant activity. Eur J Immunol 38:2085–2089
13. Sharp FA, Ruane D, Claass B, Creagh E, Harris J, Malyala P, Singh M, O'Hagan DT, Petrilli V, Tschopp J et al (2009) Uptake of particulate vaccine adjuvants by dendritic cells activates the NALP3 inflammasome. Proc Natl Acad Sci USA 106:870–875
14. Schins RP (2002) Mechanisms of genotoxicity of particles and fibers. Inhal Toxicol 14:57–78
15. Dostert C, Petrilli V, Van Bruggen R, Steele C, Mossman BT, Tschopp J (2008) Innate immune activation through Nalp3 inflammasome sensing of asbestos and silica. Science 320:674–677
16. Cassel SL, Eisenbarth SC, Iyer SS, Sadler JJ, Colegio OR, Tephly LA, Carter AB, Rothman PB, Flavell RA, Sutterwala FS (2008) The Nalp3 inflammasome is essential for the development of silicosis. Proc Natl Acad Sci USA 105:9035–9040
17. Hornung V, Bauernfeind F, Halle A, Samstad EO, Kono H, Rock KL, Fitzgerald KA, Latz E (2008) Silica crystals and aluminum salts activate the NALP3 inflammasome through phagosomal destabilization. Nat Immunol 9:847–856
18. Halle A, Hornung V, Petzold GC, Stewart CR, Monks BG, Reinheckel T, Fitzgerald KA, Latz E, Moore KJ, Golenbock DT (2008) The NALP3 inflammasome is involved in the innate immune response to amyloid-beta. Nat Immunol 9:857–865
19. Dostert C, Guarda G, Romero JF, Menu P, Gross O, Tardivel A, Suva ML, Stehle JC, Kopf M, Stamenkovic I et al (2009) Malarial hemozoin is a Nalp3 inflammasome activating danger signal. PLoS One 4:e6510
20. Shio MT, Eisenbarth SC, Savaria M, Vinet AF, Bellemare MJ, Harder KW, Sutterwala FS, Bohle DS, Descoteaux A, Flavell RA et al (2009) Malarial hemozoin activates the NLRP3 inflammasome through Lyn and Syk kinases. PLoS Pathog 5:e1000559
21. Coban C, Ishii KJ, Kawai T, Hemmi H, Sato S, Uematsu S, Yamamoto M, Takeuchi O, Itagaki S, Kumar N et al (2005) Toll-like receptor 9 mediates innate immune activation by the malaria pigment hemozoin. J Exp Med 201:19–25
22. Caicedo MS, Desai R, McAllister K, Reddy A, Jacobs JJ, Hallab NJ (2009) Soluble and particulate Co-Cr-Mo alloy implant metals activate the inflammasome danger signaling pathway in human macrophages: a novel mechanism for implant debris reactivity. J Orthop Res 27:847–854
23. Khan SR (2004) Crystal-induced inflammation of the kidneys: results from human studies, animal models, and tissue-culture studies. Clin Exp Nephrol 8:75–88
24. So A, De Smedt T, Revaz S, Tschopp J (2007) A pilot study of IL-1 inhibition by anakinra in acute gout. Arthritis Res Ther 9:R28
25. McGonagle D, Tan AL, Shankaranarayana S, Madden J, Emery P, McDermott MF (2007) Management of treatment resistant inflammation of acute on chronic tophaceous gout with anakinra. Ann Rheum Dis 66:1683–1684
26. Terkeltaub R, Tenner AJ, Kozin F, Ginsberg MH (1983) Plasma protein binding by monosodium urate crystals. Analysis by two-dimensional gel electrophoresis. Arthritis Rheum 26:775–783

27. Ramanathan VD, Badenoch-Jones P, Turk JL (1979) Complement activation by aluminium and zirconium compounds. Immunology 37:881–888
28. Ng G, Sharma K, Ward SM, Desrosiers MD, Stephens LA, Schoel WM, Li T, Lowell CA, Ling CC, Amrein MW et al (2008) Receptor-independent, direct membrane binding leads to cell-surface lipid sorting and Syk kinase activation in dendritic cells. Immunity 29:807–818
29. Zhou R, Tardivel A, Thorens B, Choi I, Tschopp J (2010) Thioredoxin-interacting protein links oxidative stress to inflammasome activation. Nat Immunol 11:136–140
30. Schroder K, Zhou R, Tschopp J (2010) The NLRP3 inflammasome: a sensor for metabolic danger? Science 327:296–300
31. Hise AG, Tomalka J, Ganesan S, Patel K, Hall BA, Brown GD, Fitzgerald KA (2009) An essential role for the NLRP3 inflammasome in host defense against the human fungal pathogen Candida albicans. Cell Host Microbe 5:487–497
32. Joly S, Ma N, Sadler JJ, Soll DR, Cassel SL, Sutterwala FS (2009) Cutting edge: Candida albicans hyphae formation triggers activation of the Nlrp3 inflammasome. J Immunol 183: 3578–3581
33. Martinon F, Burns K, Tschopp J (2002) The inflammasome: a molecular platform triggering activation of inflammatory caspases and processing of proIL-beta. Mol Cell 10:417–426
34. Petrilli V, Papin S, Dostert C, Mayor A, Martinon F, Tschopp J (2007) Activation of the NALP3 inflammasome is triggered by low intracellular potassium concentration. Cell Death Differ 14:1583–1589
35. Kahlenberg JM, Dubyak GR (2004) Mechanisms of caspase-1 activation by P2X7 receptor-mediated K+ release. Am J Physiol Cell Physiol 286:C1100–C1108
36. Lamkanfi M, Mueller JL, Vitari AC, Misaghi S, Fedorova A, Deshayes K, Lee WP, Hoffman HM, Dixit VM (2009) Glyburide inhibits the Cryopyrin/Nalp3 inflammasome. J Cell Biol 187:61–70
37. Li H, Ambade A, Re F (2009) Cutting edge: necrosis activates the NLRP3 inflammasome. J Immunol 183(3):1528–1532
38. Hsu LC, Ali SR, McGillivray S, Tseng PH, Mariathasan S, Humke EW, Eckmann L, Powell JJ, Nizet V, Dixit VM et al (2008) A NOD2-NALP1 complex mediates caspase-1-dependent IL-1beta secretion in response to Bacillus anthracis infection and muramyl dipeptide. Proc Natl Acad Sci USA 105:7803–7808
39. Cain K, Langlais C, Sun XM, Brown DG, Cohen GM (2001) Physiological concentrations of K+ inhibit cytochrome c-dependent formation of the apoptosome. J Biol Chem 276:41985–41990
40. Arlehamn CS, Petrilli V, Gross O, Tschopp J, Evans TJ (2010) The role of potassium in inflammasome activation by bacteria. J Biol Chem 285(14):10508–10518
41. Scheel O, Papavlassopoulos M, Blunck R, Gebert A, Hartung T, Zahringer U, Seydel U, Schromm AB (2006) Cell activation by ligands of the toll-like receptor and interleukin-1 receptor family depends on the function of the large-conductance potassium channel MaxiK in human macrophages. Infect Immun 74:4354–4356
42. Fay AJ, Qian X, Jan YN, Jan LY (2006) SK channels mediate NADPH oxidase-independent reactive oxygen species production and apoptosis in granulocytes. Proc Natl Acad Sci USA 103:17548–17553
43. Mariathasan S, Weiss DS, Newton K, McBride J, O'Rourke K, Roose-Girma M, Lee WP, Weinrauch Y, Monack DM, Dixit VM (2006) Cryopyrin activates the inflammasome in response to toxins and ATP. Nature 440:228–232
44. Pelegrin P, Surprenant A (2006) Pannexin-1 mediates large pore formation and interleukin-1beta release by the ATP-gated P2X7 receptor. EMBO J 25:5071–5082
45. Kanneganti TD, Lamkanfi M, Kim YG, Chen G, Park JH, Franchi L, Vandenabeele P, Nunez G (2007) Pannexin-1-mediated recognition of bacterial molecules activates the cryopyrin inflammasome independent of Toll-like receptor signaling. Immunity 26:433–443
46. Shirahama T, Cohen AS (1974) Ultrastructural evidence for leakage of lysosomal contents after phagocytosis of monosodium urate crystals. A mechanism of gouty inflammation. Am J Pathol 76:501–520

47. Harder J, Franchi L, Munoz-Planillo R, Park JH, Reimer T, Nunez G (2009) Activation of the Nlrp3 inflammasome by Streptococcus pyogenes requires streptolysin O and NF-kappa B activation but proceeds independently of TLR signaling and P2X7 receptor. J Immunol 183:5823–5829
48. Li Z, Berk M, McIntyre TM, Gores GJ, Feldstein AE (2008) The lysosomal-mitochondrial axis in free fatty acid-induced hepatic lipotoxicity. Hepatology 47:1495–1503
49. Zhao M, Antunes F, Eaton JW, Brunk UT (2003) Lysosomal enzymes promote mitochondrial oxidant production, cytochrome c release and apoptosis. Eur J Biochem 270:3778–3786
50. Willingham SB, Bergstralh DT, O'Connor W, Morrison AC, Taxman DJ, Duncan JA, Barnoy S, Venkatesan MM, Flavell RA, Deshmukh M et al (2007) Microbial pathogen-induced necrotic cell death mediated by the inflammasome components CIAS1/cryopyrin/NLRP3 and ASC. Cell Host Microbe 2:147–159
51. Felbor U, Kessler B, Mothes W, Goebel HH, Ploegh HL, Bronson RT, Olsen BR (2002) Neuronal loss and brain atrophy in mice lacking cathepsins B and L. Proc Natl Acad Sci USA 99:7883–7888
52. Gross O, Poeck H, Bscheider M, Dostert C, Hannesschlager N, Endres S, Hartmann G, Tardivel A, Schweighoffer E, Tybulewicz V et al (2009) Syk kinase signalling couples to the Nlrp3 inflammasome for anti-fungal host defence. Nature 459:433–436
53. Desaulniers P, Fernandes M, Gilbert C, Bourgoin SG, Naccache PH (2001) Crystal-induced neutrophil activation. VII. Involvement of Syk in the responses to monosodium urate crystals. J Leukoc Biol 70:659–668
54. Van Ziffle JA, Lowell CA (2009) Neutrophil-specific deletion of Syk kinase results in reduced host defense to bacterial infection. Blood 114:4871–4882
55. Bae YS, Lee JH, Choi SH, Kim S, Almazan F, Witztum JL, Miller YI (2009) Macrophages generate reactive oxygen species in response to minimally oxidized low-density lipoprotein: toll-like receptor 4- and spleen tyrosine kinase-dependent activation of NADPH oxidase 2. Circ Res 104:210–218, 221
56. Hamilton RF Jr, Wu N, Porter D, Buford M, Wolfarth M, Holian A (2009) Particle length-dependent titanium dioxide nanomaterials' toxicity and bioactivity. Part Fibre Toxicol 6:35
57. Fam AG, Schumacher HR Jr, Clayburne G, Sieck M, Mandel NS, Cheng PT, Pritzker KP (1992) A comparison of five preparations of synthetic monosodium urate monohydrate crystals. J Rheumatol 19:780–787
58. McGill NW, Hayes A, Dieppe PA (1992) Morphological evidence for biological control of urate crystal formation in vivo and in vitro. Scand J Rheumatol 21:215–219
59. Kanevets U, Sharma K, Dresser K, Shi Y (2009) A role of IgM antibodies in monosodium urate crystal formation and associated adjuvanticity. J Immunol 182:1912–1918
60. Adachi S, Yoshida S, Kawamura K, Takahashi M, Uchida H, Odagiri Y, Takemoto K (1994) Inductions of oxidative DNA damage and mesothelioma by crocidolite, with special reference to the presence of iron inside and outside of asbestos fiber. Carcinogenesis 15:753–758
61. Franchi L, Eigenbrod T, Nunez G (2009) Cutting edge: TNF-alpha mediates sensitization to ATP and silica via the NLRP3 inflammasome in the absence of microbial stimulation. J Immunol 183:792–796
62. Bauernfeind FG, Horvath G, Stutz A, Alnemri ES, MacDonald K, Speert D, Fernandes-Alnemri T, Wu J, Monks BG, Fitzgerald KA et al (2009) Cutting edge: NF-kappaB activating pattern recognition and cytokine receptors license NLRP3 inflammasome activation by regulating NLRP3 expression. J Immunol 183:787–791
63. Liu-Bryan R, Scott P, Sydlaske A, Rose DM, Terkeltaub R (2005) Innate immunity conferred by Toll-like receptors 2 and 4 and myeloid differentiation factor 88 expression is pivotal to monosodium urate monohydrate crystal-induced inflammation. Arthritis Rheum 52:2936–2946

Inflammasome Activation by Danger Signals

Pablo Pelegrín

Abstract The concept of danger signals as a crucial element to activate the immune response has risen in the last 15 years and has been opposed to the classical model of the immune response recognizing non-self patterns. Danger signals are endogenous host molecules that are *not in place*; such signals include the presence of ATP or uric acid in the extracellular space, the breakdown or release of soluble extracellular matrix components or the presence of dsDNA in the cytosol of the cell. All these danger signals have been proved to activate the release of mature interleukin (IL)-1 through caspase-1 activation. Caspase-1 activation is dependent on the assembly of an intracellular protein platform termed the inflammasome, which is formed by an NLR receptor that is responsible for detecting such danger signals. Knowledge of the involvement of inflammasome activation by danger signals in different diseases is increasing every year and has emerged as responsible for the sterile inflammatory response. This concept of danger signals activating the immune system could be the origin of the complex cytokine network which orchestrates the inflammatory response.

1 Introduction

Back in 1984, when interleukin (IL)-1 was first cloned by Dinarello [1] (Fig. 1), two remarkable characteristics were identified in its amino acid sequence, which make this protein a paradigm for cell biologists and immunologists. The first one was that IL-1 protein sequence lacks a clear signal peptide which would target it to the classical endoplasmatic reticulum–Golgi secretory pathway [1, 2]. However IL-1

P. Pelegrín
Inflammation and Experimental Surgery Group, University Hospital "Virgen de la Arrixaca"-CIBERhed, Murcia Biomedical Research Institute (IMIB)-FFIS, Carretera Madrid-Cartagena s/n, 30120 Murcia, Spain
e-mail: pablo.pelegrin@ffis.es

I. Couillin et al. (eds.), *The Inflammasomes*, Progress in Inflammation Research, DOI 10.1007/978-3-0348-0148-5_7, © Springer Basel AG 2011

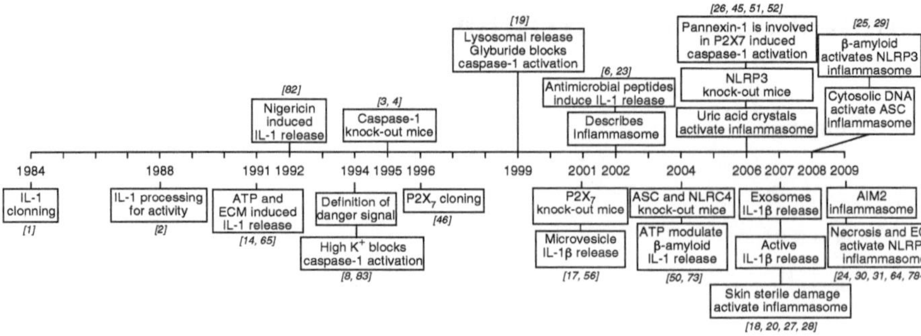

Fig. 1 Chronological key discoveries for IL-1β processing and release in response to danger signals. The number in *square brackets* is indication to the references

had a clear extracellular signalling role, raising the concept of unconventional secretion and export of proteins directly from the cytosol to the extracellular space. The second remarkable concept for a cytokine was that IL-1β needs post-translational processing to generate its mature bioactive form [2–4], introducing the concept of two regulated steps for its release (transcription and post-transcriptional processing). Nowadays, 25 years after the cloning of IL-1, both paradigms are still puzzling and are at the forefront of innate immune research. We still do not fully understand how IL-1β is processed and released to the extracellular milieu, but we understand that both mechanisms, transcription and processing, are tightly regulated [5]. In the last 8 years, major advances have been made in understanding the detailed post-transcriptional processing of IL-1β after the characterization of the inflammasome [6]. The inflammasome is a recent concept that has been quickly accepted by the scientific community, since the number of publications in which this term appears in the title has exponentially increased over the last 4 years. This is mainly due to its involvement in an increasing number of diseases, including sterile-inflammatory conditions where the immune system is activated in the absence of pathogen-derived signals [5, 7] (Table 1). In this chapter, we summarize all the available data on the inflammasome activation by danger signals.

2 Danger Signals: Basic Signals to Activate the Immune System

The concept of danger signals arose in the immunology field in 1994 after Polly Matzinger proposed a novel concept to activate the immune system called the "danger model" [8]. This model is based on the recognition of danger and destruction-associated signals by the immune system, rather than the classical distinction between the self and nonself model evolved from clonal lymphocyte selection [8, 9]. In the danger model, danger signals are self-components that are not in the

Table 1 Sterile inflammatory conditions mediated by danger signals activating the inflammasome

Condition/pathology	Specie[a]	Danger signal	Inflammasome	IL-1 involvement[b]	Refs.
Alzheimer's	m	β-Amyloid/ATP-$P2X_7$	ASC	IL-$1R^{-/-}$	[25, 57]
Bone remodelling	m	ATP-$P2X_7$/ECM	?	?	[65, 88]
CAPS	h	NLRP3 SNPs	NLRP3	Use of anakinra	[89]
Contact hypersensitivity	m	Irritants	NLRP3/ASC	IL-$1R^{-/-}$	[28, 52]
Diabetes (type 2)	h/m	High plasma glucose	NLRP3	Use of anakinra	[86, 90]
Gout and pseudogout	m	MSU	NLRP3	?	[26]
Inflammatory pain	m	ATP-$P2X_7$	?	?	[91]
Ischemic brain injury	h/m	?	?	Use of anakinra and IL-$1\alpha/\beta^{-/-}$	[92, 93]
Lung injury	m	Necrosis	NLRP3/ASC	?	[30]
Neuropathic pain	r/m	ATP-$P2X_7$	?	IL-$1\alpha/\beta^{-/-}$	[53, 91, 94]
Renal dysfunction	m	ECM/ATP-$P2X_7$	NLRP3	?	[31, 64, 95]
Rheumatoid arthritis	h/m	ATP-$P2X_7$	?	?	[54, 55]
Skin burn	m	ECM/?	NLRP3	?	[24, 27]
Tumour immunity	m	ATP-$P2X_7$	NLRP3	IL-$1R^{-/-}$	[44]

CAPS cryopyrin-associated periodic syndrome, *ECM* extracellular matrix, *MSU* monosodium urate crystals, *SNP* single nucleotide polymorphism, ? not determined-unknown

[a]Specie studied. *h* Human, *m* mouse, *r* rat

[b]IL-$1R^{-/-}$: use of IL-1 receptor-knock out mice; IL-$1\alpha/\beta^{-/-}$: use of IL-1α/β double-knock out mice, using these transgenic mice a reduced associated damage was found; use of anakinra in clinical trials or animal models was found beneficial

right place, such as normal cytosolic components that reach extracellular space under injury conditions. Professional antigen presenting cells (APCs) are primed by such danger signals to develop an appropriate immune response [9]. Danger signal patterns are recognized by *P*attern *R*ecognition *R*eceptors (PRRs) present in APCs [9]. It is well known that danger-associated molecular pattern recognition overlaps in some cases with pathogen pattern recognition, since some receptors could recognize both pathogen-associated molecular patterns (PAMPs) and danger signals. Such is the case of Toll-like receptors (TLRs) which can recognize the degraded products of some extracellular matrix components as well as bacterial and fungal compounds [10].

These host-derived danger signals were first identified as molecules derived from injured or distressed cells and tissues, and the precise nature of these signals is emerging. Now the list of danger molecules also includes signals actively released after cell activation and includes among others: host DNA, nucleotides (like ATP), heat shock proteins (Hsp), nuclear proteins (like HMGB), cytokines and breakdown products of the extracellular matrix components, such as hyaluronan [10]. In fact, the term "danger signal" could be extended to any misplaced protein or molecule in the cell or in the body. This theory explains the origin of the cytokine network, and the IL-1 family of cytokines is a clear example of such evolution. Although IL-1 cytokines play an important extracellular signalling role, they lack a classical secretion signal peptide and surprisingly they present a nuclear localization signal (like IL-1α and IL-33) [11, 12]. Therefore, early in evolution IL-1 cytokines were nuclear proteins; in fact mammalian pro-IL-1α, pro-IL-1β and pro-IL-33 localize to the nucleus, although their nuclear function remains to be elucidated [11, 12]. Under conditions of cellular damage these nuclear proteins reach the extracellular space and then signal as danger messengers in the form of early pro-inflammatory cytokines [11, 12]. Similarly, receptors for danger signals could be the first line of defence to detect such early damage-related signals; in fact IL-1 receptors and TLRs share the same intracellular transducing domain (the *T*oll */I*L-1*R* (TIR) domain) and activate the same intracellular signalling cascades [13]. This concept situates cytokines as evolved danger signals to initiate the immune response. In line with this, we could apply the same concept to PRRs and argue that each type of pathogen will have evolved its own way of binding to a specific PRR, which has a role to detect endogenous danger signals [9].

The central core of the inflammasome consists of a PRR of the NLR (*N*ucleotide-binding domain and *L*eucine-rich repeat *R*eceptor) family. Such receptors are present in the cytosol of cells and are able to recognize intracellular danger signals, supervising the sanctity of the cytosol [6, 7]. Upon activation, NLRs oligomerize and recruit other proteins by homotypic domain interaction; the most important is the adaptor protein ASC (*A*poptotic *S*peck-like protein with a *C*aspase-activating recruiting domain), together with the cysteine protease pro-caspase-1 to induce the activation of this inflammatory caspase. Active caspase-1 processes inactive pro-forms of IL-1β and IL-18, inducing the release of these active cytokines to the extracellular milieu [5–7].

3 Evolving Concepts: IL-1β Processing and Release in Response to Danger Signals

Classical biochemical experiments to study the activation of the inflammasome are based on the detection of the cleaved products produced by caspase-1 activity. The easiest and most used method is the detection of the 17 kDa mature form of IL-1β in the extracellular milieu (Fig. 2). Often, the processing of IL-1β by the inflammasome/caspase-1 pathway is tightly coupled to its release [5–7]. In these experiments, a common problem is to differentiate between active IL-1β release and the release as a consequence of cell death, since over-activation of the inflammasome leads to cell death in the form of pyroptosis or apoptosis [7]. This fact has complicated the study of IL-1β processing and release pathway, and for many years, the release of IL-1β was associated with cell death [14]. In the late 1990s cell culture, stimulation and detection techniques were refined and both pathways, IL-1β processing/release and cell death, were clearly separated [15–18]. Actually, four active release pathways for IL-1β have been identified, including lysosomal release, plasma membrane excision of microvesicles, exosome release and the direct crossing of the plasma membrane using complex unidentified release mechanisms [17–20]. All four mechanisms could coexist and their relative

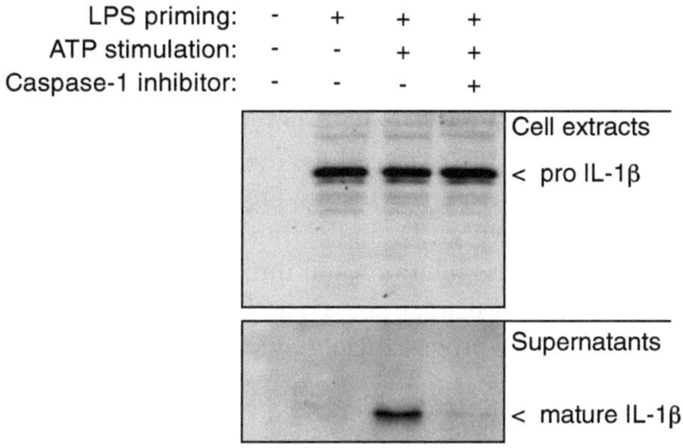

Fig. 2 Classical biochemical experiment to determine inflammasome activation. Example of IL-1β production and release in human macrophage cell line THP-1 detected by Western blot. Un-primed macrophage does not present any detectable IL-1β; stimulation with LPS (1 μg/ml for 4 h) results in the activation of NF-κB signalling and the cellular accumulation of IL-1β precursor (pro-IL-1β; 35 kDa), but does not induce its release. Further activation of the $P2X_7$ receptor-signalling pathway by extracellular ATP (5 mM for 20 min) results in the release of mature IL-1β (17 kDa). Treatment with the caspase-1 inhibitor (Ac-YVAD-AOM, 10 μM) prior to ATP stimulation completely blocks the release of mature IL-1β. Anti-IL-1β antibody used in Western blot was 3ZD (Biological Resources Branch, NCI). Experiments shown are similar to the ones in [16, 45, 48, 62] and detailed protocols could be found in these references

importance depends on the macrophage type and their activation state [17–20]. However, these IL-1β release pathways have only been studied after the activation of the inflammasome using extracellular ATP.

This is due to extracellular ATP being the first identified signal that potently activates the processing and release of IL-1β from macrophages [14], and since 1991 has been the most widely used and studied danger signal to activate the inflammasome [21] (Figs. 1 and 2). Surprisingly extracellular ATP is not recognized directly by a PRR, but is recognized by a family of plasma membrane receptors called purine P2X receptors and the different receptors of this family have evolved to recognize different concentrations of ATP in different tissues [22]. Little is known about all seven members of the P2X family, but the four best characterized receptors are associated with different distress conditions and pathologies. For example, $P2X_2$ and $P2X_3$ are involved in pain sensation in the nervous system; meanwhile, $P2X_4$ and $P2X_7$ are activated in pro-inflammatory conditions, as well as in neuropathic and inflammatory pain [22]. Among all P2X receptors, $P2X_7$ is the only one which is activated by high concentrations of extracellular ATP (in the mM range) and that specifically couples to the activation of the inflammasome. Thus, active drug discovery programmes have been launched by big pharma companies with the aim to create potent and selective $P2X_7$ receptor antagonists to use as novel anti-inflammatory drugs [21, 22].

It was not until more than 10 years after the discovery of ATP-induced IL-1β release when a new family of danger molecules were described to activate the release of mature IL-1β through the action of certain host-derived antimicrobial peptides on primed APCs [23] (Fig. 1). In the last 4 years, an increase in the number of danger signals that specifically activate the inflammasome, including uric acid crystals, skin damage by external agents, β-amyloid, cytosolic DNA, cell necrosis or degraded components of the extracellular matrix, has been described [24–31] (Fig. 1). All danger signals (except cytosolic DNA) appear to activate specifically the NLRP3 inflammasome (Fig. 3), indicating a central role for this molecule sensing potentially dangerous situations and initiating the inflammatory response.

Evolutionarily, the link of danger signals activating the inflammasome appears to be an earlier event than the processing and release of pro-IL-1β (Fig. 4). The first

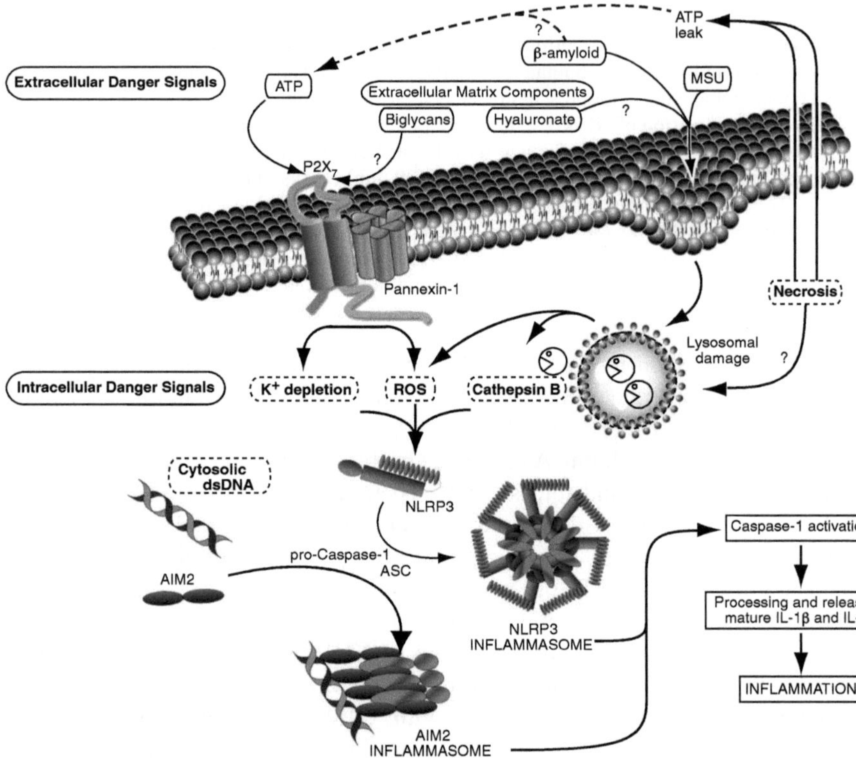

Fig. 3 Inflammasome activation by danger signals. Extracellular danger signals follow two main pathways to activate NLRP3 inflammasome: (1) $P2X_7$/pannexin-1 activation or (2) endocytosis with subsequential lysosomal damage. Both pathways convey in three main intracellular danger signals: depletion of cellular K^+, production of ROS and the presence of cathepsin B activity in the cytosol. However, AIM2 inflammasome is directly activated by the presence of cytosolic dsDNA. Formation of NLRP3 or AIM2 inflammasome requires the adaptor protein ASC and results in the activation of the cystein protease caspase-1. Active caspase-1 processes inactive pro-IL-1β and pro-IL-18 into its biologically active cytokine. Mature cytokines will be subsequently released to the extracellular milieu by unconventional secretion system and will signal inflammation. Schemes used in this figure are freely available at http://www.inflammation.ffis.es

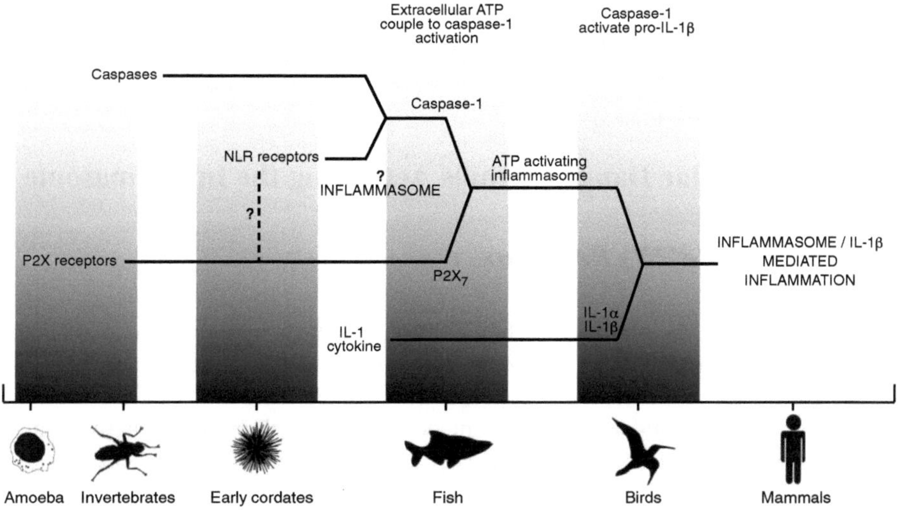

Fig. 4 Diagram illustrating evolution of NLR receptors and activation by danger signals. Purine P2Xs receptors appear early in evolution and can be found in amoeba and algae, caspases appear in invertebrates and NLR expand in early chordates. The appearance of the link between NLR receptors and the activation of caspase is not well determined, but should appear between early chordates and fish, since the presence of inflammatory caspase-1 (with a CARD domain) is present in fish. Therefore, in early chordates NLR receptors should play an unknown function independent of caspases and remain to be elucidated if early P2Xs receptor stimulation couple to NLR receptor activation. $P2X_7$ subunits are present for first time in fish where it has been demonstrated to activate fish-caspase-1; however activation of caspase-1 in these animals is independent of the release of IL-1, since early sequences of IL-1 lack a caspase-1 cleavage site. IL-1 cytokines appear early in fish as a single IL-1β and no IL-1α. Birds are the first animal group where IL-1β sequence presents a clear caspase-1 cleavage site and where IL-1α and IL-1β sequences appear. After the separation of both IL-1 sequences, pro-IL-1α was able to signal without processing; however, pro-IL-1β was inactive and requires a processing step in order to generate a bioactive form. Therefore, in the precise moment in which IL-1β sequence suffers a mutation and the caspase-1 cleavage site appears, a biologically active mature IL-1β was generated and two important signalling pathways merge together, linking inflammasome activation via danger extracellular ATP to the release of pro-inflammatory IL-1β

evidence for such a hypothesis comes from the presence of NLR proteins in early chordates and P2X receptors in amoeba. Meanwhile, cytokines appear in vertebrate genomes [32–34]. In fact, early chordates contain a larger number of NLR receptors in their genomes compared with mammalian genomes [34]. The second evidence is that the caspase-1 processing site in pro-IL-1β appears late in birds and is absent in fish, indicating that either fish pro-IL-1β is maturated by other protease or is bioactive as a precursor [33]. Furthermore, fish do not present the closely related IL-1α in their genomes. This may suggest that fish IL-1β fulfils both IL-1α and IL-1β activities in this animal group. It has been demonstrated that in fish, $P2X_7$ receptor activation was able to couple to caspase-1 activation but not to pro-IL-1β processing or release [35, 36]. Therefore, we expect to discover new roles for the inflammasome, conserved through evolution, and separated to the activation of pro-inflammatory IL-1 cytokines, and fish are the ideal model for such studies.

4 Extracellular Danger Signals Activating the Inflammasome

4.1 ATP and $P2X_7$ Receptors

In healthy tissues, the plasma membrane of living cells is impermeable to ATP and ATP remains intracellularly linked to metabolism. Thus, extracellular levels of ATP are low. However, in distressed or damage tissues extracellular ATP concentrations rise and act as a potent danger signal through the activation of purine receptors in effector cells [22, 37]. The release of ATP to the extracellular milieu could be due to a lytic-passive release coupled to cell death or to non-lytic regulated release. Different immune cell types have been described to release ATP under certain activation conditions. For example, vesicular ATP release has been reported in activated platelets, neutrophils and mast cells [38–40], whereas hemi-channel ATP release has been reported in monocytes [41]. ATP can also be released from microbial flora and pathogens [42]. In vivo, the presence of high concentrations of extracellular ATP has been demonstrated in the milieu of solid tumours [43] and the in vivo importance of the extracellular ATP activating the $P2X_7$–inflammasome pathway in the immunity against tumours has been recently demonstrated [44].

$P2X_7$ receptors are unspecific cation channels (Na^+, Ca^{2+} and K^+), causing membrane depolarization after agonist (ATP) binding. However, $P2X_7$ is an unusual ion channel and prolonged agonist activation (sec) couples to a specific cascade of secondary signals, which are important for inflammasome activation [16, 17, 22, 37, 45, 46]. $P2X_7$ receptor activation quickly leads to rearrangements of plasma membrane lipids, where phosphatidylserine is exposed in the outer leaflet of the plasma membrane [47], the opening of pannexin-1 hemichannels, which leads to cell permeabilization of molecules up to 900 Da in size [45, 48], the release of lysosomes [20], the production of reactive oxygen species (ROS) [49], the production of plasma membrane blebs [17, 47] and the excision of microvesicles containing mature IL-1β [17, 47]. All these events are fully reversible if ATP is washed from the receptor, and these $P2X_7$-reversible apoptotic events have been termed pseudo-apoptosis [47]. The dramatic drop in cytosolic K^+ levels, the pannexin-1 cell permeabilization phenomena and the production of ROS generate the ideal environment for NLRP3-inflammasome assembly [37].

The detailed mechanisms that trigger inflammasome activation are not fully understood; however, the generation of ASC- and NLRP3-deficient mice demonstrates that both proteins are important for $P2X_7$-dependent caspase-1 activation [50–52]. The high concentration of ATP needed to activate $P2X_7$ receptors in vitro has been the main obstacle to fully accept this pathway as a physiological route to activate the inflammasome. However, there is clear evidence for the involvement of $P2X_7$ receptors in different inflammatory conditions. Such evidence comes from the use of specific $P2X_7$ receptors antagonist in animal models or in clinical trials [21, 22, 53–55] and the use of mice lacking $P2X_7$, which present altered cytokine production and deficiencies in inflammatory models [22, 54, 56, 57]. All this evidence demonstrates that extracellular ATP can act as a danger signal to activate the NLRP3 inflammasome and promote caspase-1 activation and release of mature IL-1β.

There are three plausible hypotheses that could explain the activation of $P2X_7$ receptors in vivo: (1) under in vivo injured or pathological conditions $P2X_7$ could be activated by lower concentrations of ATP than the required in vitro. It has been reported that $P2X_7$ has an additional high-affinity site for ATP that responds to low concentrations (1–10 μM) of ATP [58]; (2) ATP could be locally concentrated in plasma membrane containing $P2X_7$-enriched micro-domains reaching high levels in such places [59]; (3) in injured areas extracellular levels of ATP could be high enough to activate $P2X_7$ receptors, like those found in the core of solid tumours [43, 44]. In any case, detailed in vivo studies are needed to solve the extracellular levels of ATP and $P2X_7$ receptor activation during pathogenic/injury conditions.

On the other hand, extracellular ATP levels are tightly controlled by ecto-nucleotidases that degrade ATP, thus keeping, in normal conditions, extracellular ATP levels low [60]. Under distress or damage conditions ecto-nucleotidases cannot clear all exceeding levels of released ATP, resulting in a net increase of ATP signalling. When damaged tissue is repaired and the inflammation has to resolve, extracellular ATP is cleared into two main anti-inflammatory products: adenosine and pyrophosphates (PPi) [61, 62]. Adenosine acts through adenosine receptors to raise cytosolic cAMP, known to be a potent anti-inflammatory signal [61]. Recently, we have shown that extracellular PPi is a potent NLRP3-inflammasome inhibitor, discovering one of the first drug-like small molecules with direct inflammasome inhibitory capabilities [62]. We demonstrated that bisphosphanates, a PPi drug analogue, was equally able to block casapse-1 activation by disrupting NLRP3-inflammasome assembly [62]. However, further experiments are needed to describe the inflammasome inhibitory role of PPi and analogues in animal models of inflammation.

ATP contamination or addition to macrophage preparations is a common problem that has led to falsely report some novel inflammasome activators, overlooking the fact that $P2X_7$ receptor activation lay behind such conclusions. Recently, it has been elegantly shown that human monocyte release ATP in an autocrine way, reaching concentrations high enough to activate $P2X_7$ and the inflammasome pathway [41, 63]; therefore, endotoxin priming of human monocytes was enough to detect mature IL-1β in cell supernatants. This autocrine ATP release is lost after monocyte maturation to macrophage and therefore macrophage needs two-step protocol to release IL-1β (TLR priming and $P2X_7$ activation by exogenous ATP) [41, 63]. In fact, as detailed below, some other danger molecules could be acting by inducing the autocrine release of ATP from cells and NLRP3-inflammasome activity is due to $P2X_7$ receptor activation, such is the case for biglycans, β-amyloid or necrotic cell death [31, 57, 64].

4.2 Extracellular Matrix Components

In 1991, a pioneering study demonstrate that bone particles were potent inducers of bioactive IL-1β release, describing bone collagen and hydroxyapatite as the active bone extracellular matrix components inducing IL-1β release [65]. This study leads to the idea that degradation of extracellular matrix components (ECMs) could be a direct danger signal, which activate the immune system and regulate the inflammatory response. Recently, there is evidence that the ECM exerts much more complex functions than being a mere scaffold for cells to attach [66]. There are many reports of PRR recognizing endogenous ECM breakdown products that are exposed during cellular injury and extracellular matrix remodelling. For instance, TLR2 or TLR4 have been implicated in the recognition of hyaluronan fragments, heparan sulphate, fibrinogen, fibronectin extra domain A, lung surfactant protein A and biglycans [31, 66]. However, only biglycans and hyaluronan have been shown to activate the NLRP3 inflammasome [24, 31].

Biglycans are glycoproteins of the ECM in many tissues; during tissue injury its soluble form is released from the ECM and is also secreted from activated macrophages at sites of damage and inflammation [31, 66]. It has been demonstrated that biglycans are direct ligands for TLR2 and TLR4, inducing the expression of NF-κB-dependent genes, including pro-IL-1β and NLRP3 in macrophages [31]. Biglycans are also able to activate the NLRP3/ASC inflammasome by activating purinergic $P2X_7$ receptors and in consequence caspase-1 gets active and mature IL-1β is released. The requirement of purinergic receptors in the biglycan model raises the question if this ECM component induces an autocrine release of ATP from cells and this ATP will be the mediator to activate the inflammasome. Attempts to measure ATP release from biglycan-treated cells have failed [31], but could be due to low sensitivity of the technique employed. In any case, biglycan-deficient mice displayed a reduced inflammatory infiltrate and lower levels of active caspase-1 and IL-1β in a model of non-infectious inflammatory renal injury and in LPS-induced septic shock [31], strongly supporting a pro-inflammatory role of biglycan in vivo.

Hyaluronan is a glycosaminoglycan released from the ECM after tissue injury, acting as an endogenous signal of trauma and inducing the release of chemokines [66]. Recently, it was demonstrated to be a danger signal that activates both TLR4 signalling and the NLRP3 inflammasome [24]. Hyaluronan signalling to the inflammasome was dependent on CD44 endocytosis and subsequent hyaluronidase activity in endosomes/lysosomes, where hyaluronan is hydrolyzed to small oligosaccharide fragments. These oligosaccharides are responsible for activating the NLRP3-inflammasome, thus generating active IL-1β [24].

ECM components are the first identified non-pathogenic signals that stimulate both pro-IL-1β synthesis and processing, avoiding the classical two-step protocol that requires an initial endotoxin priming step before inflammasome activation by danger signal. These findings support the hypothesis that ECM components could signal through TLR and the NLRP3/ASC inflammasome in conditions of sterile inflammation.

4.3 Uric Acid Crystals

Activation of the inflammasome by particles (including uric acid crystals) is discussed in a different chapter of this book. Therefore, I only mention the ability of uric acid crystals to act as a danger signal activating the inflammasome.

Uric acid is the final product of purine catabolism and is released by dying cells. High levels of circulating uric acid have been associated with various inflammatory diseases [67]. Extracellular uric acid reacts with free Na^+ to form monosodium urate (MSU) crystals and crystals being the biologically active structure which activates the immune system. MSU was first identified to induce adjuvanticity in vivo and later on was found to stimulate the NLRP3/ASC inflammasome to produce active IL-1β [26]. Mouse deficient in NLRP3, ASC or IL-1 receptor (IL-1R) has reduced crystal-induced peritonitis in response to MSU [26]. These findings provide new molecular mechanisms into inflammatory diseases of gout and pseudogout.

4.4 β-Amyloid

Alzheimer's disease is the leading cause of progressive dementia in humans and is characterized by the presence of senile plaques in the grey matter of the cortex, which are mainly composed of aggregates of β-amyloid protein [68]. The use of non-steroidal anti-inflammatory drugs reduces the risk, and delays the onset of the disease [69], suggesting a potential pro-inflammatory role in Alzheimer's disease pathology. In fact, IL-1β is accumulated in cells surrounding such β-amyloid plaques and is present in the cerebrospinal fluid of patients suffering this disease [70, 71].

Recently, it was demonstrated that β-amyloid is able to induce the release of IL-1β in vitro from endotoxin-primed microglia [25, 72]. β-Amyloid-induced release of IL-1β was dependent on caspase-1 activation via the NLRP3/ASC inflammasome [25]. Inflammasome activation occurs upon microglia phagocytosis of β-amyloid followed by lysosomal damage and release of cathepsin B to the cytosol [25]. In vivo, β-amyloid induces microglia migration, activation and neuronal killing; processes dependent on IL-1β, since recruitment and damage, were abrogated in mice deficient in ASC, caspase-1 or IL-1R [25].

Microglia also express high amounts of functional $P2X_7$ receptors and ATP stimulation was able to modulate the release of IL-1β induced by β-amyloid [73]. Recently, $P2X_7$ receptor was found to be a non-dispensable factor for β-amyloid-mediated microglia stimulation and IL-1β release [57]. In that sense, β-amyloid was able to induce autocrine ATP release from microglia which activate the $P2X_7$-dependent NLRP3 inflammasome in vitro [57]. Importantly, intrahippocampal injection of β-amyloid triggers IL-1β accumulation in the hippocampus of wild type but not in $P2X_7$-deficient mice, suggesting that in vivo, extracellular ATP could reach enough levels to induce the activation of $P2X_7$ receptors during Alzheimer's disease [57]. Further in vivo experiments are required to demonstrate if β-amyloid can activate NLRP3 exclusively by the lysosomal pathway, or whether there is also concomitant $P2X_7$ receptor activation.

4.5 Danger Signals During Skin Inflammation

Agents causing skin inflammation have been recently considered as danger signals or as inducers of danger signals from local cells and ECM [24, 74]. Exogenous danger signals causing sterile injury in the skin such as freezing, ultraviolet B (UVB) radiation or sensitizing chemicals [24, 27, 28, 52] have been reported to induce less tissue damage and inflammation in mice deficient in NLRP3 or ASC. Freezing temperature was able to degrade ECM and induce the release of hyaluronan and activate inflammasome via the lysosomal pathway (see above Sect. 4.2) [6, 24]. There is no current information if the other chemical or physical agents cause direct NLRP3 activation or induce the release of an associated danger signal in the skin that mediates inflammasome assembly.

Recently, it has been demonstrated that UVB radiation induces an increase of cytoplasmic Ca^{2+} required in keratinocytes for inflammasome activation and subsequent release of mature IL-1β [27]. The requirement of cytosolic Ca^{2+} for caspase-1 activation is a controversial issue; for example, ATP-mediated release of mature IL-1β from different macrophage and microglia has been variously observed to have a strict requirement for Ca^{2+} or to have little or no dependence on Ca^{2+} [16, 20, 75]. Therefore, it cannot be concluded if inflammasome activation is dependent on the intracellular Ca^{2+} levels.

Contact sensitizers or skin irritants have been found to activate, in vitro, the release of mature IL-1β from APCs and keratinocytes in an NLRP3/ASC-dependent inflammasome [28]. A skin sensitization model using trinitrophenylene chloride or trinitrochlorobenzene failed to induce ear swelling in NLRP3- or ASC-deficient mice [28, 52]. Inflammasome activation during skin sensitization is important at both the early and the late phase response of contact hypersensitivity reaction [74]. Inflammasome activation is known to be an early innate response at the level of keratinocytes, APCs, or accessory cells, such as mast cells, and therefore with an impaired early phase, late T-cell-dependent phase development is affected [74]. Before such chemicals can be considered as danger signals, which directly bind or activate the NLRP3 inflammasome, further characterization is needed to study if direct tissue damage induced by such chemicals is the actual danger signal to the inflammasome, such as the breakdown of ECM components or the release of extracellular ATP.

5 Intracellular Danger Signals Activating the Inflammasome

5.1 Cytosolic DNA

The presence of pathogen double strand (ds) DNA in the cell is a potent PAMP signal to detect the genomes or the transcription/replication of invading pathogens; classical recognition of pathogen dsDNA is mediated through TLR9 receptors [76]. However, several clinical sterile syndromes such as systemic lupus erythematosus (SLE) are characterized by auto-immunity to self-DNA, indicating that endogenous non-microbial dsDNA can also activate the immune system [77]. Endogenous host dsDNA is a potent danger signal when not in the nucleus and is present in the cytosol of the cell. Recently, the discovery of a cytosolic protein with dsDNA binding motifs which activates the innate immune system had helped to develop such idea [76].

The first report of cytosolic dsDNA activating the inflammasome came from Tschöpp's group, where they reported that any kind of dsDNA (mammalian, bacterial or viral) bigger than 250 bp in the cytosol was able to activate the release of mature IL-1β [29]. All types of dsDNA induced caspase-1 activation in an ASC-dependent manner. Surprisingly, viral and bacterial DNA activated the NLRP3 inflammasome, whereas mammalian dsDNA activated caspase-1 in an NLRP3-independent manner [29]. One year after, four different groups reported the existence of a new inflammasome, which surprisingly does not contain any NLR and plays a key role in cytosolic host dsDNA recognition and activation of caspase-1 [78–81]. Such an inflammasome is composed by the *A*bsent *in* *m*elanoma-*2* (AIM2) protein, which is expressed in the cytosol and contains an HIN200 and a pyrin domain. AIM2 binds directly to dsDNA, but not ssDNA through the C-terminal HIN200 domain resulting in AIM2 oligomerization and the recruitment of ASC through homotypic pyrin interactions. Interestingly, these results document the assembly of an inflammasome via a direct interaction between the receptor and its ligand. The subsequent recruitment and activation of caspase-1 result in the formation of an AIM2 inflammasome that mediates IL-1β maturation and release [78–81]. The AIM2 inflammasome, through ASC recruitment, also induces pyroptotic cell death [79]. Therefore, AIM2 has emerged as a key regulator of caspase-1 activation and mature IL-1β release after sensing potentially dangerous cytoplasmic DNA, responding to viral, bacterial and host dsDNA [67, 80].

5.2 *Necrosis*

Necrosis is an unexpected cell death via a damage or infection and is contrary to apoptosis, which is the natural programmed cell death. Necrosis is known to be a key activator of the innate immune response and mediates, for example, sterile inflammatory responses [9, 10, 67]. Necrosis signals alert an unexpected dangerous situation. Thus, necrosis could be considered as a danger signal to activate the inflammasome on the necrotic cell and to induce the release of necrotic-related danger signals to surrounding cells [30, 64].

It has been recently reported that necrotic cells release high amounts of ATP to the extracellular environment and activate $P2X_7$-dependent NLRP3 inflammasome in surrounding APCs [64]. During sterile renal ischemic acute tubular necrosis, NLRP3 deficiency protects animals against mortality, renal dysfunction and impairs an inflammatory response [64]. In such models, necrosis induces the release of ATP, which will act as a danger signal (see Sect. 4.1).

Another recent study has shown that necrosis induced by a specific indirubin oxime derivative was a potent inducer of caspase-1 activation and IL-1β release [30]. The activation of the inflammasome on the necrotic cell was independent of the autocrine release of uric acid or ATP, danger signals known to be released during necrosis. Indirubin oxime-induced caspase-1 activation was dependent on cathepsin B activity, suggesting lysosomal damage as a potential pathway to activate the inflammasome in response to this necrosis [30]. The activation of the inflammasome was dependent on ASC and NLRP3, but independent of NLRC4 [30]. Notably, other necrotic substances failed to activate the inflammasome [14, 30], suggesting a specific mechanism for the specific indirubin oxime derivative used in this study (7-bromoindirubin-3′-oxime), which induces NLRP3 inflammasome activation, more than a general necrosis mechanism [30]. In any case, NLRP3- or ASC-deficient mice show a reduced release of IL-1β in vivo during naphthalene-induced lung injury, which results in selective necrosis of the epithelial cell lining the airway [30].

5.3 *Decrease of Intracellular K^+, Production of ROS and Presence of Cytosolic Cathepsins*

There are three intracellular events non-dispensable to activate the *danger sensing* NLRP3 inflammasome [49, 82–86]. All of them will cover a full chapter about the regulation of the inflammasome in this book, so I only focus on the potential of such intracellular conditions as *intracellular* danger signals. The need of low intracellular K^+ to activate the release of mature IL-1β was described by Gabel's group back in 1992 [82, 83]. Excitable cells (such as neurons) uses internal K^+ depletion to repolarize membrane potential, with normal levels of K^+ quickly restored by

Na^+/K^+ pumps. However, in non-excitatory cells K^+ depletion is a death-related signal [87], which could be sensed by the NLRP3 inflammasome.

Production of excessive ROS is considered a beneficial signal to clearance invading pathogens; however, sterile production of ROS is a potent danger signal that damages cellular DNA. Overproduction of ROS has been recently implicated in the direct activation of the NLRP3 inflammasome in response to different extracellular danger signals such as ATP or MSU [49, 86].

Finally, lysosomal damage and the presence of active cathepsin B in the cytosol have been demonstrated as an indispensable step to activate the inflammasome in response to particulate extracellular danger signals such as β-amyloid or MSU [25, 85]. All three signals could be viewed as intracellular danger signal pathways that traduce the different extracellular danger environments to the intracellular activation of the NLRP3 inflammasome (Fig. 3).

6 Concluding Remarks

In summary, the inflammasome is a primary sensor for endogenous danger molecules that are *not in the right place*. Such molecules could be released after tissue or cell damage and signals toward the innate immune system to start sterile inflammation. The presence of extracellular danger molecules such as ATP or uric acid, which are released by necrotic or injured cells to the extracellular milieu is well reported [7, 26, 50, 51, 67]. Extracellular matrix degradation in response to tissue injury is another class of emerging danger signal, and release of biglycans or hyaluronan has been demonstrated to potently activate both TLR and NLR pathways, being able to induce transcription of pro-IL-1β and activation of caspase-1 [24, 31]. All these mechanisms could form the basis of sterile skin inflammation caused by damaging stimuli such as UVB or chemical sensitizers, acting at the early establishing phase of this reaction [74]. Certain pathologies are characterized by their very own particular danger signal; such is the case of Alzheimer's disease and the aggregates of β-amyloid, which potentially activates NLRP3 inflammasome in combination with $P2X_7$ receptor activation [25, 57] (Fig. 3).

Intracellular danger signals are recently emerging; novel inflammasomes are described to recognize host endogenous dsDNA that is not in the nucleus. This is the case of the AIM2 inflammasome activation in response to endogenous host cytosolic dsDNA [45, 76]. This discovery suggests the possibility for all extracellular danger signals to activate a common cytosolic mediator which will be the final NLRP3 inflammasome ligand and activator. This concept will answer the problem of a wide repertoire of different signals recognized by a sole NLR receptor (NLRP3). Common cytosolic hallmarks for danger signals inducing NLRP3 activation are described in Sect. 5.3 and include (1) the decrease of cytosolic K^+ and (2) the increase of ROS and (3) lysosomal damage [85]. These factors could make the intra-cellular environment ideal for assembly of the inflammasome and therefore could

Inflammasome Activation by Danger Signals act as intracellular danger signal pathways (Fig. 3). Recently, Tschöpp's group has discovered a Thioredoxin-interacting protein which binds to the LRR of NLRP3 in an ROS-dependent manner and allows the activation of the NLRP3 inflammasome [86]; further experiments are needed to elucidate if such protein is the key NLRP3 ligand.

Current investigations about the inflammasome and IL-1 release have found the very basis of the innate immune response. Innate immunity is all about recognizing danger signals, and this importance has been demonstrated in the pathogenesis of an increasing number of sterile inflammatory diseases, including gout, pseudogout, rheumatoid arthritis, skin damage, type 2 diabetes, Alzheimer's disease, ischemic brain injury, lung injury, renal dysfunction, bone remodelling, autoinflammatory syndromes, tumour immunity and inflammatory and neuropathic pain (Table 1). Therefore, drug-design programmes aimed to inhibit the inflammasome is a promising field expected to grow in the next decade after detailed description of the inflammasome activation mechanisms.

Acknowledgements I thank members of my lab Alberto Baroja-Mazo and Carlos M. Martinez for comments and discussions, David Brough for critically reviewing the manuscript and Sebastien Roger for helpful discussion. Work in Pelegrín's lab is supported by *instituto de salud carlos iii* (grants EMER07/049 and PS09/00120), Fundación Séneca (grant 11922/PI/09) and British Council through Alliance program (grant 09.027).

References

1. Auron PE, Webb AC, Rosenwasser LJ, Mucci SF, Rich A, Wolff SM, Dinarello CA (1984) Nucleotide sequence of human monocyte interleukin 1 precursor cDNA. Proc Natl Acad Sci USA 81:7907–7911
2. Black RA, Kronheim SR, Cantrell M, Deeley MC, March CJ, Prickett KS, Wignall J, Conlon PJ, Cosman D, Hopp TP (1988) Generation of biologically active interleukin-1 beta by proteolytic cleavage of the inactive precursor. J Biol Chem 263:9437–9442
3. Li P, Allen H, Banerjee S, Franklin S, Herzog L, Johnston C, McDowell J, Paskind M, Rodman L, Salfeld J (1995) Mice deficient in IL-1 beta-converting enzyme are defective in production of mature IL-1 beta and resistant to endotoxic shock. Cell 80:401–411
4. Kuida K, Lippke JA, Ku G, Harding MW, Livingston DJ, Su MS, Flavell RA (1995) Altered cytokine export and apoptosis in mice deficient in interleukin-1 beta converting enzyme. Science 267:2000–2003
5. Dinarello CA (2009) Immunological and inflammatory functions of the interleukin-1 family. Annu Rev Immunol 27:519–550
6. Martinon F, Burns K, Tschopp J (2002) The inflammasome: a molecular platform triggering activation of inflammatory caspases and processing of proIL-beta. Mol Cell 10:417–426
7. Martinon F, Mayor A, Tschopp J (2009) The inflammasomes: guardians of the body. Annu Rev Immunol 27:229–265
8. Matzinger P (1994) Tolerance, danger, and the extended family. Annu Rev Immunol 12:991–1045
9. Matzinger P (2002) The danger model: a renewed sense of self. Science 296:301–305
10. Bianchi ME (2007) DAMPs, PAMPs and alarmins: all we need to know about danger. J Leukoc Biol 81:1–5

11. Moussion C, Ortega N, Girard J (2008) The IL-1-like cytokine IL-33 is constitutively expressed in the nucleus of endothelial cells and epithelial cells in vivo: a novel 'alarmin'? PLoS One 3:e3331
12. Luheshi NM, Rothwell NJ, Brough D (2009) The dynamics and mechanisms of interleukin-1alpha and Beta nuclear import. Traffic 10:16–25
13. Kenny EF, O'Neill LA (2008) Signalling adaptors used by Toll-like receptors: an update. Cytokine 43:342–349
14. Hogquist KA, Nett MA, Unanue ER, Chaplin DD (1991) Interleukin 1 is processed and released during apoptosis. Proc Natl Acad Sci USA 88:8485–8489
15. Verhoef PA, Kertesy SB, Lundberg K, Kahlenberg JM, Dubyak GR (2005) Inhibitory effects of chloride on the activation of caspase-1, IL-1beta secretion, and cytolysis by the P2X7 receptor. J Immunol 175:7623–7634
16. Pelegrin P, Barroso-Gutierrez C, Surprenant A (2008) P2X7 receptor differentially couples to distinct release pathways for IL-1beta in mouse macrophage. J Immunol 180:7147–7157
17. MacKenzie AB, Wilson H, Kiss-Toth E, Dower SK, North RA, Surprenant A (2001) Rapid secretion of interleukin-1beta by microvesicle shedding. Immunity 15:825–835
18. Brough D, Rothwell N (2007) Caspase-1-dependent processing of pro-interleukin-1beta is cytosolic and precedes cell death. J Cell Sci 120:772–781
19. Andrei C, Dazzi C, Lotti L, Torrisi MR, Chimini G, Rubartelli A (1999) The secretory route of the leaderless protein interleukin 1beta involves exocytosis of endolysosome-related vesicles. Mol Biol Cell 10:1463–1475
20. Qu Y, Franchi L, Nunez G, Dubyak GR (2007) Nonclassical IL-1 beta secretion stimulated by P2X7 receptors is dependent on inflammasome activation and correlated with exosome release in murine macrophages. J Immunol 179:1913–1925
21. Pelegrin P (2008) Targeting interleukin-1 signaling in chronic inflammation: focus on P2X(7) receptor and Pannexin-1. Drug News Perspect 21:424–433
22. Surprenant A, North RA (2008) Signaling at purinergic P2X receptors. Annu Rev Physiol 71:333–359
23. Perregaux DG, Bhavsar K, Contillo L, Shi J, Gabel CA (2002) Antimicrobial peptides initiate IL-1 beta posttranslational processing: a novel role beyond innate immunity. J Immunol 168:3024–3032
24. Yamasaki K, Muto J, Taylor KR, Cogen AL, Audish D, Bertin J, Grant EP, Coyle AJ, Misaghi A, Hoffman HM et al (2009) NLRP3/cryopyrin is necessary for interleukin-1beta (IL-1beta) release in response to hyaluronan, an endogenous trigger of inflammation in response to injury. J Biol Chem 284:12762–12771
25. Halle A, Hornung V, Petzold G, Stewart C, Monks B, Reinheckel T, Fitzgerald KA, Latz E, Moore K, Golenbock D (2008) The NALP3 inflammasome is involved in the innate immune response to amyloid-beta. Nat Immunol 9:857–865
26. Martinon F, Pétrilli V, Mayor A, Tardivel A, Tschopp J (2006) Gout-associated uric acid crystals activate the NALP3 inflammasome. Nature 440:237–241
27. Feldmeyer L, Keller M, Niklaus G, Hohl D, Werner S, Beer HD (2007) The inflammasome mediates UVB-induced activation and secretion of interleukin-1beta by keratinocytes. Curr Biol 17:1140–1145
28. Watanabe H, Gaide O, Pétrilli V, Martinon F, Contassot E, Roques S, Kummer J, Tschopp J, French L (2007) Activation of the IL-1beta-processing inflammasome is involved in contact hypersensitivity. J Investig Dermatol 127:1956–1963
29. Muruve D, Pétrilli V, Zaiss A, White L, Clark S, Ross PJ, Parks R, Tschopp J (2008) The inflammasome recognizes cytosolic microbial and host DNA and triggers an innate immune response. Nature 452:103–107
30. Li H, Ambade A, Re F (2009) Cutting edge: necrosis activates the NLRP3 inflammasome. J Immunol 183:1528–1532
31. Babelova A, Moreth K, Tsalastra-Greul W, Zeng-Brouwers J, Eickelberg O, Young MF, Bruckner P, Pfeilschifter J, Schaefer RM, Groene HJ et al (2009) Biglycan: a danger signal

that activates the NLRP3 inflammasome via toll-like and P2X receptors. J Biol Chem 284:24035–24048
32. Fountain SJ, Burnstock G (2009) An evolutionary history of P2X receptors. Purinergic Signal 5:269–272
33. Bird S, Zou J, Wang T, Munday B, Cunningham C, Secombes CJ (2002) Evolution of interleukin-1beta. Cytokine Growth Factor Rev 13:483–502
34. Sodergren E, Weinstock G, Davidson E, Cameron R, Gibbs R, Angerer R, Angerer L, Arnone M, Burgess D, Burke R et al (2006) The genome of the sea urchin Strongylocentrotus purpuratus. Science 314:941–952
35. López-Castejón G, Young MT, Meseguer J, Surprenant A, Mulero V (2007) Characterization of ATP-gated P2X7 receptors in fish provides new insights into the mechanism of release of the leaderless cytokine interleukin-1β. Mol Immunol 44:1286–1299
36. López-Castejón G, Sepulcre MP, Mulero I, Pelegrin P, Meseguer J, Mulero V (2008) Molecular and functional characterization of gilthead seabream Sparus aurata caspase-1: the first identification of an inflammatory caspase in fish. Mol Immunol 45:49–57
37. Pelegrin P, Surprenant A (2009) The P2X(7) receptor-pannexin connection to dye uptake and IL-1beta release. Purinergic Signal 5:129–137
38. Eltzschig HK, Macmanus CF, Colgan SP (2008) Neutrophils as sources of extracellular nucleotides: functional consequences at the vascular interface. Trends Cardiovasc Med 18:103–107
39. Beigi R, Kobatake E, Aizawa M, Dubyak GR (1999) Detection of local ATP release from activated platelets using cell surface-attached firefly luciferase. Am J Physiol 276:C267–C278
40. Bulanova E, Bulfone-Paus S (2010) P2 receptor-mediated signaling in mast cell biology. Purinergic Signal 6(1):3–17
41. Piccini A, Carta S, Tassi S, Lasiglié D, Fossati G, Rubartelli A (2008) ATP is released by monocytes stimulated with pathogen-sensing receptor ligands and induces IL-1beta and IL-18 secretion in an autocrine way. Proc Natl Acad Sci USA 105:8067–8072
42. Atarashi K, Nishimura J, Shima T, Umesaki Y, Yamamoto M, Onoue M, Yagita H, Ishii N, Evans R, Honda K et al (2008) ATP drives lamina propria T(H)17 cell differentiation. Nature 455:808–812
43. Pellegatti P, Raffaghello L, Bianchi G, Piccardi F, Pistoia V, Di Virgilio F (2008) Increased level of extracellular ATP at tumor sites: in vivo imaging with plasma membrane luciferase. PLoS One 3:e2599
44. Ghiringhelli F, Apetoh L, Tesniere A, Aymeric L, Ma Y, Ortiz C, Vermaelen K, Panaretakis T, Mignot G, Ullrich E et al (2009) Activation of the NLRP3 inflammasome in dendritic cells induces IL-1beta-dependent adaptive immunity against tumors. Nat Med 15:1170–1178
45. Pelegrin P, Surprenant A (2006) Pannexin-1 mediates large pore formation and interleukin-1beta release by the ATP-gated P2X7 receptor. EMBO J 25:5071–5082
46. Surprenant A, Rassendren F, Kawashima E, North RA, Buell G (1996) The cytolytic P2Z receptor for extracellular ATP identified as a P2X receptor (P2X7). Science 272:735–738
47. MacKenzie AB, Young MT, Adinolfi E, Surprenant A (2005) Pseudoapoptosis induced by brief activation of ATP-gated P2X7 receptors. J Biol Chem 280:33968–33976
48. Pelegrin P, Surprenant A (2007) Pannexin-1 couples to maitotoxin- and nigericin-induced interleukin-1beta release through a dye uptake-independent pathway. J Biol Chem 282:2386–2394
49. Hewinson J, Moore SF, Glover C, Watts AG, MacKenzie AB (2008) A key role for redox signaling in rapid P2X7 receptor-induced IL-1 beta processing in human monocytes. J Immunol 180:8410–8420
50. Mariathasan S, Newton K, Monack DM, Vucic D, French D, Lee WP, Roose-Girma M, Erickson S, Dixit VM (2004) Differential activation of the inflammasome by caspase-1 adaptors ASC and Ipaf. Nature 430:213–218
51. Mariathasan S, Weiss DS, Newton K, McBride J, O'Rourke K, Roose-Girma M, Lee WP, Weinrauch Y, Monack DM, Dixit VM (2006) Cryopyrin activates the inflammasome in response to toxins and ATP. Nature 440:228–232

52. Sutterwala FS, Ogura Y, Szczepanik M, Lara-Tejero M, Lichtenberger GS, Grant E, Bertin J, Coyle AJ, Galán JE, Askenase PW et al (2006) Critical role for NALP3/CIAS1/Cryopyrin in innate and adaptive immunity through its regulation of caspase-1. Immunity 24:317–327
53. Honore P, Donnelly-Roberts D, Namovic MT, Hsieh G, Zhu CZ, Mikusa JP, Hernandez G, Zhong C, Gauvin DM, Chandran P et al (2006) A-740003 [N-(1-{[(cyanoimino) (5-quinolinylamino) methyl]amino}-2,2-dimethylpropyl)-2-(3,4-dimethoxyphenyl)acetamide], a novel and selective P2X7 receptor antagonist, dose-dependently reduces neuropathic pain in the rat. J Pharmacol Exp Ther 319:1376–1385
54. Labasi JM, Petrushova N, Donovan C, McCurdy S, Lira P, Payette MM, Brissette W, Wicks JR, Audoly L, Gabel CA (2002) Absence of the P2X7 receptor alters leukocyte function and attenuates an inflammatory response. J Immunol 168:6436–6445
55. McInnes IB, Snell NJ, Perrett JH, Parmar H, Wang MM, Astbury C. Results of a phase II clinical trial of a novel P2X7 receptor antagonist, AZD9056, in patients with active rheumatoid arthritis (CREATE study). Paper presented at ACR/ARHP Scientific Meeting, Boston, MA, 6–11 November 2007
56. Solle M, Labasi J, Perregaux DG, Stam E, Petrushova N, Koller BH, Griffiths RJ, Gabel CA (2001) Altered cytokine production in mice lacking P2X(7) receptors. J Biol Chem 276:125–132
57. Sanz JM, Chiozzi P, Ferrari D, Colaianna M, Idzko M, Falzoni S, Fellin R, Trabace L, Di Virgilio F (2009) Activation of microglia by amyloid beta requires P2X7 receptor expression. J Immunol 182:4378–4385
58. Klapperstück M, Büttner C, Schmalzing G, Markwardt F (2001) Functional evidence of distinct ATP activation sites at the human P2X(7) receptor. J Physiol 534:25–35
59. Gonnord P, Delarasse C, Auger R, Benihoud K, Prigent M, Cuif MH, Lamaze C, Kanellopoulos JM (2008) Palmitoylation of the P2X7 receptor, an ATP-gated channel, controls its expression and association with lipid rafts. FASEB J 23:795–805
60. Zimmermann H (2000) Extracellular metabolism of ATP and other nucleotides. Naunyn Schmiedebergs Arch Pharmacol 362:299–309
61. Kumar V, Sharma A (2009) Adenosine: an endogenous modulator of innate immune system with therapeutic potential. Eur J Pharmacol 616:7–15
62. Pelegrin P, Surprenant A (2009) Dynamics of macrophage polarization reveal new mechanism to inhibit IL-1beta release through pyrophosphates. EMBO J 28:2114–2127
63. Netea MG, Nold-Petry CA, Nold MF, Joosten LA, Opitz B, van der Meer JH, van de Veerdonk FL, Ferwerda G, Heinhuis B, Devesa I et al (2008) Differential requirement for the activation of the inflammasome for processing and release of IL-1{beta} in monocytes and macrophages. Blood 113:2324–2335
64. Iyer SS, Pulskens WP, Sadler JJ, Butter LM, Teske GJ, Ulland TK, Eisenbarth SC, Florquin S, Flavell RA, Leemans JC et al (2009) Necrotic cells trigger a sterile inflammatory response through the Nlrp3 inflammasome. Proc Natl Acad Sci USA 106:20388–20393
65. Pacifici R, Carano A, Santoro SA, Rifas L, Jeffrey JJ, Malone JD, McCracken R, Avioli LV (1991) Bone matrix constituents stimulate interleukin-1 release from human blood mononuclear cells. J Clin Invest 87:221–228
66. Chao W (2009) Toll-like receptor signaling: a critical modulator of cell survival and ischemic injury in the heart. Am J Physiol Heart Circ Physiol 296:H1–H12
67. Shi Y, Evans JE, Rock K (2003) Molecular identification of a danger signal that alerts the immune system to dying cells. Nature 425:516–521
68. Selkoe DJ (1994) Cell biology of the amyloid beta-protein precursor and the mechanism of Alzheimer's disease. Annu Rev Cell Biol 10:373–403
69. Breitner JC, Welsh KA, Helms MJ, Gaskell PC, Gau BA, Roses AD, Pericak-Vance MA, Saunders AM (1995) Delayed onset of Alzheimer's disease with nonsteroidal anti-inflammatory and histamine H2 blocking drugs. Neurobiol Aging 16:523–530
70. Blum-Degen D, Müller T, Kuhn W, Gerlach M, Przuntek H, Riederer P (1995) Interleukin-1 beta and interleukin-6 are elevated in the cerebrospinal fluid of Alzheimer's and de novo Parkinson's disease patients. Neurosci Lett 202:17–20

71. Simard AR, Soulet D, Gowing G, Julien JP, Rivest S (2006) Bone marrow-derived microglia play a critical role in restricting senile plaque formation in Alzheimer's disease. Neuron 49:489–502
72. Akama KT, Van Eldik LJ (2000) Beta-amyloid stimulation of inducible nitric-oxide synthase in astrocytes is interleukin-1beta- and tumor necrosis factor-alpha (TNFalpha)-dependent, and involves a TNFalpha receptor-associated factor- and NFkappaB-inducing kinase-dependent signaling mechanism. J Biol Chem 275:7918–7924
73. Rampe D, Wang L, Ringheim GE (2004) P2X7 receptor modulation of beta-amyloid- and LPS-induced cytokine secretion from human macrophages and microglia. J Neuroimmunol 147:56–61
74. Yazdi AS, Ghoreschi K, Röcken M (2007) Inflammasome activation in delayed-type hypersensitivity reactions. J Investig Dermatol 127:1853–1855
75. Brough D, Le Feuvre RA, Wheeler RD, Solovyova N, Hilfiker S, Rothwell N, Verkhratsky A (2003) Ca2+ stores and Ca2+ entry differentially contribute to the release of IL-1 beta and IL-1 alpha from murine macrophages. J Immunol 170:3029–3036
76. Vilaysane A, Muruve D (2009) The innate immune response to DNA. Semin Immunol 21:208–214
77. Decker P, Singh-Jasuja H, Haager S, Kötter I, Rammensee HG (2005) Nucleosome, the main autoantigen in systemic lupus erythematosus, induces direct dendritic cell activation via a MyD88-independent pathway: consequences on inflammation. J Immunol 174:3326–3334
78. Bürckstümmer T, Baumann C, Blüml S, Dixit E, Dürnberger G, Jahn H, Planyavsky M, Bilban M, Colinge J, Bennett KL et al (2009) An orthogonal proteomic-genomic screen identifies AIM2 as a cytoplasmic DNA sensor for the inflammasome. Nat Immunol 10:266–272
79. Fernandes-Alnemri T, Yu JW, Datta P, Wu J, Alnemri ES (2009) AIM2 activates the inflammasome and cell death in response to cytoplasmic DNA. Nature 458:509–513
80. Hornung V, Ablasser A, Charrel-Dennis M, Bauernfeind F, Horvath G, Caffrey DR, Latz E, Fitzgerald KA (2009) AIM2 recognizes cytosolic dsDNA and forms a caspase-1-activating inflammasome with ASC. Nature 458:514–518
81. Roberts TL, Idris A, Dunn JA, Kelly GM, Burnton CM, Hodgson S, Hardy LL, Garceau V, Sweet MJ, Ross IL et al (2009) HIN-200 proteins regulate caspase activation in response to foreign cytoplasmic DNA. Science 323:1057–1060
82. Perregaux DG, Barberia J, Lanzetti AJ, Geoghegan KF, Carty TJ, Gabel CA (1992) IL-1 beta maturation: evidence that mature cytokine formation can be induced specifically by nigericin. J Immunol 149:1294–1303
83. Perregaux DG, Gabel CA (1994) Interleukin-1 beta maturation and release in response to ATP and nigericin. Evidence that potassium depletion mediated by these agents is a necessary and common feature of their activity. J Biol Chem 269:15195–15203
84. Pétrilli V, Papin S, Dostert C, Mayor A, Martinon F, Tschopp J (2007) Activation of the NALP3 inflammasome is triggered by low intracellular potassium concentration. Cell Death Differ 14:1583–1589
85. Bryant C, Fitzgerald KA (2009) Molecular mechanisms involved in inflammasome activation. Trends Cell Biol 19:455–464
86. Zhou R, Tardivel A, Thorens B, Choi I, Tschopp J (2010) Thioredoxin-interacting protein links oxidative stress to inflammasome activation. Nat Immunol 11(2):136–140
87. Andersson B, Janson V, Behnam-Motlagh P, Henriksson R, Grankvist K (2006) Induction of apoptosis by intracellular potassium ion depletion: using the fluorescent dye PBFI in a 96-well plate method in cultured lung cancer cells. Toxicol In Vitro 20:986–994
88. Li J, Meyer R, Duncan RL, Turner CH (2009) P2X7 nucleotide receptor plays an important role in callus remodeling during fracture repair. Calcif Tissue Int 84:405–412
89. Goldbach-Mansky R, Dailey NJ, Canna SW, Gelabert A, Jones J, Rubin BI, Kim HJ, Brewer C, Zalewski C, Wiggs E et al (2006) Neonatal-onset multisystem inflammatory disease responsive to interleukin-1beta inhibition. N Engl J Med 355:581–592

90. Larsen CM, Faulenbach M, Vaag A, Vølund A, Ehses JA, Seifert B, Mandrup-Poulsen T, Donath MY (2007) Interleukin-1-receptor antagonist in type 2 diabetes mellitus. N Engl J Med 356:1517–1526
91. Chessell IP, Hatcher JP, Bountra C, Michel AD, Hughes JP, Green P, Egerton J, Murfin M, Richardson J, Peck WL et al (2005) Disruption of the P2X7 purinoceptor gene abolishes chronic inflammatory and neuropathic pain. Pain 114:386–396
92. Boutin H, LeFeuvre RA, Horai R, Asano M, Iwakura Y, Rothwell NJ (2001) Role of IL-1alpha and IL-1beta in ischemic brain damage. J Neurosci 21:5528–5534
93. Emsley HC, Smith CJ, Georgiou RF, Vail A, Hopkins SJ, Rothwell NJ, Tyrrell PJ, Investigators AS (2005) A randomised phase II study of interleukin-1 receptor antagonist in acute stroke patients. J Neurol Neurosurg Psychiatr 76:1366–1372
94. Honore P, Donnelly-Roberts D, Namovic M, Zhong C, Wade C, Chandran P, Zhu C, Carroll W, Perez-Medrano A, Iwakura Y et al (2009) The antihyperalgesic activity of a selective P2X7 receptor antagonist, A-839977, is lost in IL-1alphabeta knockout mice. Behav Brain Res 204 (1):77–81
95. Gonçalves R, Gabrich L, Rosário A, Takiya C, Ferreira M, Chiarini L, Persechini P, Coutinho-Silva R, Leite M (2006) The role of purinergic P2X7 receptors in the inflammation and fibrosis of unilateral ureteral obstruction in mice. Kidney Int 70:1599–1606

Toll-Like Receptors and Inflammasomes

Frank L. van de Veerdonk and Mihai G. Netea

Abstract The recognition of microorganisms is mediated by germline pattern recognition receptors (PRRs) such as Toll-like receptors (TLRs) and nucleotide-binding domain, leucine-rich repeat-containing proteins (NLRs) that interact with conserved structures of the microorganisms, the pathogen-associated molecular patterns (PAMPs). PRRs activate intracellular signals that collaborate for efficient activation of host defense. One such specific collaboration is the interaction between TLRs and cytosolic NLRs. NLRs are the triggering components of the inflammasomes, and the combination of TLR and NLR signaling results in the activation of caspase-1, followed by processing and secretion of the highly proinflammatory cytokines IL-1β and IL-18. Understanding the mechanisms of interaction between TLRs and inflammasomes is important, since defects that influence this interaction might lead to autoinflammatory diseases or impaired host defense.

List of abbreviations

PRR	Pattern recognition receptor
TLR	Toll-like receptor
NLR	Nucleotide-binding domain, leucine-rich repeat-containing protein
PAMP	Pathogen-associated molecular patterns
IL	Interleukin
TIR	Toll/IL-1 receptor
LPS	Lipopolysaccharide

F.L. van de Veerdonk (✉) • M.G. Netea
Department of Medicine, Radboud University Nijmegen Medical Center, Nijmegen, The Netherlands
and
Nijmegen Institute for Infection, Inflammation and Immunity (N4i), Nijmegen, The Netherlands
e-mail: f.veerdonk@aig.umcn.nl

I. Couillin et al. (eds.), *The Inflammasomes*, Progress in Inflammation Research,
DOI 10.1007/978-3-0348-0148-5_8,

1 Toll-Like Receptors

In 1989, Charles Janeway argued that the early phases of host defense are regulated by receptors and ligands that preceded in evolution the development of clonally distributed receptors encoded in rearranging genes and that these receptors and ligands would trigger innate or nonclonal responses [1]. The first molecules that were identified as such were the Toll-like receptors (TLRs), which are currently one of the most studied classes of pattern recognition receptors (PRRs). TLRs are named for their similarity to Toll, a receptor expressed on the membrane of the hematocyte of the fruit fly *Drosophila melanogaster*. Lemaitre et al. discovered in 1996 that Toll had an essential role in the fly's immunity to fungal infection [2]. Shortly after this discovery, the cloning of a human homologue of Toll was reported [3].

Ten different TLRs (TLR1–10) have been described in humans to date, and 12 TLRs (TLR1–9; TLR11–13) are present in mice [4]. TLRs are type I transmembrane glycoproteins and have a trimodular structure. The extracellular N-terminal domain consists of approximately 19–25 leucine-rich repeats (LRRs), and each LRR consists of 24–29 amino acids with the conserved motif "LxxLxLxxN." The intracellular C-terminal domain is known as the Toll/IL-1 receptor (TIR) domain, which shows homology with the intracellular domain of the IL-1 receptor. The TIR domain is required for the interaction and recruitment of various adaptor molecules to activate the downstream signaling pathway [5]. TLRs can form heterodimers such as TLR1-TLR2, TLR2-TLR6, and TLR4-MD2 or homodimers such as TLR2-TLR2 or TLR3-TLR3 after association with their ligands [6]. TLRs are expressed in distinct cellular compartments; TLR1, TLR2, TLR4, TLR5, and TLR6 are expressed on the cell surface, whereas TLR3, TLR7, TLR8, and TLR9 are expressed intracellularly in the endosomes.

TLRs recognize pathogen-associated molecular patterns (PAMPs) derived from various microorganisms. LPS from Gram-negative bacteria is recognized by TLR4, diacyl or triacyl lipopeptides from bacteria are recognized by TLR2/1 and TLR2/6 respectively, bacterial flagellin by TLR5, single-stranded RNA from viruses is recognized by TLR7 and TLR8, and genomic DNA rich in unmethylated CpG DNA and malarial hemozoin are recognized by TLR9. Recognition of these ligands by TLRs results in the activation of signaling cascades that lead to the induction of genes involved in inflammatory responses.

2 TLR-Induced IL-1β and IL-18 Production Is Controlled by the Inflammasome

Next to direct activation of several other proinflammatory cyokines and chemokines, activation of TLRs as described above results in intracellular accumulation of the inactive proforms of IL-1β and IL-18. In contrast to other cytokines that are readily secreted from the cell, pro-IL-1β and pro-IL-18 do not contain a signal peptide to direct the process of secretion, and they need cleavage by the cysteine protease caspase-1 in order to result in the secretion of the active molecule. It has been therefore postulated that the release of active IL-1 and IL-18 needs two signals: (1) the first signal is provided by TLR engagement (or a similar pathway) and results in gene transcription and pro-IL-1β accumulation, a process sometimes described as "priming"; and (2) a second signal through a nucleotide-binding domain, leucine-rich repeat-containing proteins (NLRs) activates the inflammasome, which results in caspase-1 activation and subsequently processing of the TLR induced pro-IL-1β into mature bioactive IL-1β. Together these two signals result in inflammasome-dependent processing of mature IL-1β, with TLRs providing the substrate for the inflammasome. Dual triggering of TLR and NLR is believed to be necessary for the tight regulation of proinflammatory cytokines such as IL-1β. Several interactions between TLR stimulation/signaling and caspase-1 activation have been described (Table 1). Although TLR signaling is crucial for priming and regulating the transcription of pro-IL-1β that needs to be processed in order to become active, it is still controversial whether there is a direct role for TLR signaling in the activation of the inflammasome. Interestingly, it has

Table 1 The interaction of TLR stimulation and caspase-1

Toll-like receptor	Ligand	Interaction with caspase-1
TLR2	Bacterial lipoproteins (BLPs)	BLPs activates caspase-1 through TLR2 [8]
TLR1/2	Alkane particles	Alkane particles also activate caspase-1 which leads to a two component response and results in a synergistic proinflammatory reaction [9]
TLR1/2 and TLR2/6	Unknown ligand(s) of *Francisella tularensis*	Caspase-1 controls TLR2-mediated IL-1β induced by live *F. tularensis* [10]
TLR2/6	Unknown ligand from *Mycobacterium tuberculosis*	Caspase-1 controls TLR2-mediated IL-1β induced by *M. tuberculosis* [11]
TLR2 TLR4	Malp2 *Escherichia coli* LPS	Caspase-1 regulates TLR2 and TLR4 signaling via an effect on Mal [7]
TLR2 and TLR4	Biglycan	The interaction of biglycan with TLR2/TLR4 and P2X7 receptor induces caspase-1 activation and release of mature IL-1β [12]
TLR4	*E. coli* LPS	TLR4-mediated TRIF-dependent caspase-1 activation [13]
TLR4	Pneumolysin	TLR4 recognizes pneumolysin, which induces caspase-1-dependent IL-1β [14]
TLR4	*Helicobacter pylori* LPS	*H. pylori* LPS directly induces caspase-1 through the TLR4 pathway [15]
TLR7 and TLR8	R848	Activation of caspase-1 is a downstream signaling event after stimulation TR7/8 [16]
TLR9	Free DNA released from apoptotic hepatocytes	Activation of both caspase-1 and TLR9 is essential for acetaminophen-induced liver injury [17]

been reported that caspase-1 is able to cleave one of the TLR adaptor molecules, Mal/TIRAP, which activates the signaling pathways of TLR2 and TLR4, and NF-κB activation induced by TLR2 and TLR4 was blocked in the presence of a caspase-1 inhibitor [7]. This suggests a direct role for caspase-1 in the regulation of TLR signaling.

3 TLRs Can Directly Induce Inflammasome-Dependent IL-1β Secretion in Monocytes, But Not in Macrophages

Despite the progress made in understanding the regulation of IL-1β processing, controversy surrounded the capacity of TLR ligands such as LPS or lipopeptides to activate caspase-1 resulting in the release of active IL-1β. Based on defective responses of the monocyte-like leukemia cell line THP-1 to LPS stimulation, it has been suggested LPS by itself is ineffective as a stimulator of IL-1β release [18]. It was therefore proposed that a second signal such as MDP or ATP is required to process and release IL-1β. This second signal activates the inflammasome and subsequently induces the activation of caspase-1 [19]. However, this model is derived from studies that investigated THP-1 cells [18] and primary mouse macrophages [20], and is in contradiction with studies demonstrating that IL-1β is released from monocytes after single stimulation with a TLR ligand, such as purified LPS, lipopeptides, or lipoteichoic acid [21, 22].

These inconsistencies have recently been resolved by a study showing that the synthesis and release of IL-1β differ between human monocytes and macrophages [23]. Monocytes were found to have constitutively activated caspase-1. Therefore, a single stimulation event with a TLR ligand leads to gene transcription and pro-IL-1β synthesis that is processed by the constitutively active caspase-1, leading to the release of active IL-1β (Fig. 1). In contrast, macrophages need two distinct signals: one signal that induces transcription and translation and a second signal that activates caspase-1. These two signals will finally result in IL-1β processing and secretion [23] (Fig. 1). It appears therefore that the rate-limiting effect of IL-1β production in monocytes is represented by transcription, while both transcription and processing control IL-1β production in the macrophage. Although caspase-1 is constitutively activated in human monocytes, it is still dependent on inflammasome components [23]. A crucial functional aspect in relation to activation of the inflammasome relates to the release of endogenous ATP by monocytes that will induce IL-1β secretion through P2X7, while macrophages completely lack the capacity to release ATP [24].

The demonstration that monocytes have a constitutive activation of caspase-1 uncouples caspase-1 activation from PAMP recognition in human primary monocytes. This likely represents an adaptation of the monocyte to its respective environment. Circulating monocytes function in the surveillance of an essentially pathogen-free environment, so they must respond promptly to any danger signal. In contrast, macrophages are confined to an environment (e.g., alveolar space and

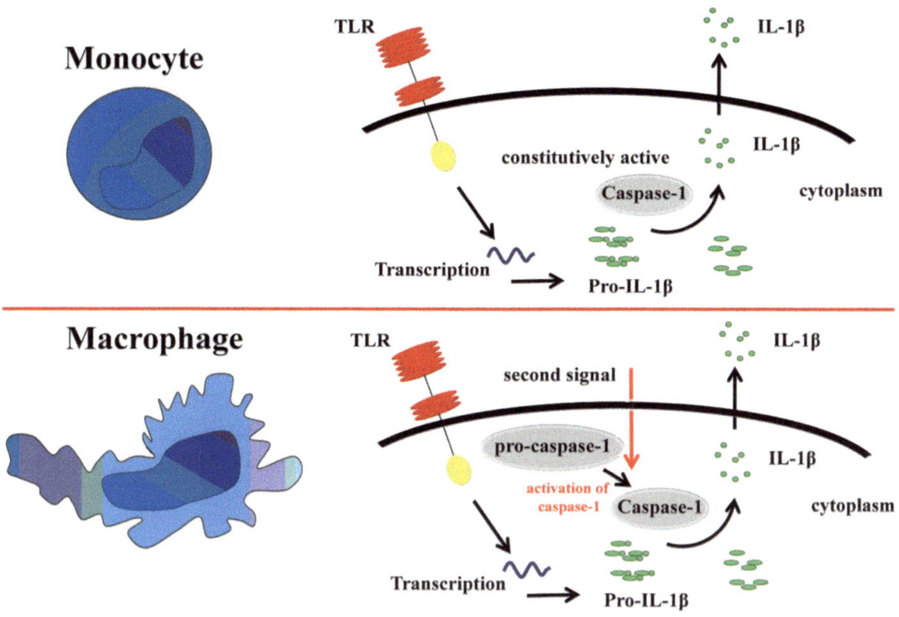

Fig. 1 Differential IL-1β secretion pathways in monocytes and macrophages. Caspase-1 is constitutively active in monocytes, and these cells release mature IL-1β after single stimulation with a TLR ligand. In contrast, macrophages need two signals: one signal such as a TLR-ligand that induces transcription and a second signal that induces inflammasome activation and IL-1β processing and secretion

mucosal surfaces) in which they are constantly exposed to microbial stimuli and danger signals. A sensitive response of macrophages to release IL-1β for each encounter with such stimuli would result in chronic deleterious inflammatory reactions. Thus, repeated bouts of inflammation are likely reduced by the requirement of a second signal for the activation of the inflammasome and release of active IL-1β. Such a second signal would be available at the site of infection, trauma, or necrosis where ATP levels are elevated due to cell lysis and can trigger the P2X7 receptor [25]. In addition, second signals can come from the cathelicidin-derived peptide LL37 from infiltrating neutrophils [26], or the release of bacterial toxins [27].

4 Functional Consequences of Defective TLR–Inflammasome Interaction

Understanding the interplay between TLRs and inflammasomes is important, since defects that influence this interaction can result in autoinflammatory diseases. Some of the best-known diseases in this group include familial Mediterranean fever

(FMF) [28], and cryopyrin-associated periodic syndromes also known as cryopyrinopathies, which include familial cold autoinflammatory syndrome (FCAS) [29], Muckle–Wells syndrome (MWS) [30], and neonatal onset multisystem inflammatory disease (NOMID) [31]. Blood monocytes from patients with these disorders produce significantly more IL-1β after TLR stimulation. This is a relevant pathophysiological condition since treatment of these patients with IL-1 receptor antagonist has been reported to improve clinical outcome. Furthermore, monocytes from patients with familial Mediterranean fever who have mutations in pyrin release more IL-1β upon TLR stimulation than cells from control subjects, suggesting a failure to suppress the activation of caspase-1 [28].

Defects between the interplay of TLRs and the inflammasome/caspase-1 activation might also have impact on host defense. This is highlighted by LPS models performed in caspase-1−/− mice, which demonstrate that mice defective in inflammasome function were protected against LPS-induced shock [32, 33]. This clearly demonstrates the impact of TLR and inflammasome interaction on host defense mechanisms.

5 Beyond the Inflammasomes

Despite the reported relevance of inflammasome activation in experimental models of inflammation, certain in vivo infection models in mice deficient in inflammasome components show intriguing results that question the importance of the inflammasome. In clear contrast with the LPS models where caspase-1−/− mice are protected [32, 33], caspase-1−/− mice are not protected in a model of turpentine-induced sterile inflammation that is highly dependent on IL-1β [34, 35]. Therefore, it appears that caspase-1 and inflammasome activation is important in some, but not all, types of IL-1β-driven inflammation [36].

When it was discovered that the cleavage of the procytokine is needed for activation, the question arose whether other enzymes apart from caspase-1 would also be capable of processing pro-IL-1β. Indeed, subsequent studies have identified neutrophil-derived serine proteases, such as proteinase-3 (PR3), elastase, and cathepsin-G, as enzymes that can process pro-IL-1β into its active fragments [37, 38]. Furthermore, proteinase-3 can also process pro-IL-18 [39]. The crucial role played by neutrophil-dependent, inflammasome-independent activation of pro-IL-1β has been elegantly confirmed recently by the group of Karin et al. [40].

The inflammasome-independent activation of pro-IL-1β in situations when neutrophils are the major cell population in the inflammatory infiltrate can explain many of the puzzling observations described above. Inflammatory infiltrates during arthritis consist of both macrophages and neutrophils. Indeed, it has been observed that there is a minimal role for caspase-1 during acute inflammation of arthritis. During this phase of inflammation, which is characterized by an overwhelming neutrophil infiltrate, serine proteases such as PR3 play an important role for the processing of pro-IL-1β [41]. In contrast, during the chronic phase of arthritis caspase-1 seems to have a more significant effect [41]. This could be explained by the fact that in the chronic phase macrophages, in which caspase-1 rather than PR3 cleaves pro-IL-1β, are the main component of the infiltrate in arthritis.

6 Conclusions

Much information regarding the mechanisms of signaling and activation of innate immunity has been gained during the past few years, and this has greatly contributed to a better understanding of the host immune response. Exciting areas of advance were represented by the discovery of TLRs, which recognize pathogen-associated molecular patterns, and the description of the inflammasomes leading to processing and activation of cytokines of the IL-1 family. An accumulating body of evidence has brought to light a concept that these two inflammatory pathways have strong interactions during infection and inflammation, which results in a balanced immune response. One of the most important effects of this interaction has been unraveled in the requirement for double stimulation with TLR ligands (inducing gene transcription) and NLR agonists (activate caspase-1) to release mature IL-1β from macrophages. This is different from the IL-1β activation in monocytes that can release active IL-1β upon stimulation with TLR ligands alone due to constitutively active caspase-1. The relevance of this difference is apparent when one considers the role of IL-1β for the host response to exogenous challenges. These new insights into the interplay between TLRs and inflammasomes have contributed to a better understanding of autoinflammatory processes and host defense mechanisms and have provided important clues in the mechanisms responsible for the regulation of IL-1β.

References

1. Janeway CA Jr (1989) Approaching the asymptote? Evolution and revolution in immunology. Cold Spring Harb Symp Quant Biol 54(Pt 1):1–13
2. Lemaitre B, Nicolas E, Michaut L, Reichhart J-M, Hoffmann JA (1996) The dorsoventral regulatory gene cassette Spaetzle/Toll/Cactus controls the potent antifungal response in Drosophila adults. Cell 86:973–983
3. Medzhitov R, Preston-Hurlburt P, Janeway CA Jr (1997) A human homologue of the Drosophila Toll protein signals activation of adaptive immunity. Nature 388:394–397
4. O'Neill LA (2006) How Toll-like receptors signal: what we know and what we don't know. Curr Opin Immunol 18:3–9
5. Kawai T, Akira S (2006) TLR signaling. Cell Death Differ 13:816–825
6. Choe J, Kelker MS, Wilson IA (2005) Crystal structure of human toll-like receptor 3 (TLR3) ectodomain. Science (New York, NY) 309:581–585
7. Miggin SM, Palsson-McDermott E, Dunne A, Jefferies C, Pinteaux E, Banahan K, Murphy C, Moynagh P, Yamamoto M, Akira S, Rothwell N, Golenbock D, Fitzgerald KA, O'Neill LA

(2007) NF-kappaB activation by the Toll-IL-1 receptor domain protein MyD88 adapter-like is regulated by caspase-1. Proc Natl Acad Sci USA 104:3372–3377
8. Aliprantis AO, Yang RB, Weiss DS, Godowski P, Zychlinsky A (2000) The apoptotic signaling pathway activated by Toll-like receptor-2. EMBO J 19:3325–3336
9. Maitra R, Clement CC, Scharf B, Crisi GM, Chitta S, Paget D, Purdue PE, Cobelli N, Santambrogio L (2009) Endosomal damage and TLR2 mediated inflammasome activation by alkane particles in the generation of aseptic osteolysis. Mol Immunol 47:175–184
10. Li H, Nookala S, Bina XR, Bina JE, Re F (2006) Innate immune response to Francisella tularensis is mediated by TLR2 and caspase-1 activation. J Leukoc Biol 80:766–773
11. Kleinnijenhuis J, Joosten LA, van de Veerdonk FL, Savage N, van Crevel R, Kullberg BJ, van der Ven A, Ottenhoff TH, Dinarello CA, van der Meer JW, Netea MG (2009) Transcriptional and inflammasome-mediated pathways for the induction of IL-1beta production by Mycobacterium tuberculosis. Eur J Immunol 39:1914–1922
12. Babelova A, Moreth K, Tsalastra-Greul W, Zeng-Brouwers J, Eickelberg O, Young MF, Bruckner P, Pfeilschifter J, Schaefer RM, Grone HJ, Schaefer L (2009) Biglycan, a danger signal that activates the NLRP3 inflammasome via toll-like and P2X receptors. J Biol Chem 284:24035–24048
13. Imamura M, Tsutsui H, Yasuda K, Uchiyama R, Yumikura-Futatsugi S, Mitani K, Hayashi S, Akira S, Taniguchi S, Van Rooijen N, Tschopp J, Yamamoto T, Fujimoto J, Nakanishi K (2009) Contribution of TIR domain-containing adapter inducing IFN-beta-mediated IL-18 release to LPS-induced liver injury in mice. J Hepatol 51:333–341
14. Shoma S, Tsuchiya K, Kawamura I, Nomura T, Hara H, Uchiyama R, Daim S, Mitsuyama M (2008) Critical involvement of pneumolysin in production of interleukin-1alpha and caspase-1-dependent cytokines in infection with Streptococcus pneumoniae in vitro: a novel function of pneumolysin in caspase-1 activation. Infect Immun 76:1547–1557
15. Basak C, Pathak SK, Bhattacharyya A, Mandal D, Pathak S, Kundu M (2005) NF-kappaB- and C/EBPbeta-driven interleukin-1beta gene expression and PAK1-mediated caspase-1 activation play essential roles in interleukin-1beta release from Helicobacter pylori lipopolysaccharide-stimulated macrophages. J Biol Chem 280:4279–4288
16. Hurst J, Prinz N, Lorenz M, Bauer S, Chapman J, Lackner KJ, von Landenberg P (2009) TLR7 and TLR8 ligands and antiphospholipid antibodies show synergistic effects on the induction of IL-1beta and caspase-1 in monocytes and dendritic cells. Immunobiology 214:683–691
17. Imaeda AB, Watanabe A, Sohail MA, Mahmood S, Mohamadnejad M, Sutterwala FS, Flavell RA, Mehal WZ (2009) Acetaminophen-induced hepatotoxicity in mice is dependent on Tlr9 and the Nalp3 inflammasome. J Clin Invest 119:305–314
18. Martinon F, Agostini L, Meylan E, Tschopp J (2004) Identification of bacterial muramyl dipeptide as activator of the NALP3/cryopyrin inflammasome. Curr Biol 14:1929–1934
19. Martinon F, Tschopp J (2005) NLRs join TLRs as innate sensors of pathogens. Trends Immunol 26:447–454
20. Kanneganti TD, Lamkanfi M, Kim YG, Chen G, Park JH, Franchi L, Vandenabeele P, Nunez G (2007) Pannexin-1-mediated recognition of bacterial molecules activates the cryopyrin inflammasome independent of Toll-like receptor signaling. Immunity 26:433–443
21. Dinarello CA, Cannon JG, Wolff SM (1986) Tumor necrosis factor (cachectin) is an endogenous pyrogen and induces production of interleukin-1. J Exp Med 163:1433–1450
22. Dinarello CA, Ikejima T, Warner SJC, Orencole SF, Lonnemann G, Cannon JG, Libby P (1987) Interleukin 1 induces interleukin 1. I. Induction of interleukin 1 in rabbits in vivo and in human mononuclear cells in vitro. J Immunol 139:1902–1910
23. Netea MG, Nold-Petry CA, Nold MF, Joosten LA, Opitz B, van der Meer JH, van de Veerdonk FL, Ferwerda G, Heinhuis B, Devesa I, Funk CJ, Mason RJ, Kullberg BJ, Rubartelli A, van der Meer JW, Dinarello CA (2009) Differential requirement for the activation of the inflammasome for processing and release of IL-1beta in monocytes and macrophages. Blood 113:2324–2335

24. Ferrari D, Chiozzi P, Falzoni S, Hanau S, Di Virgilio F (1997) Purinergic modulation of interleukin-1 beta release from microglial cells stimulated with bacterial endotoxin. J Exp Med 185:579–582
25. Ferrari D, Pizzirani C, Adinolfi E, Lemoli RM, Curti A, Idzko M, Panther E, Di Virgilio F (2006) The P2X7 receptor: a key player in IL-1 processing and release. J Immunol 176:3877–3883
26. Elssner A, Duncan M, Gavrilin M, Wewers MD (2004) A novel P2X7 receptor activator, the human cathelicidin-derived peptide LL37, induces IL-1 beta processing and release. J Immunol 172:4987–4994
27. Mariathasan S, Weiss DS, Newton K, McBride J, O'Rourke K, Roose-Girma M, Lee WP, Weinrauch Y, Monack DM, Dixit VM (2006) Cryopyrin activates the inflammasome in response to toxins and ATP. Nature 440:228–232
28. Chae JJ, Wood G, Masters SL, Richard K, Park G, Smith BJ, Kastner DL (2006) The B30.2 domain of pyrin, the familial Mediterranean fever protein, interacts directly with caspase-1 to modulate IL-1beta production. Proc Natl Acad Sci USA 103:9982–9987
29. Hoffman HM, Rosengren S, Boyle DL, Cho JY, Nayar J, Mueller JL, Anderson JP, Wanderer AA, Firestein GS (2004) Prevention of cold-associated acute inflammation in familial cold autoinflammatory syndrome by interleukin-1 receptor antagonist. Lancet 364:1779–1785
30. Hawkins PN, Lachmann HJ, Aganna E, McDermott MF (2004) Spectrum of clinical features in Muckle-Wells syndrome and response to anakinra. Arthritis Rheum 50:607–612
31. Aksentijevich I, Nowak M, Mallah M, Chae JJ, Watford WT, Hofmann SR, Stein L, Russo R, Goldsmith D, Dent P, Rosenberg HF, Austin F, Remmers EF, Balow JE Jr, Rosenzweig S, Komarow H, Shoham NG, Wood G, Jones J, Mangra N, Carrero H, Adams BS, Moore TL, Schikler K, Hoffman H, Lovell DJ, Lipnick R, Barron K, O'Shea JJ, Kastner DL, Goldbach-Mansky R (2002) De novo CIAS1 mutations, cytokine activation, and evidence for genetic heterogeneity in patients with neonatal-onset multisystem inflammatory disease (NOMID): a new member of the expanding family of pyrin-associated autoinflammatory diseases. Arthritis Rheum 46:3340–3348
32. Kuida K, Lippke JA, Ku G, Harding MW, Livingston DJ, Su MS-S, Flavell RA (1995) Altered cytokine export and apoptosis in mice deficient in interleukin-b converting enzyme. Science (New York, NY) 267:2000–2003
33. Li P, Allen H, Banerjee S, Franklin S, Herzog L, Johnston C, McDowell J, Paskind M, Rodman L, Salfield J, Towne E, Tracey D, Wardwell S, Wei F-Y, Wong W, Kamen R, Seshadri T (1995) Mice deficient in IL-1b-converting enzyme are defective in production of nature IL-1-beta and resistant to endotoxic shock. Cell 80:401–411
34. Fantuzzi G, Ku G, Harding MW, Livingston DL, Sipe JD, Kuida K, Flavell RA, Dinarello CA (1997) Response to local inflammation of IL-1beta converting enzyme-deficient mice. J Immunol 158:1818–1824
35. Labow M, Shuster D, Zetterstrom M, Nunes P, Terry R, Cullinan EB, Bartfai T, Solorzano C, Moldawer LL, Chizzonite R, McIntyre KW (1997) Absence of IL-1 signaling and reduced inflammatory response in IL-1 type I receptor-deficient mice. J Immunol 159:2452–2461
36. Fantuzzi G, Dinarello CA (1996) The inflammatory response in interleukin-1beta-deficient mice: comparison with other cytokine-related knock-out mice. J Leukoc Biol 59:489–493
37. Coeshott C, Ohnemus C, Pilyavskaya A, Ross S, Wieczorek M, Kroona H, Leimer AH, Cheronis J (1999) Converting enzyme-independent release of tumor necrosis factor alpha and IL-1beta from a stimulated human monocytic cell line in the presence of activated neutrophils or purified proteinase 3. Proc Natl Acad Sci USA 96:6261–6266
38. Dinarello CA (1996) Biologic basis for interleukin-1 in disease. Blood 87:2095–2147
39. Sugawara S, Uehara A, Nochi T, Yamaguchi T, Ueda H, Sugiyama A, Hanzawa K, Kumagai K, Okamura H, Takada H (2001) Neutrophil proteinase 3-mediated induction of bioactive IL-18 secretion by human oral epithelial cells. J Immunol 167:6568–6575
40. Greten FR, Arkan MC, Bollrath J, Hsu LC, Goode J, Miething C, Goktuna SI, Neuenhahn M, Fierer J, Paxian S, Van Rooijen N, Xu Y, O'Cain T, Jaffee BB, Busch DH, Duyster J, Schmid

RM, Eckmann L, Karin M (2007) NF-kappaB is a negative regulator of IL-1beta secretion as revealed by genetic and pharmacological inhibition of IKKbeta. Cell 130:918–931
41. Joosten LA, Netea MG, Fantuzzi G, Koenders MI, Helsen MM, Sparrer H, Pham CT, van der Meer JW, Dinarello CA, van den Berg WB (2009) Inflammatory arthritis in caspase 1 gene-deficient mice: contribution of proteinase 3 to caspase 1-independent production of bioactive interleukin-1beta. Arthritis Rheum 60:3651–3662

Inflammasome in the Skin

Olivier Gaide

Abstract The skin, our largest organ, protects us from our environment and is therefore constantly subjected to microbial, chemical, and physical insults. Not surprisingly, it is endowed with the capacity to detect these events and respond accordingly, alerting the immune system when needed, often through the inflammasome. In this chapter, we propose a synthetic view of the role and implication of the inflammasome in the skin. We will consider first the molecular events leading to its activation and the consequences of IL-1β secretion, and then lessons from the bedside, reviewing diseases that are inflammasome dependent and their effect on the skin.

1 The Skin as an Immune Organ

Once considered merely a physical barrier, the skin is now known to play a major role in the induction and maintenance of local and systemic immunity [1]. It protects us from our environment and is therefore constantly subjected to microbial, chemical, and physical insults that ultimately result in the generation of nonself antigens. Accordingly, it has the capacity to trigger innate and adaptive immune responses [2]. Cutaneous cells express a wide variety of danger sensors, including Toll-like receptors (TLR) and NOD-like receptors (NLR), and can secrete numerous cytokines, chemokines, and antimicrobial peptides [2]. The importance of its immunological role is highlighted by the fact that several vaccine and immune assays (tuberculin test, pathergy reaction) are done in the skin.

O. Gaide (✉)
Department of Dermatology-Venereology, Geneva University, Rue Michel-Servet 1, 1211 Geneva, Switzerland
and
Department of Pathology-Immunology, Geneva University, Rue Michel-Servet 1, 1211 Geneva, Switzerland
e-mail: oliver.gaide@unibe.ch

I. Couillin et al. (eds.), *The Inflammasomes*, Progress in Inflammation Research,
DOI 10.1007/978-3-0348-0148-5_9, © Springer Basel AG 2011

The skin is composed of a surprisingly high number of different cells, and most seem to play an important immunological role. The epidermis, under normal conditions, is mainly composed of keratinocytes, but also contains Langerhans cells, melanocytes, merkel cells, and few memory B and T cells. Beneath the epidermis, the dermis contains a very significant number of memory T cells, dendritic cells (DC) of various types, macrophages, mast cells, and more. To set things into perspective, it is important to know that the skin contains more memory T cells than the blood. At this time, most of these cells have been shown to modulate the inflammatory and immune response [3, 4]. Deleting one type of more of these cells may lead to compensatory mechanisms affecting other cell populations, thus resulting in partially contradictory data [5]. As a result, the exact contribution of each cell type is difficult to assess in a given inflammatory or immune reaction. In any case, the type of response generated at a given danger signal is likely to result, in vivo, from the interplay of each independent stimulatory or inhibitory signal given by the different cell types residing in the skin. The study of the individual role of each of these cells will benefit greatly from the study of conditional inflammasome deficient mice, which are currently being studied by different research groups.

2 Danger Sensing in the Skin

Keratinocytes represent by far the largest cell population in the epidermis. Its top most layer is actually composed only of keratinocytes. Numerous saprophytic bacteria, fungi, and parasites colonize this layer, without eliciting any inflammatory reaction (in humans). Interestingly, these keratinocytes express few danger sensors [6]. The situation is completely different in the more basal layers, where keratinocytes are known to express several danger sensors and an array of cytokines and are an important source of innate immune danger signals [1, 4]. Keratinocytes express a variety of danger sensors, including TLR and NLR [4, 7]. These signals elicit a local inflammation that influence the primary immune response through its effect on the skin resident DC, the draining lymph nodes (LN) milieu and the dermal microvasculature, thus recruiting and modifying memory effector T cells [2]. Epidermal and dermal DC are critical for capturing antigens and allowing their delivery to skin draining LN, where these DC can activate naïve or central memory T cells [8, 9]. Alternatively, Langerhans cells and the few B and T cells found in the epidermis also express various danger sensors and can act as antigen presenting cells (APC). This is also the case of melanocytes. As stated earlier, the relative contribution of each cell types is still unknown. However, it is tempting to speculate that keratinocytes should play a key role in early detection of pathogen and physical or chemical insults, as they cover most of the mass and surface of the epidermis.

The situation is even more complex in the dermis, where the inflammatory cell flow varies tremendously. However, any breach of the barrier resulting in the penetration of pathogen or chemicals below the dermis results in strong inflammatory signals and there is little doubt that danger can be detected in the dermis as well.

3 The Inflammasome in the Skin

The skin has been recognized long ago as an important source of cytokines, including IL-1β. Delayed type hypersensitivity (DTH) is a paradigm of the processes leading to the adaptive immune response. In this type of immune response, nonself antigens are first taken up by APC and presented to naïve T cells, resulting in their activation (provided they carry the suitable T cell receptor) and the creation of a repertoire of active T cells that can, upon renewed antigen encounter, mount a rapid and potent protective immune response. One of the most studied forms of DTH is contact hypersensitivity (CHS). In this reaction, a chemical is painted on the skin, penetrates the epidermis, reacts with proteins to create neo-antigens that can stimulate naïve T cells, and results in a life-lasting immune reaction upon new exposure to the chemical. Efficient antigen presentation, as well as the later recruitment of primed lymphocytes to the skin, was recognized early as being dependent on the secretion of IL-1β originating in the skin itself [10, 11]. IL-1β was also demonstrated to play a key role in several diseases that affect predominantly the skin. Examples, range from acne and diabetes to Behçet disease and skin cancer. However, the mechanism responsible for the control of this cytokine remained unclear for decades. This changed rapidly after Fabio Martinon and colleagues identified the inflammasome as the major machinery controlling its activation [12]. In this chapter, we present the studies that have allowed a better characterization of the role of the inflammasome in the skin.

3.1 Expression of the Inflammasome Machinery in the Skin

The importance of IL-1β in several skin diseases (see below) has led us to postulate the existence of a functional inflammasome in the skin. We therefore initiated studies aiming at characterizing the role of proteins of the NALP family in the skin. We found that that NALP1, NALP3, ASC-1, caspase-1, caspase-5, pro-IL-1β, and IL-18 are present at the protein level in primary keratinocytes and HaCaT cells (an untransformed keratinocyte-derived cell line), a finding that was confirmed by Feldmeyer et al. [13]. Moreover, NALP8, 10-11, and 13-14 could also be detected at the mRNA level [4]. We also found that keratinocytes can, when exposed to several skin irritants, activate and secrete IL-1β in a caspase-1-dependent manner [4]. This was the first demonstration that nonprofessional immune cells contain the molecular components necessary for inflammasome assembly, and that an innate immune receptor other than a TLR was active on epithelial cells. We also found expression of several NALP proteins in Langerhans cells (LC) and melanocytes (our unpublished observations). However, we could not yet find the trigger that leads to a significant activation and secretion of significant levels of IL-1β in these cells.

3.2 The NALP3 Inflammasome and Skin Sensitizers

Repeated exposure to certain skin irritants induces CHS [14]. This immunological process can be divided into a sensitization phase and an elicitation phase. Sensitization depends on antigen uptake by skin-resident APC and their migration to draining LN where T cell priming occurs. An irritant effect of the antigen is essential at this stage, as is the presence of a functional caspase-1, IL-1β, and IL-18 [10, 15, 16]. Elicitation occurs after challenge with a relevant hapten for the primed T cells and is not dependent on caspase-1 or an irritant effect of the chemical agent. We and others have recently demonstrated that sensitization with several skin sensitizers, such as di-nitro-chloro-benzene (DNCB) is dependent on the NALP3 inflammasome, thus identifying the molecular link between the irritant effect of the hapten and its capacity to induce a potent immune response [4, 17].

3.3 Absence of Danger Leads to Treg Mediated Tolerance, While Danger Signaling Favors Th1 and Th17

It seems very unlikely that mammals should have evolved sensors to sensitizing chemicals, but it is very likely that these molecules are strong sensitizers because they happen to activate the inflammasome. We have explored this hypothesis and found that the level of inflammasome activation in vitro correlates with the sensitizing potency of a given hapten in vivo, and that it can be further modulated by providing concomitant activators/inhibitors of the NALP3 inflammasome [18]. The cytokine expression pattern of CD4+ T lymphocytes defines four main subtypes: T helper 1 are important in antiviral response; T helper 2 ensure parasite elimination and play a major role in the development of asthma, hay fever, and atopic dermatitis; T helper 17 play target bacteria infected cells and are implicated in autoimmune and autoinflammatory disorders; finally, regulatory T cells (Treg) oppose the activity of other T cells and act to maintain immune reaction in balance. We could show that mice lacking either NALP3 or ASC in the skin fail to mount an efficient T cell response to contact sensitizers [4]. More interestingly, using a model of adoptive transfer, we found that the absence of inflammasome activation renders the mice resistant to subsequent sensitization [18]. This tolerization is hapten specific as sensitization and challenge with a structurally unrelated antigen could still elicit a potent immune response. Depletion of T regulatory cells restored the capacity of mice to be sensitized, strongly suggesting that in the absence of danger signals, the skin instructs the immune system to tolerate antigens present in the skin. This provides the first explanation for the apparent tolerance of the immune system to neoplastic cells, which often start to develop in the absence of danger signals [18].

Di-nitro-thyocyano-benzene (DNTB) is a chemical that does not activate the IL-1β signaling pathway, yet is an efficient source of epitopes that cross-react with those arising from the very potent sensitizer di-nitro-fluoro-benzene (DNFB). Using these molecular tools, we could show that concomitant danger signals turn DNTB into a strong sensitizer, and that this correlates with the expression, within the draining LN, of a cytokine environment (IL-6, IL-12, and TGF-β) that favors the development of Th1 and Th17 [18]. We believe that these findings hint to the molecular basis underlying the clinical efficacy of DNCB as a booster of the antitumoral activity in melanoma patients [19].

It is important to note here that not all sensitizers activate the inflammasome. DNCB, TNCB, DNFB, and urushiol, the active component of poison ivy, all do. However, oxazolone is a very potent sensitizer that is not affected by the absence of ASC or NALP3 (our unpublished observation). We also failed to observe a significant secretion of IL-1β when we exposed keratinocytes to weak sensitizers such as nickel and fragrances, although this may be due to the lack of sensitivity of our assays. Taken together, our data suggest that potent activators of the skin inflammasome provide danger signals that ensure a rapid immune response and leads in every case to a clinically detectable CHS. On the other hand, weak activators will depend on additional immunological events to activate naïve T cells. This explains why certain molecules such as urushiol and DNCB will sensitize all, when only selected patients become allergic to nickel or fragrances.

3.4 *Th2 Response Modulated by the Inflammasome*

More recently, we have studied the role of the thymic stromal lymphopoietin (TSLP) in relation to the inflammasome. This cytokine is expressed at high level in the skin and is known to direct a T helper 2 cell fate [20]. It is believed that the TSLP gene is under the control of an NF-κB responsive element that is under direct control of the IL-1 receptor (IL-1R). As the inflammasome controls the activation of IL-1β, we decided to study its impact on TSLP expression. We used a model of Th2 sensitization, in which mice are expose to fluorescein-iso-thyo-cyanate (FITC) dissolved in acetone and dibutyl-phtalate. We found that ASC deficient mice developed a full response to FITC. This strongly suggests that, contrary to the current belief, TSLP is not under the control of IL-1 and the inflammasome. Moreover, we observed that dibutyl-phtalate is sufficient to induce a strong expression of TSLP. This TSLP level is sufficient to induce a robust Th2 response to DNTB, instead of the Treg population it normally elicits (Schuepbach et al., manuscript in preparation). Taken together, our results indicate that the skin contains a series of functional danger sensors that control the local and distant expression of cytokines that in turn direct the T cell fate.

3.5 NALP3 and UV

The inflammatory role of IL-1β following acute sun exposure has been described long ago [21]. We, as an example, had observed that UV irradiation resulted in a caspase-1-dependent processing and release of IL-1β from cultured keratinocytes, strongly suggesting an involvement of the inflammasome in UV sensing [4]. An implication of the NALP1 and NALP3 inflammasomes was later corroborated in vitro [13] and the in vivo relevance confirmed in a mouse model of acute sunburn, where caspase-1 was found to be essential for inflammatory cell recruitment in the skin [13]. The authors of this study suggested that the increase in calcium levels after UV irradiation may act as a danger signal activating the inflammasome, since they found that caspase-1 and IL-1β activation are calcium dependent in keratinocytes [13]. An alternative explanation is that UV irradiation induces reactive oxygen species formation, which acts as danger signal leading to inflammasome activation through thioredoxin-interacting protein [22]. The observation that the oxidative state of certain contact sensitizers affects their ability to activate the inflammasome in vitro (our own unpublished data) and relate to their sensitizing potential in vivo [23] and is in accordance with this hypothesis.

4 Human Skin Diseases and the Inflammasome

4.1 NALP1 and Vitiligo

Vitiligo is characterized by acquired skin depigmentation due to local loss of melanocytes, which are the sole pigment producing cells in the skin. Its etiology is still not fully understood, but several lines of evidence suggest an autoimmune origin. For example, autoantibodies binding to melanocyte-derived peptides [24] and skin infiltrating T cells [25] can be detected in vitiligo patients, suggesting a dual involvement of the humoral and cellular response. Whether these antibodies and cells are the cause or the consequence of the melanocyte cell death remains to be determined. More importantly, familial vitiligo, which accounts for one third of patients with vitiligo, is associated with other autoimmune diseases, including diabetes, autoimmune thyroid disease, rheumatoid arthritis, Addison's disease, and systemic lupus erythematosus [26].

A locus for familial vitiligo was recently identified on chromosome 17, where the gene coding for NALP1 is positioned [27]. Specific mutations within NALP1 could be linked with vitiligo alone or with an autoimmune disease. The functional effect of these NALP1 variants has not been studied yet. Since NALP3, NOD2, and possibly NALP5 allow a strong priming of the adaptive immune response [4, 12, 28], it is possible that the NALP1 mutations promote IL-1β and IL-18 secretion, which have been demonstrated to favor CD25+ effector T cell over regulatory T cell growth, thus supporting the development of autoreactive T cells

[29]. Interestingly, microtrauma plays a role in the development of vitiligo, but the underlying mechanisms are still unknown [30]. The idea that microtrauma may trigger NALP1 inflammasome-mediated autoreactive T cell activation in patients with predisposing mutations is an attractive hypothesis that remains to be investigated.

4.2 NALP3 and Inherited Autoinflammatory Disorders

4.2.1 Cryopyrinopathies or CAPS

Mutations in the gene encoding NALP3 are associated with a clinical continuum of autoinflammatory diseases referred to, collectively, as cryopyrinopathies or cryopyrin-associated periodic syndromes (CAPS) (NALP3 is also known as cryopyrin) [31]. These disorders present early in life and are transmitted in an autosomal dominant fashion, although there are several sporadic cases reported, which have been linked to de novo mutations. Heterozygous mutations in NALP3 induce single amino acid substitutions primarily located in the NACHT domain, which are thought to result in unchecked inflammasome activity. The mildest of the phenotypes is known as familial cold autoinflammatory syndrome (FCAS) and is characterized by recurrent episodes of fever, urticaria-like skin rash, and joint pain triggered by mild generalized cold exposure. Muckle–Wells syndrome (MWS) is characterized by more severe and chronic symptoms, and patients develop progressive sensory hearing loss and renal amyloidosis. Finally, the most severe of the cryopyrinopathies is known as chronic infantile neurologic cutaneous and articular syndrome (CINCA) also known as neonatal-onset multisystem inflammatory disease (NOMID), and results in inflammatory damage to the joints and the central nervous system. The genotype–phenotype correlation is somewhat limited but the online compilation of the mutations and their phenotype represents a great guide to CAPS biology [32].

Blocking of IL-1β signaling by neutralizing antibodies or the IL-1R antagonist (IL-1Ra) brings spectacular relief in these patients [31, 33–35], demonstrating the central role of this cytokine in the cutaneous and systemic symptoms characterizing these diseases. It also illustrates the importance of maintaining a tight control over inflammasome activation and IL-1β secretion.

All forms of CAPS display some form of cutaneous involvement. This illustrates the sensitivity of the skin to the inflammasome activity. Does this also point to a direct effect of the NALP3 mutations in the keratinocytes? The question remains unanswered for the moment. In fact, mast cells are currently believed to be the major source of IL-1β secretion in these diseases [36]. Mast cells are an integral component of the skin immune cell population that normally resides in the dermis. They are direct targets of IL-1β and secrete histamine and bradykinine, among others, which induce the urticaria-like skin lesions observed in CAPS patients. However, this does not rule out that additional cells expressing the inflammasome participate to the mast cell activation. This is particularly the case for epithelial cells in the cochlea (the cochlea is affected in MWS and CINCA patients, leading to the hearing loss), where mast cells are normally not found, but could also be the case for other epithelial cells such as keratinocytes. Moreover, skin lesions of CAPS patients also contain a neutrophilic infiltration atypical for hives, probably resulting from the direct chemoattractant effect of IL-1β, IL-18, and IL-6.

4.2.2 SoJIA and AOSD and FMF

Several other genetic diseases involving the skin are associated with the inflammasome, even though they are not associated with mutations in NALP proteins so far [31, 37]. Systemic onset juvenile idiopathic arthritis (SoJIA) and adult onset Still's disease (AOSD) represent two autoinflammatory syndromes that can have skin lesions indiscernible from those observed in cryopyrinopathies. The cause of both diseases still has to be determined, but response to IL-1β blocking drugs strongly suggests an involvement of the inflammasome. Familial Mediterranean fever (FMF) is due to mutations in the gene that codes for pyrin, a proposed inhibitor of the inflammasome [31]. Alterations in pyrin are believed to result in excessive IL-1β secretion due to loss of this inhibitory function. It is interesting to note that colchicine has been used to treat patients with FMF since the antiquity [38], and is also a classical therapy of gout, a disease that was recently associated with the NALP3 inflammasome [39].

4.2.3 PAPA Syndrome

Pyogenic arthritis, pyoderma gangrenosum, and acne (PAPA) is another genetic disease strongly involving the skin, as its name suggests. PAPA has similarities to other autoinflammatory diseases, including fever bouts and an exaggerated tissue infiltration by neutrophils, which are attracted by IL-1β and IL-18 [31]. It is due to mutations in the proline–serine–threonine phosphatase-interacting protein-1 (PSTPIP1). This protein binds pyrin and is believed, when mutated, to relieve pyrin's inhibitory activity on the secretion of IL-1β, although PSTPIP1 has also been suspected to form an inflammasome-like structure itself [40, 41]. Although the disease is excessively rare, it points toward a potentially important role of the inflammasome in acne, which affects 90% of the general population.

4.2.4 Acne

The pathophysiology of acne is currently thought to result from an abnormal inflammatory response to *Propionibacterium acnes*, but also to the presence of ruptured microcyst (known as comedone or white/blackheads). Interestingly, acne is known to be a multifactor disease with genetic involvement. Indeed, high clinical concordance in monozygotic twins and similar acne levels in acne prone families are routinely observed. Whether *P. acnes* and keratin from the ruptured cyst activate the inflammasome, and whether polymorphism in inflammasome components is responsible for the intensity of the inflammatory response in acne, is currently being investigated. Pyoderma gangrenosum, known by French dermatologists as phagedenic ulcer, is characterized by an explosive yet sterile neutrophilic infiltration of the skin resulting in its digestion, leading to spectacular ulcers that only respond to inflammatory drugs, such as steroids. In this still poorly understood disease, any insult to the skin, and especially surgical debridement by clinicians who failed to recognize the affliction, results in an immediate and dramatic neutrophilic infiltration and worsening of the ulcer. Since neutrophils are highly effectively recruited to sites of IL-1β and IL-18 secretion, one can speculate on an important role of the inflammasome in this disease as well.

4.2.5 Behçet's Disease

Several evidences suggest that Behçet disease is another autoinflammatory syndrome with skin lesions that may implicate the inflammasome. First, IL-1β has been implicated in its development [42, 43]. Second, Behçet is characterized by a fascinating clinical finding called the pathergy reaction. Patients suffering from the disease respond to a needle puncture on the forearm by forming a pustule at the wound site within 1–2 days. This pustule is due to the recruitment of neutrophils to skin wounds and may even occur spontaneously, much alike pyoderma gangrenosum. Third, IL-1R antagonists has been reported to alleviate the symptoms of a Behçet patient resistant to several potent immunosuppressive or immunomodulating drugs [44].

4.2.6 Crohn's Disease

Crohn's disease is a chronic relapsing inflammation of the digestive tract characterized by a markedly increased IL-1β secretion in the mucosa [45], but also skin lesions that overlap, from a clinical point of view, with the lesions of Behçet and pyoderma-gangrenosum. Mutations in the NOD2 gene were shown to be strongly associated with Crohn's disease [46]. More recently, several single nucleotide polymorphisms in NALP3, associated with hyperproduction of IL-1β and decreased NLRP3 expression, were recently identified in certain Crohn patients [47]. Hence, development of this multifactorial disease may depend on the synergistic activation of NOD2 and NALP3.

4.3 NALP3 and Gout

Gout and pseudogout are autoinflammatory diseases characterized by inflammation in the joint induced by two microcrystals, monosodium urate (MSU) and calcium pyrophosphate dihydrate (CPPD), respectively [39]. Both crystals are thought to act as "danger signals" released by dying cells that release their DNA [48] and were recently found to activate the NALP3 inflammasome [39]. Little known is the fact that the skin is affected by gout, just like joints are. In fact, crystal deposits in the skin are responsible for the occurrence of tophi, a hallmark of gout, which are big crystal agglomerates surrounded by an intense inflammatory reaction often leading to skin ulceration and expulsion of the crystals. Since MSU-induced inflammation is abolished in IL-1R deficient mice, IL-18 does not seem to play an important role in vivo [39]. The importance of IL-1β in the pathology of gout is also highlighted by promising preliminary studies in humans, showing that IL-1 blockade by its natural inhibitor induced a rapid and efficient response in patients that could not tolerate or had failed standard anti-inflammatory gout therapies [49, 50].

4.4 NALP3, Vaccine, and the Skin

Immunization through the skin is a very effective means of generating a protective systemic immune response. Skin draining LN provide an environment that drives activated T cells to acquire a skin homing phenotype as they express cutaneous lymphocyte antigen (CLA) that binds to skin microvasculature-expressed E-selectin [49, 51]. This should in theory preclude the use of skin vaccination to achieve global immunity. However, the Kupper team recently showed that subsets of proliferating T cells can leave the skin draining LN and migrate to other LN where they acquire different homing receptors (matching the region they migrated to) [52]. As an example, MMR and pneumococcal vaccines are routinely injected in the skin.

Efficient vaccination in the skin most often requires the use of adjuvants, molecules that acts as danger signals and enhance the recipient's immune response, be it a T or B lymphocyte response. Freund's adjuvant, which is such an immune booster, consists of inactivated and dried mycobacteria in water in oil emulsion. Not surprisingly, this mix is rich in muramyl-dipeptide and signals notably through NALP3. However, Freund's adjuvant is not used in clinical setting. In fact, the only human approved adjuvant is "alum," actually either crystalline aluminum oxy-hydroxide or amorphous aluminum hydroxyphosphate [53]. Alum induces IL-1β and IL-18 release in an NALP3 and caspase-1-dependent way in vitro and in vivo [54–57]. The degree to what NALP3 is genuinely essential in vivo is however still debated, as mice studies have generated partially contradictory results and may not be good reflections of what happens in the human skin [53, 58, 59].

Variola major (smallpox), however, remains the paradigm of skin-based vaccination. Variola indeed claimed hundreds of millions lives before it could be eradicated by the inoculation of the vaccinia virus (VV) into an epidermis disrupted by a bifurcated needle, a process known as "skin scarification." This vaccination strategy is so potent that it does not require any adjuvant. Liu et al. have recently shown that it outclasses the intradermal and intramuscular injections routes [60]. This is likely to be due to the activation in the skin of several danger pathways, which have only been recently identified, and include the NALP3 inflammasome itself [61].

5 Downstream Signaling Pathways Controlled by the Inflammasome and Their Role in the Skin

There is evidence to suggest that NLRs, such as NOD1, NOD2, and NALP3 directly activate NF-κB, independently of capase-1 [3]. However, the in vivo relevance of these findings remains to be determined, whereas there is ample evidence that caspase-1 is under the direct control of the inflammasome. Since the major effect of caspase-1 is mediated through its target cytokines, it is reasonable to suspect that inflammasome is also implicated in the pathogeny of diseases due to deregulated IL-1β, IL-18, and possibly IL-33 activation. One should note here that several other proteases have been proposed to cleave IL-1β and IL-18, including mast cell chymases, but only in vitro [62], unlike caspase-1 [63–65].

IL-1β and IL-18 have been the subject of several excellent reviews providing insights into the pleiotropic effects of these cytokines [66–68]. IL-1β affects virtually every tissue, initiating or amplifying a wide variety of processes mainly associated with innate immunity and host response to environmental insults (bacteria, UV, irritating chemicals), but also controlling tumor angiogenesis and neuron activity. In the skin, IL-1β has been reported to play a role in Behçet's disease, acne, cryopyrinopathies, SoJIA and AOSD, PAPA, as described earlier, but also in melanoma and psoriasis [37, 69].

IL-18 was formerly known as interferon-γ inducing factor, but it also has many other direct effects, including the regulation of leukocyte migration and activity. It is highly expressed in the skin, especially in suprabasal keratinocytes. However, the role of IL-18 in the skin is less well studied, and its role in vivo remains to be discovered.

IL-33 is a recently identified member of the IL-1 family that has been proposed to be cleaved by caspase-1 [70], although the in vivo relevance of this finding is limited [71]. IL-33 is associated with the production of T helper 2-associated cytokines, which play a key role in the development of atopic dermatitis [70]. This complex disease is also characterized by an increase in the risk of developing HSC as well as increased Th1 cytokine production, which are under partial control of the inflammasome [18]. Whether the inflammasome is truly responsible for diverging Th1 vs. Th2 responses remains to be determined. However, one can postulate the expression of the different pro-cytokines may vary in time, thus providing one signal or another, depending on the context. However, the role of IL-33 in the skin, if any, still has to be demonstrated.

IL-1 family cytokines exert their function through receptors that belong to the TLR-IL-1R superfamily [72] and signal through both NF-κB and MAPK. Activation of the IL-1R, IL-18R, and ST2R (IL-33 receptor) results in the release of a number of additional cytokines, such as IL-6, IFNγ, and IL-4, respectively [72, 73]. These cytokines are not regulated uniquely by interleukins, but can be readily detected after the injection of recombinant interleukins. Hence, it may prove difficult to dissect the direct effect of IL-1β, IL-18, and IL-33 independently of the effect of downstream cytokines, and it may be reasonable to think of IL-6 and other downstream cytokines as inflammasome-regulated cytokines.

6 Therapeutic Tools Targeting the Inflammasome and the Skin

6.1 Inhibition of the Inflammasome

Currently, the only widely available drug directed at this pathway is anakinra (Amgen Kineret®, which is a recombinant soluble form of IL-1 receptor antagonist (IL-1Ra) [31, 33–35]. It has a short half-life (about 12 h) and therefore requires daily self-administered subcutaneous injections. As underlined by several very successful studies in CAPS and SOJIA patients, this systemic injection has a spectacular effect on skin inflammation [31, 35]. Additional drugs targeting the IL-1 pathway will soon be available. A recombinant fusion protein known as IL-1 TRAP (Regeneron IL-1 TRAP) and a fully human anti-IL-1β antibody (Novartis ACZ885) are undergoing clinical trials (CAPS, phase III). Both molecules, which have a much longer half-life, display systemic effects that include skin healing.

Therapeutic substances can also be delivered locally to the skin, through topical creams or ointments, but topical inhibitors of the inflammasome still have to reach the bedside. A proof of concept of this clinical potential is no longer necessary. Indeed, we have shown that the local application of a caspase-1 inhibitor (zYVAD) could block T cell activation in the context of contact dermatitis. However, the extremely short half-life (20 min) and irritant effect of zYVAD are serious limitations that prevent its clinical use [74]. Recently, Vertex has developed zYVAD-peptido-mimetic molecules, which were selected for their high specificity and activity [74]. Their activity has been demonstrated both in vivo and in vitro [75] but due to liver toxicity, they may find their true clinical potential as topical drugs (which will not be the case for larger molecules described earlier). Finally, geldanamycin and its derivatives, which influence the function of heat-shock proteins, such as HSP-90, and thus the activity of the inflammasome [76], are currently undergoing clinical trials in cancer patients and represent interesting candidates for topical inhibition of the inflammasome, as their size should allow trans-epidermal delivery.

6.2 *Activation of the Inflammasome*

Systemic activation of the inflammasome will result in the secretion of high levels of IL-1β and are very likely to induce intolerable fever bouts. It therefore has a very limited clinical potential. Topical activators of the inflammasome, on the contrary, may have therapeutic uses. We have recently shown that contact sensitizers such as DNCB are potent activators of the inflammasome. This small lipid-soluble molecule can be applied directly on the skin to modulate the adaptive immune response. Indeed, DNCB is currently used for the treatment of alopecia areata, an autoimmune disease causing hair loss [77]. DNCB has also been used to boost host vs. tumor immune reactions, in malignancies such as melanoma, where it is considered a safe and promising treatment [19, 77]. Interestingly, TLR7 triggering by imiquimod, which is used to boost the adaptive-immune response against HPV viruses and basal cell carcinoma, has also been used to treat melanoma [78]. The convergence of TLR and NLR signaling pathways may suggest that inflammasome activators could be used for vaccination and virus/tumor rejection.

7 Conclusion

The skin contains immune and nonimmune cells, most of which express the inflammasome. These cells can react to several stimuli that activate the inflammasome and results in IL-1β secretion both in vitro and in vivo. On the other hand, all diseases associated with mutations in NALP1 or NALP3 may affect the skin. Taken together, these data show the importance of the skin as a source and target of the inflammasome activity.

References

1. Kupper TS, Fuhlbrigge RC (2004) Immune surveillance in the skin: mechanisms and clinical consequences. Nat Rev Immunol 4:211–222
2. Murphy JE, Robert C, Kupper TS (2000) Interleukin-1 and cutaneous inflammation: a crucial link between innate and acquired immunity. J Invest Dermatol 114:602–608
3. Martinon F, Tschopp J (2005) NLRs join TLRs as innate sensors of pathogens. Trends Immunol 26:447–454
4. Watanabe H, Gaide O, Petrilli V, Martinon F, Contassot E, Roques S, Kummer JA, Tschopp J, French LE (2007) Activation of the IL-1beta-processing inflammasome is involved in contact hypersensitivity. J Invest Dermatol 127:1956–1963
5. Kaplan DH, Kissenpfennig A, Clausen BE (2008) Insights into Langerhans cell function from Langerhans cell ablation models. Eur J Immunol 38:2369–2376
6. Baker BS, Ovigne JM, Powles AV, Corcoran S, Fry L (2003) Normal keratinocytes express Toll-like receptors (TLRs) 1, 2 and 5: modulation of TLR expression in chronic plaque psoriasis. Br J Dermatol 148:670–679
7. Miller LS (2008) Toll-like receptors in skin. Adv Dermatol 24:71–87

8. Lenz A, Heine M, Schuler G, Romani N (1993) Human and murine dermis contain dendritic cells. Isolation by means of a novel method and phenotypical and functional characterization. J Clin Invest 92:2587–2596
9. Stingl G, Tamaki K, Katz SI (1980) Origin and function of epidermal Langerhans cells. Immunol Rev 53:149–174
10. Grabbe S, Steinert M, Mahnke K, Schwartz A, Luger TA, Schwarz T (1996) Dissection of antigenic and irritative effects of epicutaneously applied haptens in mice. Evidence that not the antigenic component but nonspecific proinflammatory effects of haptens determine the concentration-dependent elicitation of allergic contact dermatitis. J Clin Invest 98:1158–1164
11. Griffiths CE, Dearman RJ, Cumberbatch M, Kimber I (2005) Cytokines and Langerhans cell mobilisation in mouse and man. Cytokine 32:67–70
12. Martinon F, Agostini L, Meylan E, Tschopp J (2004) Identification of bacterial muramyl dipeptide as activator of the NALP3/cryopyrin inflammasome. Curr Biol 14:1929–1934
13. Feldmeyer L, Keller M, Niklaus G, Hohl D, Werner S, Beer HD (2007) The inflammasome mediates UVB-induced activation and secretion of interleukin-1beta by keratinocytes. Curr Biol 17:1140–1145
14. Grabbe S, Schwarz T (1998) Immunoregulatory mechanisms involved in elicitation of allergic contact hypersensitivity. Immunol Today 19:37–44
15. Allenby CF, Basketter DA (1993) An arm immersion model of compromised skin (II). Influence on minimal eliciting patch test concentrations of nickel. Contact Dermatitis 28: 129–133
16. Cumberbatch M, Scott RC, Basketter DA, Scholes EW, Hilton J, Dearman RJ, Kimber I (1993) Influence of sodium lauryl sulphate on 2,4-dinitrochlorobenzene-induced lymph node activation. Toxicology 77:181–191
17. Sutterwala FS, Ogura Y, Szczepanik M, Lara-Tejero M, Lichtenberger GS, Grant EP, Bertin J, Coyle AJ, Galan JE, Askenase PW et al (2006) Critical role for NALP3/CIAS1/Cryopyrin in innate and adaptive immunity through its regulation of caspase-1. Immunity 24:317–327
18. Watanabe H, Gehrke S, Contassot E, Roques S, Tschopp J, Friedmann PS, French LE, Gaide O (2008) Danger signaling through the inflammasome acts as a master switch between tolerance and sensitization. J Immunol 180:5826–5832
19. Terheyden P, Kortum AK, Schulze HJ, Durani B, Remling R, Mauch C, Junghans V, Schadendorf D, Beiteke U, Junger M et al (2007) Chemoimmunotherapy for cutaneous melanoma with dacarbazine and epifocal contact sensitizers: results of a nationwide survey of the German Dermatologic Co-operative Oncology Group. J Cancer Res Clin Oncol 133: 437–444
20. Yoo J, Omori M, Gyarmati D, Zhou B, Aye T, Brewer A, Comeau MR, Campbell DJ, Ziegler SF (2005) Spontaneous atopic dermatitis in mice expressing an inducible thymic stromal lymphopoietin transgene specifically in the skin. J Exp Med 202:541–549
21. Granstein RD, Sauder DN (1987) Whole-body exposure to ultraviolet radiation results in increased serum interleukin-1 activity in humans. Lymphokine Res 6:187–193
22. Zhou R, Tardivel A, Thorens B, Choi I, Tschopp J (2010) Thioredoxin-interacting protein links oxidative stress to inflammasome activation. Nat Immunol 11:136–140
23. Pickard C, Louafi F, McGuire C, Lowings K, Kumar P, Cooper H, Dearman RJ, Cumberbatch M, Kimber I, Healy E et al (2009) The cutaneous biochemical redox barrier: a component of the innate immune defenses against sensitization by highly reactive environmental xenobiotics. J Immunol 183:7576–7584
24. Naughton GK, Eisinger M, Bystryn JC (1983) Antibodies to normal human melanocytes in vitiligo. J Exp Med 158:246–251
25. Le Poole IC, van den Wijngaard RM, Westerhof W, Das PK (1996) Presence of T cells and macrophages in inflammatory vitiligo skin parallels melanocyte disappearance. Am J Pathol 148:1219–1228
26. Sehgal VN, Srivastava G (2006) Vitiligo: auto-immunity and immune responses. Int J Dermatol 45:583–590

27. Jin Y, Mailloux CM, Gowan K, Riccardi SL, LaBerge G, Bennett DC, Fain PR, Spritz RA (2007) NALP1 in vitiligo-associated multiple autoimmune disease. N Engl J Med 356: 1216–1225
28. Riedl SJ, Li W, Chao Y, Schwarzenbacher R, Shi Y (2005) Structure of the apoptotic protease-activating factor 1 bound to ADP. Nature 434:926–933
29. O'Sullivan BJ, Thomas HE, Pai S, Santamaria P, Iwakura Y, Steptoe RJ, Kay TW, Thomas R (2006) IL-1 beta breaks tolerance through expansion of CD25+ effector T cells. J Immunol 176:7278–7287
30. Gauthier Y, Cario-Andre M, Lepreux S, Pain C, Taieb A (2003) Melanocyte detachment after skin friction in non lesional skin of patients with generalized vitiligo. Br J Dermatol 148: 95–101
31. Ting JP, Kastner DL, Hoffman HM (2006) CATERPILLERs, pyrin and hereditary immunological disorders. Nat Rev Immunol 6:183–195
32. Hoffman HM (2010) The registry of NLRP3 sequence variants http://fmf.igh.cnrs.fr/ISSAID/infevers/
33. Hawkins PN, Lachmann HJ, McDermott MF (2003) Interleukin-1-receptor antagonist in the Muckle-Wells syndrome. N Engl J Med 348:2583–2584
34. Hoffman HM, Rosengren S, Boyle DL, Cho JY, Nayar J, Mueller JL, Anderson JP, Wanderer AA, Firestein GS (2004) Prevention of cold-associated acute inflammation in familial cold autoinflammatory syndrome by interleukin-1 receptor antagonist. Lancet 364:1779–1785
35. Goldbach-Mansky R, Dailey NJ, Canna SW, Gelabert A, Jones J, Rubin BI, Kim HJ, Brewer C, Zalewski C, Wiggs E et al (2006) Neonatal-onset multisystem inflammatory disease responsive to interleukin-1beta inhibition. N Engl J Med 355:581–592
36. Nakamura Y, Kambe N, Saito M, Nishikomori R, Kim YG, Murakami M, Nunez G, Matsue H (2009) Mast cells mediate neutrophil recruitment and vascular leakage through the NLRP3 inflammasome in histamine-independent urticaria. J Exp Med 206:1037–1046
37. McCulloch CA, Downey GP, El-Gabalawy H (2006) Signalling platforms that modulate the inflammatory response: new targets for drug development. Nat Rev Drug Discov 5:864–876
38. Kallinich T, Haffner D, Niehues T, Huss K, Lainka E, Neudorf U, Schaefer C, Stojanov S, Timmann C, Keitzer R et al (2007) Colchicine use in children and adolescents with familial Mediterranean fever: literature review and consensus statement. Pediatrics 119:e474–e483
39. Martinon F, Petrilli V, Mayor A, Tardivel A, Tschopp J (2006) Gout-associated uric acid crystals activate the NALP3 inflammasome. Nature 440:237–241
40. Shoham NG, Centola M, Mansfield E, Hull KM, Wood G, Wise CA, Kastner DL (2003) Pyrin binds the PSTPIP1/CD2BP1 protein, defining familial Mediterranean fever and PAPA syndrome as disorders in the same pathway. Proc Natl Acad Sci USA 100:13501–13506
41. Yu JW, Fernandes-Alnemri T, Datta P, Wu J, Juliana C, Solorzano L, McCormick M, Zhang Z, Alnemri ES (2007) Pyrin activates the ASC pyroptosome in response to engagement by autoinflammatory PSTPIP1 mutants. Mol Cell 28:214–227
42. Katayama T, Tachinami K, Ishiguro M, Kubota Y (1994) The relation between Behcet's disease and interleukin-1 beta production. Nippon Ganka Gakkai Zasshi 98:197–201
43. Yosipovitch G, Shohat B, Bshara J, Wysenbeek A, Weinberger A (1995) Elevated serum interleukin 1 receptors and interleukin 1B in patients with Behcet's disease: correlations with disease activity and severity. Isr J Med Sci 31:345–348
44. Botsios C, Sfriso P, Furlan A, Punzi L, Dinarello CA (2008) Resistant Behcet disease responsive to anakinra. Ann Intern Med 149:284–286
45. Mamula P, Markowitz JE, Baldassano RN (2003) Inflammatory bowel disease in early childhood and adolescence: special considerations. Gastroenterol Clin North Am 32:967–995, viii
46. Eckmann L, Karin M (2005) NOD2 and Crohn's disease: loss or gain of function? Immunity 22:661–667
47. Villani AC, Lemire M, Fortin G, Louis E, Silverberg MS, Collette C, Baba N, Libioulle C, Belaiche J, Bitton A et al (2009) Common variants in the NLRP3 region contribute to Crohn's disease susceptibility. Nat Genet 41:71–76

48. Shi Y, Evans JE, Rock KL (2003) Molecular identification of a danger signal that alerts the immune system to dying cells. Nature 425:516–521
49. So A, De Smedt T, Revaz S, Tschopp J (2007) A pilot study of IL-1 inhibition by anakinra in acute gout. Arthritis Res Ther 9:R28
50. McGonagle D, Tan AL, Shankaranarayana S, Madden J, Emery P, McDermott MF (2007) Management of treatment resistant inflammation of acute on chronic tophaceous gout with anakinra. Ann Rheum Dis 66:1683–1684
51. von Andrian UH, Mackay CR (2000) T-cell function and migration. Two sides of the same coin. N Engl J Med 343:1020–1034
52. Liu L, Fuhlbrigge RC, Karibian K, Tian T, Kupper TS (2006) Dynamic programming of CD8 + T cell trafficking after live viral immunization. Immunity 25:511–520
53. Spreafico R, Ricciardi-Castagnoli P, Mortellaro A (2010) The controversial relationship between NLRP3, alum, danger signals and the next-generation adjuvants. Eur J Immunol 40:638–642
54. Li H, Nookala S, Re F (2007) Aluminum hydroxide adjuvants activate caspase-1 and induce IL-1beta and IL-18 release. J Immunol 178:5271–5276
55. Kool M, Petrilli V, De Smedt T, Rolaz A, Hammad H, van Nimwegen M, Bergen IM, Castillo R, Lambrecht BN, Tschopp J (2008) Cutting edge: alum adjuvant stimulates inflammatory dendritic cells through activation of the NALP3 inflammasome. J Immunol 181:3755–3759
56. Eisenbarth SC, Colegio OR, O'Connor W, Sutterwala FS, Flavell RA (2008) Crucial role for the Nalp3 inflammasome in the immunostimulatory properties of aluminium adjuvants. Nature 453:1122–1126
57. Hornung V, Bauernfeind F, Halle A, Samstad EO, Kono H, Rock KL, Fitzgerald KA, Latz E (2008) Silica crystals and aluminum salts activate the NALP3 inflammasome through phagosomal destabilization. Nat Immunol 9:847–856
58. Franchi L, Nunez G (2008) The Nlrp3 inflammasome is critical for aluminium hydroxide-mediated IL-1beta secretion but dispensable for adjuvant activity. Eur J Immunol 38: 2085–2089
59. McKee AS, Munks MW, MacLeod MK, Fleenor CJ, Van Rooijen N, Kappler JW, Marrack P (2009) Alum induces innate immune responses through macrophage and mast cell sensors, but these sensors are not required for alum to act as an adjuvant for specific immunity. J Immunol 183:4403–4414
60. Liu L, Zhong Q, Tian T, Dubin K, Athale SK, Kupper TS (2010) Epidermal injury and infection during poxvirus immunization is crucial for the generation of highly protective T cell-mediated immunity. Nat Med 16:224–227
61. Delaloye J, Roger T, Steiner-Tardivel QG, Le Roy D, Knaup Reymond M, Akira S, Petrilli V, Gomez CE, Perdiguero B, Tschopp J et al (2009) Innate immune sensing of modified vaccinia virus Ankara (MVA) is mediated by TLR2-TLR6, MDA-5 and the NALP3 inflammasome. PLoS Pathog 5:e1000480
62. Dinarello CA (1998) Interleukin-1 beta, interleukin-18, and the interleukin-1 beta converting enzyme. Ann N Y Acad Sci 856:1–11
63. Kuida K, Lippke JA, Ku G, Harding MW, Livingston DJ, Su MS, Flavell RA (1995) Altered cytokine export and apoptosis in mice deficient in interleukin-1 beta converting enzyme. Science 267:2000–2003
64. Li P, Allen H, Banerjee S, Franklin S, Herzog L, Johnston C, McDowell J, Paskind M, Rodman L, Salfeld J et al (1995) Mice deficient in IL-1 beta-converting enzyme are defective in production of mature IL-1 beta and resistant to endotoxic shock. Cell 80:401–411
65. Ghayur T, Banerjee S, Hugunin M, Butler D, Herzog L, Carter A, Quintal L, Sekut L, Talanian R, Paskind M et al (1997) Caspase-1 processes IFN-gamma-inducing factor and regulates LPS-induced IFN-gamma production. Nature 386:619–623
66. Dunne A, O'Neill LA (2003) The interleukin-1 receptor/Toll-like receptor superfamily: signal transduction during inflammation and host defense. Sci STKE 2003:re3

67. Dinarello CA (2004) Infection, fever, and exogenous and endogenous pyrogens: some concepts have changed. J Endotoxin Res 10:201–222
68. Boraschi D, Dinarello CA (2006) IL-18 in autoimmunity: review. Eur Cytokine Netw 17: 224–252
69. Numerof RP, Asadullah K (2006) Cytokine and anti-cytokine therapies for psoriasis and atopic dermatitis. BioDrugs 20:93–103
70. Schmitz J, Owyang A, Oldham E, Song Y, Murphy E, McClanahan TK, Zurawski G, Moshrefi M, Qin J, Li X et al (2005) IL-33, an interleukin-1-like cytokine that signals via the IL-1 receptor-related protein ST2 and induces T helper type 2-associated cytokines. Immunity 23:479–490
71. Palmer G, Lipsky BP, Smithgall MD, Meininger D, Siu S, Talabot-Ayer D, Gabay C, Smith DE (2008) The IL-1 receptor accessory protein (AcP) is required for IL-33 signaling and soluble AcP enhances the ability of soluble ST2 to inhibit IL-33. Cytokine 42:358–364
72. Boraschi D, Tagliabue A (2006) The interleukin-1 receptor family. Vitam Horm 74:229–254
73. Dinarello CA (2005) An IL-1 family member requires caspase-1 processing and signals through the ST2 receptor. Immunity 23:461–462
74. Cornelis S, Kersse K, Festjens N, Lamkanfi M, Vandenabeele P (2007) Inflammatory caspases: targets for novel therapies. Curr Pharm Des 13:365–383
75. Stack JH, Beaumont K, Larsen PD, Straley KS, Henkel GW, Randle JC, Hoffman HM (2005) IL-converting enzyme/caspase-1 inhibitor VX-765 blocks the hypersensitive response to an inflammatory stimulus in monocytes from familial cold autoinflammatory syndrome patients. J Immunol 175:2630–2634
76. Mayor A, Martinon F, De Smedt T, Petrilli V, Tschopp J (2007) A crucial function of SGT1 and HSP90 in inflammasome activity links mammalian and plant innate immune responses. Nat Immunol 8:497–503
77. Holzer AM, Kaplan LL, Levis WR (2006) Haptens as drugs: contact allergens are powerful topical immunomodulators. J Drugs Dermatol 5:410–416
78. Berman B, Perez OA, Zell D (2006) Immunological strategies to fight skin cancer. Skin Therapy Lett 11:1–7

Cancer, Inflammasomes, and Adjuvanticity

Lionel Apetoh

Abstract Cancer progression is associated with chronic inflammation and the dampening of antitumor immune responses. Although proinflammatory cytokines such as interleukin-1β (IL-1β) have been proposed to be involved in the initial development of cancer, there is emerging evidence that these proinflammatory molecules can also act as potent adjuvants of T-cell-mediated immune responses. The inflammasomes have been characterized as multimeric complexes leading to the release of the proinflammatory cytokines IL-1β, IL-18, and IL-33 after the detection of danger signals that can be released from stressed or dying cells. Strikingly, the activation of the NLRP3 inflammasome was recently shown to be instrumental in the initiation of an immune anticancer response that was required for the success of chemotherapy. In this chapter, I review the current knowledge on the links between the inflammasomes and cancers, with a focus on the role of the NLRP3 inflammasome in anticancer therapies.

1 Introduction

Mutant DNA sequences leading to altered protein signaling pathways within cells can possibly trigger their transformation that eventually results in tumor formation and in the development of cancer [1]. While it has been clearly established that the intrinsic molecular properties of the tumor cells will determine their aggressiveness and the course of the cancer [2, 3], numerous reports have underlined that the tumor microenvironment plays a key role in the progression of this disease [4–6]. Cancers

L. Apetoh (✉)
INSERM, U866, Dijon, France
and
INSERM AVENIR Team, Dijon, France
and
Centre Georges François Leclerc, Dijon, France
e-mail: lionel.apetoh@yahoo.fr

I. Couillin et al. (eds.), *The Inflammasomes*, Progress in Inflammation Research,
DOI 10.1007/978-3-0348-0148-5_10,

are indeed composed of multiple cell types, such as fibroblasts and epithelial cells, cells that form blood and lymphatic vasculature, and innate and adaptive immune cells [7]. The actual impact of an activation of the immune system in the setting of a tumor remains intensely debated. Recent studies in both mouse models and humans have underlined the importance of immune cells during cancer progression. It has, for instance, been proposed that the chronic activation of immune effector cells by tumor cells triggers the release of proinflammatory cytokines, chemokines, Reactive Oxygen Species (ROS), histamine, and other bioactive mediators that generate an inflammatory milieu that favors tumor cell survival and thus further supports tumor progression [8, 9]. On the other hand, the concept of tumor immunosurveillance suggests that immune effector cells are instrumental in preventing tumor outgrowth [10, 11]. Innate effectors such as gamma delta T-cells [12], natural killer cells (NK) [13], dendritic cells (DC) [14], and natural killer T-cells (NKT) [15], as well as members of adaptive immune responses (B- and T-cells) have indeed been described to restrain tumor progression [16]. The contribution of innate and adaptive immune responses to the success of anticancer therapies has also been recently explored in mice and humans suffering from cancer [17, 18]. Beyond their direct cytotoxic effects on tumor cells, radiotherapeutic and chemotherapeutic regimens have been reported to trigger the release of danger signals from tumor cells that in turn supported the induction of a strong antitumor immune response [17, 19]. These studies underlined the paradoxical roles of the immune system during tumor progression and provided impetus to investigate the molecular mechanisms leading to the development of an inflammatory response in cancer.

The inflammasomes were recently described as molecular complexes leading to caspase-1 cleavage, allowing for the release of the proinflammatory cytokines IL-1β, IL-18, and IL-33 (reviewed in [20]). The central importance of these complexes in promoting tissue inflammation has been underlined in several autoimmune diseases, including rheumatoid arthritis and autoinflammatory syndromes (reviewed in [21]). Lately, several findings have revealed that the nacht domain-, leucine-rich repeat-, and pyrin domain (PYD)-containing protein 3 (NLRP3) inflammasome were able to detect danger signals emitted from dying cells such as ATP and uric acid [22, 23]. The links between the inflammasomes and cancer progression and therapy are progressively unraveled. Here, I discuss recent findings regarding the involvement of the inflammasomes in the control of cancer development focusing on the functional consequences of NLRP3 activation in T-cell-mediated anticancer responses.

2 Inflammasomes, Inflammation, and Cancer Formation

2.1 Cancer Cells Enhance Their Aggressiveness Through the Inflammasomes Activation

The association between inflammation and cancer has been exemplified in many clinical studies. It has been established that patients suffering from inflammatory bowel disease have a higher risk than healthy individuals for the development of colorectal cancer [24]. Similarly, it has been proposed that the severity and the duration of inflammatory diseases in the respiratory tract were associated with an increased cancer formation risk [25]. In mice, the in vivo administration of IL-1β preceding the injection of melanoma cancer cells has been shown to dramatically enhance their metastatic potential [26]. At the cellular level, the importance of the inflammasome in tumor progression is progressively being unraveled. It has recently been reported that the NLRP3 inflammasome is constitutively active in human melanoma cancer cells, resulting in the cleavage of caspase-1 and the spontaneous secretion of IL-1β [27]. The release of IL-1β from tumor cells had two major functional consequences that favored cancer progression. First, the autocrine action of IL-1β on tumor cells triggered their secretion of other proinflammatory mediators such as IL-6, IL-8, and MCP-1. In addition, tumor-derived IL-1β enhanced macrophage chemotaxis and promoted in vitro angiogenesis. Importantly, the release of IL-1β from tumor cells could be prevented by pretreating melanoma cells with inhibitors of caspases-1 and -5 or IL-1 receptor blockade [27], suggesting that the inhibition of the inflammasome signaling may hinder the growth of melanoma cancer cells.

In addition to acting on the tumor microenvironment to promote the invasiveness of tumor cells, molecules of the inflammasome have also been proposed to favor tumor progression through cell-autonomous effects. Absent in melanoma 2 (AIM2), a member of the interferon-inducible HIN-200 protein family, has previously been shown to detect cytoplasmic DNA and interact with ASC through its pyrin domain to activate caspase-1 [28–30]. The role of AIM2 within the tumor microenvironment has not yet been established. Nevertheless, one study performed using human colon cancer cell concludes that AIM2 mediates reduction of cell proliferation by cell cycle arrest, thereby conferring an invasive phenotype in colon cancer cells [31]. While the molecular mechanisms involved in this process remain to be defined, this study suggests that individual members of the inflammasomes participate in the intrinsic cell transformation process.

2.2 *Inflammasomes in the Early Development of Cancer*

The role of the inflammasome in the development of malignancies has also been suggested in several mouse models. Dostert et al. have brought up evidence that point to the involvement of the NLRP3 inflammasome in the development of mesotheliomas induced by the inhalation of asbestos dust [32]. In their study, the authors have indeed shown that human macrophages sense asbestos and silica particles through the NLRP3 inflammasome, which then triggers IL-1β secretion. During particle phagocytosis, macrophages were indeed shown to release reactive oxygen species, which in turn activated the NLRP3 inflammasome [32]. These findings were relevant in vivo since in a model of asbestos inhalation, NLRP3-deficient mice showed diminished recruitment of inflammatory cells to the lungs and lower proinflammatory cytokine production [32]. These results thus imply that the inhibition of the inflammasome activity may inhibit the development of malignant mesotheliomas after exposure to asbestos.

The role of IL-1β has also recently been explored in the process of carcinogenesis. Using a mouse model where the administration of 3-methylcholantrene (3-MCA) induces tumor development, Krelin et al. have reported that IL-1β was critically involved in the early steps of carcinogenesis [33]. The authors have shown that the administration of 3-MCA resulted in a much lower incidence of tumor formation in IL-1β-deficient mice compared to WT hosts and further demonstrated that IL-1β-driven inflammation enhances the invasiveness of 3-MCA-induced tumors. In an elegant study, Tu et al. have also reported that the stomach-specific expression of human IL-1β in transgenic mice leads to spontaneous gastric inflammation and cancer [34]. These results suggest that neutralization of IL-1β (or caspase-1 activity) may prove useful in the treatment of human neoplasms. In this regard, it is interesting to note that the clinical efficacy of the anticancer agent thalidomide, which is currently implemented for the treatment of multiple myeloma, has been proposed to be mediated through its ability to inhibit caspase-1 [35]. In a UVB-induced skin inflammation model, Keller et al. indeed show that the thalidomide-mediated inhibition caspase-1 leads to reduced inflammation in mice. The authors propose that thalidomide might similarly inhibit tumor-induced inflammation, possibly impairing tumor progression. While the relevance of these results in cancer therapy has yet not been demonstrated at this stage, these findings raise the hypothesis that the efficacy of some anticancer agents might be mediated by their capacity to reduce chronic inflammation induced by tumor cells.

3 Inflammasomes and the Promotion of T-Cell Immune Responses

3.1 The NLRP3 Inflammasome as a Regulator of the Adjuvant Effect of Alum

Even if an excessive immune activation may lead to the development of chronic inflammation, which can be detrimental to the host, the release of proinflammatory mediators has also been shown to be instrumental in eliciting T-cell immune responses against infections. While the initiation of an immune response to an antigen is primarily dictated by signaling events emanating from the T-cell receptor (signal 1) and costimulatory molecules (signal 2), the release of the proinflammatory cytokines IL-1β and IL-12 by DC have been shown to be essential for full T-cell activation (signal 3) [36]. Therefore, vaccine adjuvants – exogenic preparations that boost the immune response to achieve protective immunity – are routinely used to enhance the immune to a defined antigen. Adjuvants have indeed been reported to enhance the expression of costimulatory molecules and the cytokine secretion of DC, leading to an increased T-cell activation. Most of the current adjuvants currently used were shown to mediate their effect through the activation of Toll-like receptor (TLR) signaling. However, several recent studies have unraveled the ability of inflammasomes to detect specific adjuvants. Glenny et al. first reported the adjuvant effect of Alum in 1926 [37]. Despite this early discovery, the actual mode of action of Alum has long remained unknown. While the ability of adjuvants containing pathogen-associated molecular patterns to trigger the activation DC fully relied on the activation of TLR signaling, the adjuvant effect of Alum was not dependent on TLRs. It was later found that the NLRP3 inflammasome was involved in the immunomodulatory properties of aluminum adjuvants [38, 39]. Alum was indeed shown in vitro to activate caspase-1, leading to IL-1β and IL-18 secretion [40]. Alum-induced caspase-1 activation was then demonstrated to strictly rely on the NLRP3 inflammasome [38, 41]. Consistently, mice deficient in NLRP3, ASC, or caspase-1 failed to mount a significant antibody response to antigen administration with alum adjuvants [38, 39]. Therefore, these findings emphasize that the activation of the NLRP3 inflammasome not only triggers local inflammation by innate immune cells but also favors the generation of adaptive immune responses.

3.2 *Proinflammatory Cytokines and T-Cell Responses*

Effector cytokines released after the activation of the inflammasome have been reported to support T-cell immune responses. IL-18, initially identified as IFN-γ inducing factor, has indeed been shown to support IFN-γ production from CD4+ Th1 cells, T-cells, B-cells, and NK cells [42]. Similarly, IL-1β has also been described to synergize with IL-12 to enhance IFN-γ production from human T-cells [43]. IFN-γ production from activated T-cells has already been shown to play a crucial role in cancer progression. It was indeed shown that mice lacking recombination-activating gene 2 (RAG2) and the signaling pathway of IFN-γR or STAT-1 are more likely to develop spontaneous and chemically induced tumors, suggesting that T-cell-dependent antitumor responses stave off tumor development [16]. These findings from the literature thus suggested that the cytokines released after the activation of the inflammasome could possibly drive anticancer immune responses. We therefore investigated the putative links between the activation of the inflammasomes in innate immune cells, the release of proinflammatory cytokines, the enhancement of IFNγ secretion from T-cells, and anticancer responses.

4 The Efficacy of Anticancer Therapies Relies on the NLRP3 Inflammasome

4.1 *Immunogenic Cell Death and the Release of Danger Signals*

The hypothesis of an immune surveillance of cancer is based on the ability of the immune system to discriminate self from foreign antigens. In this model, tumor-specific antigens are thus identified as nonself and thus trigger an immune response. In 1994, Polly Matzinger balanced this prevailing viewpoint by introducing a novel concept where the activation of the immune response is promoted by the detection of danger signals from stressed or dying cells [44]. Several danger signals such as uric acid and ATP have afterward been shown to trigger dendritic cell activation and thus support adaptive immune responses [45, 46]. We have previously reported that while the activity of anticancer agents is frequently evaluated through their ability to directly kill cancer cells, anticancer therapies were also triggering T-cell-mediated anticancer responses that were indispensable to obtain a potent antitumor effect in vivo [17, 47]. The release of High Mobility Group Box 1 (HMGB1) from dying tumor cells and its interaction with TLR4 on DC was required for the crosspresentation of tumor antigens and the activation of T-cell responses [17]. In line with these results, the injection of dying tumor cells killed with anticancer agents such as anthracyclines or oxaliplatin into the footpad of mice triggered the activation of IFN-γ producing $CD8^+$ draining lymph node cells in a TLR4-dependent manner [17]. However, the concomitant administration of HMGB1 with live tumor cells into mice was not able to trigger an anticancer immune response in vivo, suggesting that additional danger signals released from dying tumor cells were actually responsible for their immunogenicity.

4.2 The NLRP3 Inflammasome in T-Cell-Mediated Anticancer Responses

The NLRP3 inflammasome has been shown to be instrumental in the detection of danger signals ([22, 23]). To evaluate the putative involvement of the NLRP3 inflammasome in anticancer therapy, we implanted EL4 thymoma cells into either immunocompetent mice or mice deficient in key molecules forming the NLRP3 inflammasome and we then monitored the effect of the anticancer agent oxaliplatin on tumor growth. While oxaliplatin featured a potent antitumor activity in immunocompetent hosts, tumor growth was only marginally affected by oxaliplatin administration in NLRP3-deficient mice [48]. We similarly showed that oxaliplatin lost its tumoricidal activity in mice deficient in the NLRP3 molecule adaptor apoptosis-associated speck-like protein containing a caspase recruitment domain (ASC) or caspase-1. In line with those findings, the administration of an anti-IL-1β antibody to immunocompetent tumor-bearing mice along with oxaliplatin greatly diminished its therapeutic action. We further confirmed that the neutralization of IL-1β in vivo was also detrimental for the antitumor effects of anthracyclines against CT26 colon cancer and the spontaneous MCA-induced tumor MCA2 [48]. Overall, our results suggested that each of the individual members that assemble into the NLRP3 inflammasome complex was required for an optimal antitumor response after the administration of either anthracyclines or oxaliplatin. To further explore the role of the NLRP3 inflammasome in anticancer immune responses elicited by chemotherapy, live or dying treated tumor cells treated with oxaliplatin in vitro were injected into the footpad of immunocompetent or NLRP3-, Casp-1-, or IL-1R1-deficient mice and we assessed the IFN-γ production from $CD8^+$ T-cells after ex vivo reactivation of draining lymph node cells with specific tumor antigens. In this setting, while immunocompetent mice mounted a robust anticancer response as measured by the secretion of IFN-γ from $CD8^+$ T-cells, mice deficient for NLRP3 or Casp-1 or unable to respond to IL-1 in vivo failed to do so [48]. These results emphasized the role of the inflammasome and that of IL-1 signaling in the immunogenicity of tumor cell death triggered by anthracyclines and oxaliplatin.

4.3 The Release of ATP as an Essential Step for the Efficacy of Chemotherapy

We investigated whether danger signals released by tumor cells may account for the activation of the inflammasome in vivo. ATP was shown to trigger IL-1β secretion through the activation of the NLRP3 inflammasome [22, 49]. In this regard, it is interesting to note that anthracyclines have previously been reported to lower the cellular ATP concentration of cardiac myocytes and to promote the release of adenine molecules, events that occurred concomitantly with cell death [50]. Recently, the group headed by Guido Kroemer demonstrated that multiple cell death inducers such as mitomycin C, oxaliplatin, cisplatin, thapsigargin, and mitoxanthrone were triggering the extracellular release of ATP from tumor cells [51]. It is worth noting that since all these anticancer agents act through different mechanisms, this work suggests that the release of ATP is a general feature of tumor cell death. We thus hypothesized that extracellular ATP released from dying tumor cells might activate the inflammasome and amplify anticancer responses. To test this assumption, we first investigated the putative role of ATP in the immunogenicity of tumor cell death by administrating tumor cells that had been incubated with the ATP-degrading enzyme apyrase or with inhibitors of ATP synthesis such as antimycin A plus deoxyglucose (A/D) or the oxidative phosphorylation uncoupler 2,4-dinitrophenol (DNP). Strikingly, the depletion of ATP from dying cells dramatically compromised their ability to prime antigen-specific T-cells for IFN-γ production in vivo [48]. These results strongly indicated that ATP release from dying tumor cells was essential for the development of an anticancer response.

Given that ATP has been reported to bind to the purinergic receptor P2X7 [52], we then assessed the ability of P2X7-deficient mice to develop an anticancer immune response in vivo upon administration of oxaliplatin. Consistently, we noticed that the injection of oxaliplatin to EL4-bearing P2X7-deficient mice failed to prevent tumor outgrowth and that the inoculation of dying tumor cells into P2X7-deficient mice was unable to trigger the expansion of specific IFN-γ-producing CD8

T-cells [48]. We had previously reported that the DCs of the host were critical in supporting the generation of a T-cell-mediated anticancer response in vivo following the administration of dying tumor cells [17, 47]. In our recent study, we reported that the expression of *P2rx7* by the DC of the host was required for the promotion of antitumor responses. The adoptive transfer of bone marrow-derived $P2rx7^{+/+}$ (but not $P2rx7^{-/-}$) DC loaded with dying tumor cells could indeed restore the specific $CD8^{+}$ T-cell antitumor response in P2X7-deficient mice. When looking into the consequences of an activation of DC by dying tumor cells through the P2X7 receptor, it turned out that dying tumor cells promoted the release of IL-1β from DC. The secretion of IL-1β was clearly dependent on the activation of inflammasome as $Nlrp3^{-/-}$ or $Asc^{-/-}$ DC failed to secrete IL-1β (but not IL-12p40) after incubation with dying tumor cells.

4.4 IL-1β Secretion from Dendritic Cells Enhances $CD8^{+}$ T-Cell Anticancer Activity

It has been previously proposed that the sequential activation of the inflammasome by TLR-mediated signaling and a damage-associated molecular pattern (DAMP) was required to achieve a potent release of IL-1β [49]. This hypothesis was supported by initial reports showing that the initial activation of the inflammasome in macrophages with LPS solely leads to the formation of pro-IL-1β that is then processed into mature IL-1β upon the detection of a DAMP [49]. In agreement with this model, we found that the TLR4 ligand HMGB1 and ATP, which are both released from dying tumor cells, synergize for inducing IL-1β secretion [48]. Accordingly, the addition of oxidized ATP (which blocks the interaction of ATP with P2X7) or of an anti-HMGB1 antibody in a dendritic cell culture impaired their ability to secrete IL-1β after coculture with dying tumor cells [48].

Our results suggested that the release of ATP by dying tumor cells is sensed through P2X7 receptors on DC, thereby triggering the activation of the NLRP3 inflammasome that leads to the secretion of IL-1β and the promotion of CD8 T-cell anticancer responses. As underlined previously in this chapter, IL-1β has been reported to act as an adjuvant of T-cell-based immune responses. We therefore hypothesized that IL-1β could act as an adjuvant of antitumor responses by enhancing the priming of anticancer $CD8^{+}$ T-cells. To test this, we first loaded EG7 cells, which are EL4 cells expressing ovalbumin, on DC from either immunocompetent or caspase-1-deficient host and tested their respective ability to induce the polarization of naïve TCR-transgenic OT-1 $CD8^{+}$ T-cells (which specifically recognize the OVA peptide presented by DC). Unlike wild-type DCs, which were efficient at promoting IFN-γ secretion from OT-1 cells, caspase-1-deficient DCs were poor stimulators of T-cell IFN-γ production. Similarly, the neutralization of IL-1β activity using either IL-1Ra or an anti-IL-1β antibody decreased the ability of

WT DC to activate OT-1 cells in this setting, suggesting that IL-1β was required to obtain fully differentiated T-cells [48]. To directly assess the impact of IL-1β on T-cells, we isolated mouse $CD3^+$ $CD8^+$ T-cells and activated these with anti-CD3 and anti-CD28 in the presence of IL-1β. As previously proposed for human T-cells [43], we noted that IL-1β enhanced IFN-γ production from mouse T-cells. These results prompted us to evaluate the ability of IL-1β to restore an efficient anticancer response in mice that lacked key elements of the inflammasome. In line with our hypothesis, the failure of *P2rx7*-, *nlrp3*-, and *casp1*-deficient mice to mount an OVA-specific CD8 T-cell response after immunization with oxaliplatin-treated EG7 cells could be overcome by the addition of recombinant IL-1β [48]. Our results thus indicate that the activation of the inflammasome by dying tumor cells is essential for the full differentiation of effector T-cells and the achievement of strong antitumor responses after chemotherapeutic treatment with anthracyclines or oxaliplatin.

4.5 Clinical Relevance of the P2X7-Mediated Activation of the Inflammasome

We eventually investigated the relevance of the ATP/P2X7 pathway in anticancer therapies in humans. The Glu496Ala polymorphism of P2X7 has been associated with a decreased ability of human PBMC to release IL-1β after stimulation with LPS [53]. We thus decided to investigate the impact of the loss-of-function polymorphism on the capacity of patients suffering from cancer to respond to chemotherapy. Our results from a retrospective study on a cohort of 225 women treated with anthracyclines for breast cancer indicate that patients harboring the variant form of P2X7 had an accelerated disease course. This observation emphasizes the potential clinical relevance of the P2X7/inflammasome/IL-1β pathway in humans.

5 Concluding Remarks

Chronic inflammation has been shown to favor cancer development and progression. However, we found that the release of proinflammatory mediators occurring after the administration of anticancer agents was beneficial for anticancer responses. We have identified tumor-derived ATP as a novel danger signal that accounts for the immunogenicity of tumor cell death. We propose a model where ATP would bind to P2X7 receptors on DC, stimulate the aggregation/activation of the NLRP3/ASC/Casp-1 inflammasome, and stimulate the proteolytic maturation of caspase-1, which finally results in IL-1β release (Fig. 1). IL-1β then acts as adjuvant of the anticancer immune response triggered by anthracyclines and oxaliplatin by

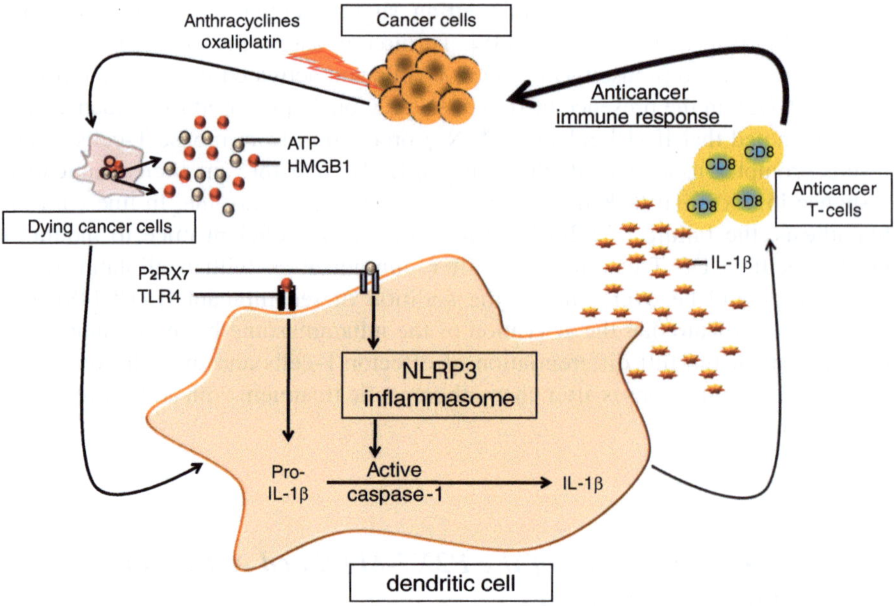

Fig. 1 The NLRP3 inflammasome in the promotion of anticancer immune responses triggered by anthracyclines and oxaliplatin. Upon treatment with cytotoxic agents as anthracyclines or oxaliplatin, dying tumor cells release the danger signals HMGB1 and ATP that respectively bind to the TLR4 and P2X7 receptor on dendritic cells. This sequence of events triggers the assembly of members of the NLRP3 inflammasome and ultimately leads to the release of IL-1β, which then in turn will reinforce the IFN-γ production from T-cells and thus contribute to the success of anticancer therapeutic agents

enhancing the IFN-γ-production from tumor antigen-specific $CD8^+$ T-cells. Importantly, these results were corroborated in humans, suggesting that the in vivo modulation of the activity of the inflammasomes might be beneficial in anticancer therapy. These findings thus prompt for a better molecular characterization of the inflammasomes in order to design specific pharmacological molecules that would be implemented in the clinic as safe and effective adjuvants of anticancer responses.

Acknowledgements I would like to thank Dr François Ghiringhelli for critical reading of this review.

Lionel Apetoh received support from the Agence Nationale de la Recherche (ANR-10-PDOC-014-01).

References

1. Hanahan D, Weinberg RA (2000) The hallmarks of cancer. Cell 100:57–70
2. van't Veer LJ, Dai H, van de Vijver MJ, He YD, Hart AA, Mao M, Peterse HL, van der Kooy K, Marton MJ, Witteveen AT et al (2002) Gene expression profiling predicts clinical outcome of breast cancer. Nature 415:530–536

3. Shedden K, Taylor JM, Enkemann SA, Tsao MS, Yeatman TJ, Gerald WL, Eschrich S, Jurisica I, Giordano TJ, Misek DE et al (2008) Gene expression-based survival prediction in lung adenocarcinoma: a multi-site, blinded validation study. Nat Med 14:822–827
4. Kim S, Takahashi H, Lin WW, Descargues P, Grivennikov S, Kim Y, Luo JL, Karin M (2009) Carcinoma-produced factors activate myeloid cells through TLR2 to stimulate metastasis. Nature 457:102–106
5. Karnoub AE, Dash AB, Vo AP, Sullivan A, Brooks MW, Bell GW, Richardson AL, Polyak K, Tubo R, Weinberg RA (2007) Mesenchymal stem cells within tumour stroma promote breast cancer metastasis. Nature 449:557–563
6. Colmone A, Amorim M, Pontier AL, Wang S, Jablonski E, Sipkins DA (2008) Leukemic cells create bone marrow niches that disrupt the behavior of normal hematopoietic progenitor cells. Science 322:1861–1865
7. Liotta LA, Kohn EC (2001) The microenvironment of the tumour-host interface. Nature 411:375–379
8. Coussens LM, Raymond WW, Bergers G, Laig-Webster M, Behrendtsen O, Werb Z, Caughey GH, Hanahan D (1999) Inflammatory mast cells up-regulate angiogenesis during squamous epithelial carcinogenesis. Genes Dev 13:1382–1397
9. de Visser KE, Eichten A, Coussens LM (2006) Paradoxical roles of the immune system during cancer development. Nat Rev Cancer 6:24–37
10. Dunn GP, Bruce AT, Ikeda H, Old LJ, Schreiber RD (2002) Cancer immunoediting: from immunosurveillance to tumor escape. Nat Immunol 3:991–998
11. Koebel CM, Vermi W, Swann JB, Zerafa N, Rodig SJ, Old LJ, Smyth MJ, Schreiber RD (2007) Adaptive immunity maintains occult cancer in an equilibrium state. Nature 450:903–907
12. Gao Y, Yang W, Pan M, Scully E, Girardi M, Augenlicht LH, Craft J, Yin Z (2003) Gamma delta T cells provide an early source of interferon gamma in tumor immunity. J Exp Med 198:433–442
13. Borg C, Terme M, Taieb J, Menard C, Flament C, Robert C, Maruyama K, Wakasugi H, Angevin E, Thielemans K et al (2004) Novel mode of action of c-kit tyrosine kinase inhibitors leading to NK cell-dependent antitumor effects. J Clin Invest 114:379–388
14. Taieb J, Chaput N, Menard C, Apetoh L, Ullrich E, Bonmort M, Pequignot M, Casares N, Terme M, Flament C et al (2006) A novel dendritic cell subset involved in tumor immunosurveillance. Nat Med 12:214–219
15. Crowe NY, Smyth MJ, Godfrey DI (2002) A critical role for natural killer T cells in immunosurveillance of methylcholanthrene-induced sarcomas. J Exp Med 196:119–127
16. Shankaran V, Ikeda H, Bruce AT, White JM, Swanson PE, Old LJ, Schreiber RD (2001) IFNgamma and lymphocytes prevent primary tumour development and shape tumour immunogenicity. Nature 410:1107–1111
17. Apetoh L, Ghiringhelli F, Tesniere A, Obeid M, Ortiz C, Criollo A, Mignot G, Maiuri MC, Ullrich E, Saulnier P et al (2007) Toll-like receptor 4-dependent contribution of the immune system to anticancer chemotherapy and radiotherapy. Nat Med 13:1050–1059
18. Apetoh L, Mignot G, Panaretakis T, Kroemer G, Zitvogel L (2008) Immunogenicity of anthracyclines: moving towards more personalized medicine. Trends Mol Med 14:141–151
19. Apetoh L, Ghiringhelli F, Tesniere A, Criollo A, Ortiz C, Lidereau R, Mariette C, Chaput N, Mira JP, Delaloge S et al (2007) The interaction between HMGB1 and TLR4 dictates the outcome of anticancer chemotherapy and radiotherapy. Immunol Rev 220:47–59
20. Ogura Y, Sutterwala FS, Flavell RA (2006) The inflammasome: first line of the immune response to cell stress. Cell 126:659–662
21. Church LD, Cook GP, McDermott MF (2008) Primer: inflammasomes and interleukin 1beta in inflammatory disorders. Nat Clin Pract Rheumatol 4:34–42
22. Duncan JA, Bergstralh DT, Wang Y, Willingham SB, Ye Z, Zimmermann AG, Ting JP (2007) Cryopyrin/NALP3 binds ATP/dATP, is an ATPase, and requires ATP binding to mediate inflammatory signaling. Proc Natl Acad Sci USA 104:8041–8046

23. Martinon F, Petrilli V, Mayor A, Tardivel A, Tschopp J (2006) Gout-associated uric acid crystals activate the NALP3 inflammasome. Nature 440:237–241
24. Itzkowitz SH, Yio X (2004) Inflammation and cancer IV. Colorectal cancer in inflammatory bowel disease: the role of inflammation. Am J Physiol Gastrointest Liver Physiol 287:G7–G17
25. Borm PJ, Driscoll K (1996) Particles, inflammation and respiratory tract carcinogenesis. Toxicol Lett 88:109–113
26. Vidal-Vanaclocha F, Alvarez A, Asumendi A, Urcelay B, Tonino P, Dinarello CA (1996) Interleukin 1 (IL-1)-dependent melanoma hepatic metastasis in vivo; increased endothelial adherence by IL-1-induced mannose receptors and growth factor production in vitro. J Natl Cancer Inst 88:198–205
27. Okamoto M, Liu W, Luo Y, Tanaka A, Cai X, Norris DA, Dinarello CA, Fujita M (2010) Constitutively active inflammasome in human melanoma cells mediating autoinflammation via caspase-1 processing and secretion of interleukin-1{beta}. J Biol Chem 285(9):6477–6488
28. Burckstummer T, Baumann C, Bluml S, Dixit E, Durnberger G, Jahn H, Planyavsky M, Bilban M, Colinge J, Bennett KL et al (2009) An orthogonal proteomic-genomic screen identifies AIM2 as a cytoplasmic DNA sensor for the inflammasome. Nat Immunol 10:266–272
29. Fernandes-Alnemri T, Yu JW, Datta P, Wu J, Alnemri ES (2009) AIM2 activates the inflammasome and cell death in response to cytoplasmic DNA. Nature 458:509–513
30. Hornung V, Ablasser A, Charrel-Dennis M, Bauernfeind F, Horvath G, Caffrey DR, Latz E, Fitzgerald KA (2009) AIM2 recognizes cytosolic dsDNA and forms a caspase-1-activating inflammasome with ASC. Nature 458:514–518
31. Patsos G, Germann A, Gebert J, Dihlmann S (2010) Restoration of absent in melanoma 2 (AIM2) induces G2/M cell cycle arrest and promotes invasion of colorectal cancer cells. Int J Cancer 126(8):1838–1849
32. Dostert C, Petrilli V, Van Bruggen R, Steele C, Mossman BT, Tschopp J (2008) Innate immune activation through Nalp3 inflammasome sensing of asbestos and silica. Science 320:674–677
33. Krelin Y, Voronov E, Dotan S, Elkabets M, Reich E, Fogel M, Huszar M, Iwakura Y, Segal S, Dinarello CA et al (2007) Interleukin-1beta-driven inflammation promotes the development and invasiveness of chemical carcinogen-induced tumors. Cancer Res 67:1062–1071
34. Tu S, Bhagat G, Cui G, Takaishi S, Kurt-Jones EA, Rickman B, Betz KS, Penz-Oesterreicher M, Bjorkdahl O, Fox JG et al (2008) Overexpression of interleukin-1beta induces gastric inflammation and cancer and mobilizes myeloid-derived suppressor cells in mice. Cancer Cell 14:408–419
35. Keller M, Sollberger G, Beer HD (2009) Thalidomide inhibits activation of caspase-1. J Immunol 183:5593–5599
36. Curtsinger JM, Schmidt CS, Mondino A, Lins DC, Kedl RM, Jenkins MK, Mescher MF (1999) Inflammatory cytokines provide a third signal for activation of naive CD4+ and CD8+ T cells. J Immunol 162:3256–3262
37. Glenny AT, Pope CG, Waddington H, Wallace U (1926) Immunological notes XVII to XXIV. J Pathol 29:31–40
38. Eisenbarth S, Colegio O, O'Connor W, Sutterwala F, Flavell R (2008) Crucial role for the Nalp3 inflammasome in the immunostimulatory properties of aluminium adjuvants. Nature 453:1122–1126
39. Kool M, Petrilli V, De Smedt T, Rolaz A, Hammad H, van Nimwegen M, Bergen IM, Castillo R, Lambrecht BN, Tschopp J (2008) Cutting edge: alum adjuvant stimulates inflammatory dendritic cells through activation of the NALP3 inflammasome. J Immunol 181:3755–3759
40. Li H, Nookala S, Re F (2007) Aluminum hydroxide adjuvants activate caspase-1 and induce IL-1beta and IL-18 release. J Immunol 178:5271–5276
41. Franchi L, Nunez G (2008) The Nlrp3 inflammasome is critical for aluminium hydroxide-mediated IL-1beta secretion but dispensable for adjuvant activity. Eur J Immunol 38: 2085–2089

42. Dinarello CA (1999) IL-18: A TH1-inducing, proinflammatory cytokine and new member of the IL-1 family. J Allergy Clin Immunol 103:11–24
43. Tominaga K, Yoshimoto T, Torigoe K, Kurimoto M, Matsui K, Hada T, Okamura H, Nakanishi K (2000) IL-12 synergizes with IL-18 or IL-1beta for IFN-gamma production from human T cells. Int Immunol 12:151–160
44. Matzinger P (1994) Tolerance, danger, and the extended family. Annu Rev Immunol 12:991–1045
45. Shi Y, Evans JE, Rock KL (2003) Molecular identification of a danger signal that alerts the immune system to dying cells. Nature 425:516–521
46. Idzko M, Hammad H, van Nimwegen M, Kool M, Willart MA, Muskens F, Hoogsteden HC, Luttmann W, Ferrari D, Di Virgilio F et al (2007) Extracellular ATP triggers and maintains asthmatic airway inflammation by activating dendritic cells. Nat Med 13:913–919
47. Casares N, Pequignot M, Tesniere A, Ghiringhelli F, Roux S, Chaput N, Schmitt E, Hamai A, Hervas-Stubbs S, Obeid M et al (2005) Caspase-dependent immunogenicity of doxorubicin-induced tumor cell death. J Exp Med 202:1691–1701
48. Ghiringhelli F, Apetoh L, Tesniere A, Aymeric L, Ma Y, Ortiz C, Vermaelen K, Panaretakis T, Mignot G, Ullrich E et al (2009) Activation of the NLRP3 inflammasome in dendritic cells induces IL-1beta-dependent adaptive immunity against tumors. Nat Med 15:1170–1178
49. Di Virgilio F (2007) Liaisons dangereuses: P2X(7) and the inflammasome. Trends Pharmacol Sci 28:465–472
50. Singh Y, Ulrich L, Katz D, Bowen P, Krishna G (1989) Structural requirements for anthracycline-induced cardiotoxicity and antitumor effects. Toxicol Appl Pharmacol 100:9–23
51. Martins I, Tesniere A, Kepp O, Michaud M, Schlemmer F, Senovilla L, Seror C, Metivier D, Perfettini JL, Zitvogel L et al (2009) Chemotherapy induces ATP release from tumor cells. Cell Cycle 8:3723–3728
52. Surprenant A, Rassendren F, Kawashima E, North RA, Buell G (1996) The cytolytic P2Z receptor for extracellular ATP identified as a P2X receptor (P2X7). Science 272:735–738
53. Sluyter R, Shemon AN, Wiley JS (2004) Glu496 to Ala polymorphism in the P2X7 receptor impairs ATP-induced IL-1 beta release from human monocytes. J Immunol 172:3399–3405

Inflammasome and Lung Pathologies

Bernhard Ryffel, Dieudonnée Togbe, and Isabelle Couillin

Abstract Nonregulated inflammation in the lung often leads to chronic inflammatory diseases such as asthma, chronic obstructive pulmonary disease (COPD), and pulmonary fibrosis. Interleukin-1α (IL-1α) and interleukin-1β (IL-1β) are potent mediators of innate immunity and IL-1β is processed by large multiprotein complexes called inflammasomes. Using a mice model of lung inflammatory diseases, it was shown that acute cigarette smoke exposure induces inflammasome-dependent IL-1β maturation and ASC-associated protein complex is necessary to lung emphysema. Moreover, bleomycin- or crystalline silica particle-induced pulmonary inflammation and fibrosis are Nlrp3 inflammasome dependent. Nlrp3-containing inflammasome appears to be an essential mediator for allergic lung inflammation. Pulmonary exposure to nanomaterials may lead to pulmonary inflammation and fibrosis. Inhalation of nano-TiO_2 provokes lung inflammation which is strongly dependent on IL-1α.

Nlrp3 inflammasome appears as a corner stone in pulmonary inflammatory diseases, but growing evidences suggest that IL-1α plays an important role through the common IL-1 receptor.

B. Ryffel
Molecular Immunology and Embryology, UMR6218, University of Orleans and CNRS, 3B rue de la Férollerie, 45071 Orleans, France

D. Togbe
Molecular Immunology and Embryology, UMR6218, University of Orleans and CNRS, 3B rue de la Férollerie, 45071 Orleans, France
and
ArtImmune SAS, Orleans, France

I. Couillin (✉)
Molecular Immunology and Embryology, UMR6218, University of Orleans and CNRS, 3B rue de la Férollerie, 45071 Orleans, France
and
KeyObs SAS, Orleans, France
e-mail: couillin@cnrs-orleans.fr

I. Couillin et al. (eds.), *The Inflammasomes*, Progress in Inflammation Research,
DOI 10.1007/978-3-0348-0148-5_11, © Springer Basel AG 2011

1 Introduction

Inflammation is a fundamental immune mechanism of defense to infection, noninfectious tissue, or cell damage that allows pathogens eradication, repair of damaged tissue with recovery to normal conditions. However, nonregulated inflammation often leads to abnormal situation where the inflammatory response worsens infectious, auto-inflammatory, and auto-immune diseases. Lung inflammation is an initial trigger of different chronic pulmonary diseases such as asthma, COPD and lung fibrosis where infiltrated immune cells producing immune mediators such as chemokines and cytokines induce the adaptive immune response, eventually leading to tissue destruction or abnormal repair.

1.1 Pulmonary Inflammation and Inflammasomes

Lung inflammation is characterized by the production of pro-inflammatory cytokines of the interleukin-1 family in particular interleukin-1α (IL-1α), interleukin-1β (IL-1β), and interleukin-18 (IL-18) which are potent mediators of innate immunity. IL-1β is produced by activated macrophages and dendritic cells, inducing the production of chemokines or cytokines such as TNF and IL-6, or proteases such as matrix metalloproteinases (MMPs) with neutrophil recruitment and proliferation of resident cells in particular fibroblasts [1–3], features that are closely associated with chronic inflammation and structural changes observed in the lungs of patients with severe lung disease [3]. IL-1α, IL-1β, and IL-18 are highly inflammatory cytokines and dysregulation of their expression can lead to severe pathology. They are produced as pro-forms, which are further processed upon activation. IL-1α is active both as pro-IL-1α and as cleaved fragment. Processing pathway has still to be characterized in detail even if the involvement of the Ca^{2+}-dependent protease calpain was reported [4]. IL-1α is an autocrine growth factor and drive sterile inflammation, in particular in response to dying cells [5]. In contrast to IL-1α processing and activation, the mechanisms mediating pro-IL-1β and pro-IL-18 maturation are well characterized. Pro-IL-1β and pro-IL-18 are processed by the protease caspase-1, that itself is auto-activated through caspase-1 clustering within large multiprotein complexes called inflammasomes [6]. The best-characterized inflammasome is the Nod-like receptor (NLR) pyrin domain containing three inflammasome or Nalp3 inflammasome which consists of three main components, the NLR-family protein, Nlrp3, the pro-caspase-1 and the apoptosis speck-like protein containing a CARD (ASC) adapter, which bridges interactions between the proteins [7]. Following auto-activation via inflammasome assembly, caspase-1 cleaves IL-1β, to its biological active form and then is secreted. Since mature IL-1β is very potent, its synthesis and secretion are tightly regulated. A first signal triggers synthesis of pro-IL-1β by transcriptional induction, whereas a second signal leads to inflammasome oligomerization, caspase-1 auto-activation, and caspase-1-dependent cleavage and release of the biologically active, mature IL-1β. This second signal is triggered by a group of chemically and biologically unrelated pathogen-associated molecular patterns (PAMPs) or danger-associated molecular patterns (DAMPs) [8]. The Nlrp3 inflammasome can initiate pathological inflammatory reactions in response to endogenous or exogenous danger signals such as uric acid crystals [9], extracellular ATP [10], UVB irradiation [11], and the environmental pollutants, silica and asbestos [12, 13]. IL-1α, IL-1β, or IL-1 receptor antagonist (IL-1Ra), acting as a natural antagonist of IL-1α/β by trapping IL-1R1 molecules, bind to a cellular receptor complex consisting of IL-1R1 and IL-1 receptor accessory protein (IL-1RAcp) [14].

Nlrp3 inflammasome was identified in orphan auto-inflammatory diseases. The different genetic disorders were all due to different mutations in the Nlrp3 gene leading to constitutive activation of the Nlrp3 inflammasome and excessive production of IL-1β [15]. Nlrp3 inflammasome activation appears as the corner stone of many inflammatory diseases [16–18]. There is a spectrum of auto-inflammatory diseases which correspond to mutations in the nlrp3 gene leading to constitutive activation of the Nlrp3 inflammasome and production of IL-1β. Inflammatory diseases are also due to excessive activation of the Nlrp3 inflammasome by exogenous environmental particles such as silica, asbestos [13] or by endogenous molecules called endogenous danger signal or metabolic danger signal such as uric acid in gout arthritis [9], fibrillar peptide amyloid-beta in the pathogenesis of Alzheimer's disease [19] or excessive glucose in type 2 diabetes mellitus (T2DM) [20].

1.2 Sterile Pulmonary Inflammatory Diseases

Around 300 million patients are currently suffering from asthma in the world. Most patients have moderate symptoms that are controlled by anti-inflammatory medication, mainly inhaled corticosteroids and β adrenergic agonists. But long-term use of these drugs is associated with side-effects that highlight the need for additional therapeutics. Moreover, a minority of patients have severe asthma that is refractory to current therapy.

Chronic obstructive pulmonary disease (COPD) refers to chronic bronchitis and emphysema, a pair of two commonly co-existing diseases of the lungs in which the airways become obstructed. This leads to a limitation of the flow of air to and from the alveolar space causing shortness of breath. In contrast to asthma, the limitation of airflow is poorly reversible and usually gets progressively worse over time. COPD is caused by noxious particles or gas, most commonly from tobacco smoking, which triggers an abnormal inflammatory response in the lung. The inflammatory response in the larger airways is known as chronic bronchitis. In the alveoli, the inflammatory response causes destruction of the lung tissues, a process known as emphysema. The natural course of COPD is characterized by occasional sudden worsening of symptoms called acute exacerbations, most of which are caused by infections or air pollution. Important management strategies are smoking cessation, vaccinations, rehabilitation, and drug therapy. Some patients go on to long-term oxygen therapy or lung transplantation. Worldwide, COPD ranked as the sixth leading cause of death in 1990. It is projected to be the fourth leading cause of death worldwide by 2030 due to an increase in smoking rates and demographic changes in many countries. COPD is the fourth leading cause of death in the United States, and the economic burden of COPD in the United States in 2007 was $42.6 billion in health care costs and lost productivity.

Pulmonary fibrosis literally means lung (pulmonary) scarring (fibrosis). The lung scarring occurs in the tissue of the lung called the interstitium, which supports the structures of the lung (air sacs/alveoli). There are an estimated 130–200 related diseases called interstitial lung disease (ILD) that are similar in characteristics and may result in fibrosis. Pulmonary fibrosis causes the lung septae to thicken and become stiff. Scarring inhibits oxygen from entering the blood stream. The term ILD is used to distinguish these diseases from obstructive airway diseases. Idiopathic pulmonary fibrosis (IPF) is the most common and the more severe form of interstitial lung disease. By definition, the term is used only when the cause of the fibrosis is unknown ("idiopathic") [21, 22]. The actual incidence is unknown. It is estimated that 50,000 new cases are diagnosed annually. More than 200,000 people in the United States suffer from this disease. There are at least five million cases worldwide. It affects both men and women, with a slightly higher incidence in men. The average onset age is 40–70 years, but the disease can occur at any age. Although uncommon, IPF does occur in children as young as 3 years of age. Interstitial lung disease has been diagnosed in children less than 1 year of age. IPF has no specific geographical distribution; it is found in equal proportions in urban and rural environments. After 30 years of investigation and research, the exact causes of pulmonary fibrosis continue to elude us. In addition, we still do not have an effective treatment for this devastating disease.

A history of smoking has been associated with an increased risk of idiopathic pulmonary. Indeed, despite IPF was described for a long time without known origin, it appears now that 75% of IPF patients are cigarette smokers or ex-cigarette smokers [23]. The course is generally unpredictable. Although there is no consensus on the optimal management of IPF, no satisfactory treatment exists at present except lung transplantation but only for young people. Inflammation is now recognized as an important component of fibrotic disease and in particularly of exacerbation events of the disease. It is urgent to identify novel therapies for and raising awareness of pulmonary fibrosis, an often fatal lung disease and also to improving the quality of life for those afflicted with this illness through.

For these two devastating diseases, lung emphysema and lung fibrosis, there is no current therapy. After exposure to cigarette smoke or toxics, pulmonary injury results in inflammatory and remodeling responses leading normally to tissue repair. But in case of recurrent injury, if there is an excess of tissue proteases (MMPs), chronic damage leads to absence of repair with alveolar wall destruction or emphysema whereas excess of protease inhibitors (TIMPs) leads to excessive repair with collagen deposition in blue, fibroblast proliferation and fibrosis (Fig. 1).

Pulmonary exposure to nanomaterials, increasingly used for manufacturing industrial items, may constitute a new health hazard. Nanoparticles are single

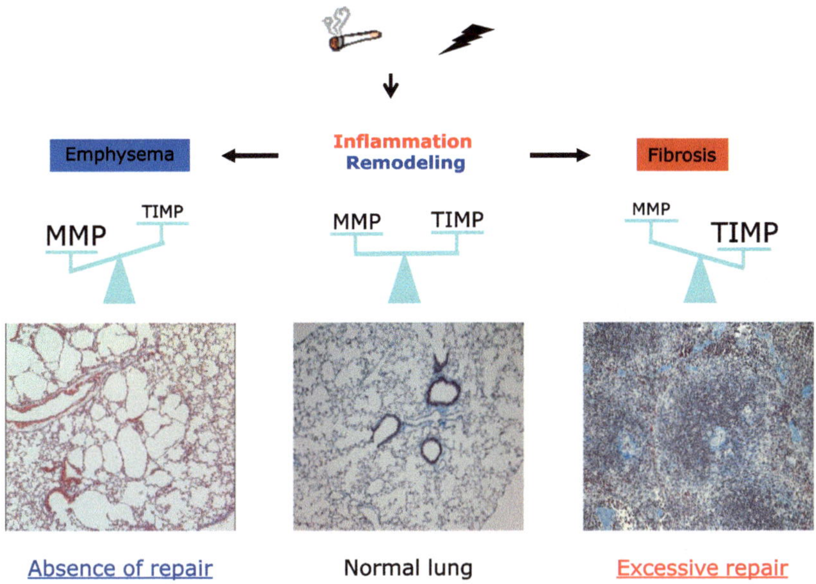

Fig. 1 Proteases/antiproteases imbalance leads to emphysema or fibrosis. Pulmonary emphysema and lung fibrosis are two devastating chronic inflammatory diseases without current therapy. After exposure to cigarette smoke or toxics, pulmonary injury results in inflammatory and remodeling responses resulting normally to tissue repair. But in case of recurrent injury, if there is an excess of tissue proteases (MMPs), chronic damage leads to absence of repair with alveolar wall destruction or emphysema whereas excess of inhibitors of proteases (TIMPs) leads to excessive repair with collagen deposition in *blue*, fibroblast proliferation, and fibrosis

particles with a diameter less than 100 nm [24]. In particular metallic TiO_2 nanoparticles are used in pain, paper and cosmetics. Repeated exposition to TiO_2 nanoparticles may lead to pulmonary inflammation and fibrosis. Indeed an augmented frequency of respiratory diseases in workers exposed to TiO_2 was reported in the late 1980s [25]. Supporting this observation, several recent studies show that pulmonary uptake of TiO_2 nanoparticles worsened respiratory diseases such as pulmonary fibrosis or lipopolysaccharide (LPS)-induced lung inflammation in vivo [26, 27].

The importance of Nlrp3 inflammasome and/or IL-1 in the physiopathology of pulmonary inflammatory diseases, such as asthma, silicosis, chronic obstructive pulmonary disease and lung fibrosis, is presented here.

2 Cigarette Smoke Exposure, Emphysema and Inflammasome

Chronic obstructive pulmonary disease (COPD) is a major cause of morbidity among pulmonary diseases with high mortality [28]. COPD is defined as a disease state characterized by poorly reversible chronic inflammatory response with progressive loss of lung function commonly as a result of cigarette smoking [29]. In the bronchoalveolar lavage (BAL) fluid from COPD patients, an increase of pro-inflammatory cytokines and chemokines including TNF-α and IL-8 has been reported, and these mediators may play an important role in establishing and maintaining the inflammatory condition, characterized by high local neutrophilia [1]. Cigarette smoke is well known to induce chronic inflammation and destruction of alveolar septae, with the loss of surface area for gas exchange and to loss of elasticity known as emphysema [30]. The mechanisms leading to these changes after lung exposure to cigarette smoke are not completely understood. Emphysema may be due to a relative excess of cell-derived proteases, mainly serine proteases such as neutrophil elastase and matrix metalloproteinases (MMPs) that degrade the connective tissue of the lung, and to a relative paucity of anti-proteolytic defenses.

2.1 Cigarette Smoke Exposure

In a mice model of acute cigarette smoke exposure (two cigarettes twice daily for 3 days), we showed that airway inflammation is dependent on Toll-like receptor 4 and IL-1R1 signaling with a significant reduction in neutrophil efflux in the bronchoalveolar space and pulmonary parenchyma of TLR4-, MyD88-, and IL-1R1-deficient mice. Cigarette smoke condensate (CSC) induced a macrophage pro-inflammatory response in vitro, which was dependent on MyD88, IL-1R1, and TLR4 signaling, but not attributable to LPS. Airway inflammation may result from TLR4/MyD88 activation by heat shock protein 70 which was induced in the airways upon smoke exposure. CSC-activated macrophages were able to release mature IL-1 in the presence of ATP, whereas CSC alone promoted the TLR4/MyD88 signaling-dependent production of pro-IL-1 implicating cooperation between TLRs and the inflammasome. This work showed that acute cigarette exposure results in LPS-independent TLR4 activation, leading to pro-IL-1β production and in inflammasome-dependent maturation of IL-1β, crucial for cigarette smoke-induced inflammation leading to chronic obstructive pulmonary disease with emphysema [31].

2.2 Pulmonary Emphysema

To investigate the molecular mechanisms of emphysema, rodent models of elastase-induced emphysema have been established. For example, the intratracheal administration of porcine pancreatic elastase (PPE) induces acute inflammation with destruction of the alveolar wall. Elastase causes a rapid degradation of extracellular matrix and tissue cells with the release of a broad variety of breakdown products and inflammatory mediators. Therefore, the elastase/PPE-induced model of emphysema allows deciphering of the role of several critical factors in the pathogenesis of emphysema downstream of protease activation such as the role of the oxidative stress, cytokines, and chemokines [32]. Using the elastase-induced lung tissue injury model, we reported that inflammation and alveolar destruction occur in the absence of TLR signaling using gene-deficient mice. IL-1β was induced upon elastase administration in the lung, and most importantly IL-1R1- and MyD88-deficient mice have reduced inflammation and emphysema. We found that uric acid was released upon lung injury activating the apoptosis-associated speck-like protein containing a caspase recruitment domain (ASC)-containing Nlrp inflammasome. Inflammation and IL-1β production were reduced in the absence of ASC but not of Nlrp3 suggesting involvement of another inflammasome than the Nlrp3 inflammasome. We showed also that neutralization of IL-1 with the IL-1 receptor antagonist (IL-1Ra) or antibodies reduced lung inflammation and emphysema pointing on the important role for IL-1 [33].

It will be interesting to further characterize the ASC-associated protein complex leading to IL-1β maturation inflammasome involved.

3 Inflammasome and Idiopathic Pulmonary Fibrosis

Idiopathic pulmonary fibrosis (IPF) is a progressive and fatal interstitial pneumonitis of unknown cause, characterized by sequential episodes of acute lung injury with subsequent scarring, leading ultimately to respiratory failure and no effective therapy is available [23]. IPF is characterized by alveolar epithelial injury and hyperplasia, inflammatory cell accumulation, fibroblast proliferation, deposition of extracellular matrix with fibrotic lesions. Bleomycin-induced inflammation and fibrosis represents an experimental model for IPF. Bleomycin (BLM) induces oxidative stress, DNA damage and apoptosis of alveolar macrophages and epithelial cells leading to chemokine and pro-inflammatory cytokine secretion, inflammatory cell recruitment and models in part interstitial fibrosis [34, 35]. Tissue injury and acute inflammatory responses in human and mice models lead to abnormal repair process with dysregulation of the balance between MMPs and their inhibitors, the tissue inhibitors of metalloproteinases (TIMPs), resulting in degradation of extracellular matrix components and accumulation of collagen with progression to pulmonary fibrosis [36]. Chronic inflammation and profibrotic cytokine network activate proliferation of fibroblasts and myofibroblasts producing MMPs and TIMPs resulting in disruption of alveolar structure and loss of lung function [37]. Circulating fibrocytes derived from bone marrow progenitor cells also contribute to the pathogenesis of pulmonary fibrosis [38, 39]. The molecular mechanisms leading to lung injury and inflammation are largely unknown. The degradation of the extracellular matrix components, such as hyaluronan, collagens may be involved in triggering of the inflammatory response as endogenous danger signals.

3.1 Essential Role for IL-1β, Nlrp3 Inflammasome

We showed that the IL-1 receptor 1 (IL-1R1) and Toll/IL-1R (TIR) signal transduction adaptor myeloid differentiation protein 88 (MyD88) are required for BLM-induced acute inflammation and fibrosis. BLM-induced KC, IL-6 secretions and neutrophil recruitment in the lung are IL-1R1 and MyD88 dependent, but TLR2, TLR4 and IL-18R independent. Exogenous rIL-1β recapitulates BLM-induced inflammation features suggesting that blocking IL-1 may be an effective therapy in pulmonary interstitial diseases. In conclusion, we showed that lung IL-1β production and inflammation in response to BLM are dependent upon Nlrp3, ASC, and caspase-1 demonstrating that BLM-induced lung pathology depends on the Nlrp3 inflammasome and IL-1R1/MyD88 signaling, and IL-1 represents a critical effector of pathology and therapeutic target of chronic lung inflammation and fibrosis [40].

3.2 Uric Acid as a Danger Signal

Then investigating the danger signals involved in pulmonary fibrosis, we showed that lung injury depends on the Nlrp3 inflammasome which is triggered by uric acid locally produced in the lung upon BLM-induced DNA damage and degradation. Reduction of uric acid levels using the inhibitor of uric acid synthesis allopurinol or uricase leads to a decrease in BLM-induced IL-1β production, lung inflammation, repair, and fibrosis. Local administration of exogenous uric acid crystals recapitulates lung inflammation and repair which depend on the Nlrp3 inflammasome, MyD88 and IL-1R1 pathways, and Toll-like receptor (TLR) 2 and TLR4 for optimal inflammation, but are independent of the IL-18 receptor. Uric acid released from injured cells constitutes a major endogenous danger signal which activates the Nlrp3 inflammasome leading to IL-1β production. Reducing uric acid tissue levels represents a novel therapeutic approach to control IL-1β production and chronic inflammatory lung pathology [33].

3.3 Extracellular ATP as a Danger Signal

In an attempt to identify other danger signals, we addressed the role of extracellular adenosine tri-phosphate (eATP) in pulmonary inflammation and fibrosis. ATP was quantified in bronchoalveolar lavage fluid (BALF) of controls and patients with idiopathic pulmonary fibrosis. We showed that fibrotic patients have elevated ATP content in BALF in comparison to control individuals. In mice, we reported early increase of eATP levels in BALF upon bleomycin administration. Modulation of eATP levels with the ATP-degrading enzyme apyrase greatly reduced BLM-induced inflammatory cell recruitment, lung IL-1β and TIMP-1 productions, while administration of ATP-γS, a stable ATP derivative, enhanced inflammation. P2X7 receptor-deficient mice presented a dramatically reduced lung inflammation, with reduced fibrosis markers such as lung collagen content and matrix remodeling proteins TIMP-1 and MMP-9. The acute inflammation depends on a functional pannexin-1 hemichannel protein. In vitro, ATP is released by pulmonary epithelial cells upon BLM-induced stress and this is partly dependent on the presence of functional P2X7 receptor and pannexin-1 hemichannel [41]. This work has shown that ATP released from BLM-injured lung cells constitutes a major endogenous danger signal which engages the P2X7 receptor/pannexin-1 axis leading to IL-1β maturation and lung fibrosis.

3.4 A Role for IL-1α and IL-17

Unpublished data indicate that mice deficient for IL-1α or IL-1β presented reduced pulmonary inflammation to BLM suggesting that IL-1α plays also an important role through IL-1R1 and probably in a manner independent of Nlrp3 inflammasome. Further experiments need to be performed to understand the role of IL-1α in pulmonary inflammation and fibrosis. BLM or IL-1β-induced lung injury increased expression of early IL-23p19 and IL-17. Early IL-17A and IL-17F are produced preferentially by $ROR\gamma t^+$ γδT cells and to a lesser extent by $CD4\alpha\beta^+$ T cells, but not by iNKT cells. Moreover, IL-23p19 and IL-17A expressions or IL-17RA signaling are necessary to pulmonary TGF-β1 production, collagen deposition and evolution to fibrosis (unpublished observations).

Our findings demonstrate downstream of Nlrp3 activation, the existence of an early IL-1β–IL-23–IL-17A axis leading to pulmonary inflammation and fibrosis and identify innate IL-23 and IL-17A as interesting drug targets for IL-1β driven lung pathology (Fig. 2).

4 Inflammasome and Silicosis

Prolonged exposure to crystalline silica particles (SiO_2) in occupational and environmental settings induces chronic lung inflammation, which may progress to fibrosis, i.e., silicosis [42]. Despite existing standards in the workplace, silicosis remains a prevalent health problem throughout the world. Insufficient information about the mechanisms that drive the expansion of inflammatory cells and collagen accumulation in silicosis has severely limited the development of effective therapeutic strategies. Exposure to excessive concentrations of SiO_2 was shown to be associated with the development of acute or chronic silicosis. Chronic silicosis is characterized by the formation of nodules with collagen-rich content and inflammatory cell infiltration. The development of silicotic nodules is initiated by

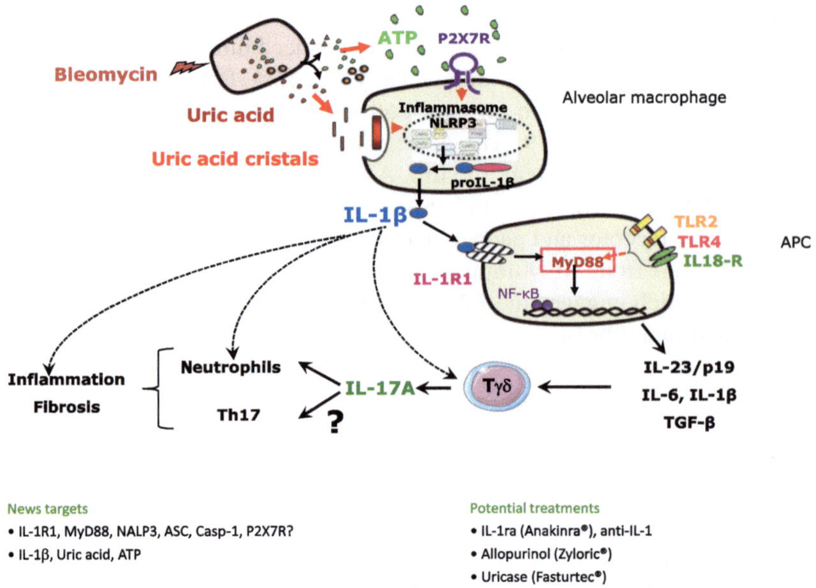

Fig. 2 Mechanisms upstream and downstream of IL-1β. Bleomycin-induced injury of lung cells, probably epithelial cells, leads to lung IL-1β production by alveolar macrophages and probably dendritic cells, inflammation and remodeling which are dependent on the Nlrp3 receptor, the ASC adaptor and the Casp-1 effector molecules. Upstream of IL-1β, pulmonary bleomycin administration induces the release of uric acid in the lung which represents a danger/stress signal likely generated from dying lung cells upon injury. Local concentration increase probably induces uric acid crystallization. Uric acid crystals activate assembly of the Nlrp3 inflammasome resulting in IL-1β maturation and secretion inflammation and remodeling. TLR2 and/or TLR4 may be involved in crystal-induced production of pro-IL-1β or in uric acid crystals uptake by alveolar macrophages and/or resident cells. ATP released from BLM-injured lung cells constitutes also a major endogenous danger signal which engages the P2X7 receptor/pannexin-1 axis leading to IL-1β maturation. Downstream of IL-1β, we showed the existence of an early IL-1β–IL-23–IL-17A axis leading to pulmonary inflammation and fibrosis: antigen-presenting cells produce IL-23 resulting in early IL-17A production by γδ T cells. This early IL-17A probably induces the production of IL-17 by Th17 lymphocytes leading to chronic lung inflammation and fibrosis

aggregates of alveolar macrophages laden with SiO_2 cleared from the alveoli. In acute silicosis, nodules are poorly developed and fibrosis is mainly interstitial with alveolar spaces filled with SiO_2. IL-1, the crucial mediator of inflammatory responses, has been implicated in the pathophysiology of human and experimental silicosis [43, 44]. In the lung of experimental animals, SiO_2 induce the production of IL-1 [45, 46], and IL-1 gene knockout mice develop a reduced response to SiO_2 [47]. In vitro studies showed that Nlrp3 inflammasome sense SiO_2 as a danger signal and subsequently induces the release of IL-1 contributing to explain the mechanism of SiO_2 toxicity. SiO_2 was shown to activate the Nlrp3 inflammasome in macrophages and ASC−/− and Nlp3−/− mice have been reported to develop reduced lung responses with SiO_2 administration [13, 48]. Phagocytosis of SiO_2 by macrophages leads to lysosomal rupture and leakage of lysosomal content into the cytosol, and this lysosomal degradation and loss of cathepsin B are associated with caspase-1 activation and IL-1 secretion. SiO_2-induced ROS production also seems required for the activation of Nalp3 [12, 13]. In vivo, Hornung et al. [13] treated IL-1R1- and MyD88-deficient mice and their corresponding congeners with SiO_2 and observed that the early influx of neutrophils was almost abolished in the deficient strains. Cassel et al. [48] have shown that ASC- and Nalp3-deficient mice developed reduced lung pathology in response to SiO_2. They observed that both knockout strains showed a reduced infiltration of the alveoli by inflammatory cells. Collagen deposition appeared less abundant in the lung of both deficient strains than in wild-type animals. These results demonstrate a role for the Nlrp3 inflammasome and IL-1β maturation in inflammation to silicosis.

5 Role of Nlrp3 Inflammasome in Asthma

Allergic inflammatory response is characterized by activation of Th2 pathway, which is initiated by the allergen uptake and processing by professional antigen-presenting cells (APCs) that present selected peptides on MHC class II molecules to naive T cells, followed by B-cell activation to generate IgE antibodies specific for common environmental allergens. Cytokines attributed to a Th2 immune response are IL-4, IL-5, IL-13 and IL-9 [49–51]. IL-1, which regulates multiple aspects of immune and inflammatory responses [52, 53], was known to promote mast cell activation and production of Th2 cytokines suggesting that IL-1 functions in the development of allergic diseases [54–56]. IL-1R1 plays critical role for the development of Th2 allergic lung inflammation induced with low doses of ovalbumin (OVA) in the absence of alum adjuvant [57]. Elevated concentrations of IL-1β have been found in the BALF from asthmatic patients [58]. Recent evidence on Nlrp3 inflammasome pointed on this crucial role in experimental allergic asthma induced in the presence of alum adjuvant [59]. Likewise, Nlrp3 inflammasome activation by aluminum hydroxide (alum) was shown to be responsible for the adjuvant effect of alum leading to IL-1β maturation and humoral adaptive immune response [59, 60]. Mice deficient for Nlrp3, ASC, and caspase-1 failed to mount a significant antibody response to antigen administration with alum adjuvant [59]. Extracellular ATP is an important mediator in the pathogenesis of allergic asthma. Elevated ATP levels are found in the airways of patient with asthma and COPD and in BALF from OVA-sensitized and challenged mice [61, 62]. In addition, neutralizing extracellular ATP or blocking airway P2 receptors inhibits allergic lung inflammation in mice. Endogenous danger signals such as ATP may activate the Nlrp3 inflammasome in asthma, leading to IL-1 production. Nevertheless, the role of Nlrp3 inflammasome in allergic asthma in the absence of aluminum adjuvant which correspond to the natural occurring disease remains unclear. A recent work in our laboratory demonstrate that Nlrp3 inflammasome is required for allergen-specific Th2 cell activation in the absence of alum adjuvant and confirm that IL-1R1 signaling is critical for the development of mouse model of allergic airways inflammation [71]. This study report that Nlrp3 deficiency causes impaired Th2 response and allergic lung inflammation with lower eosinophil influx in the airways and lower pulmonary Th2 cytokine and chemokine productions. These effects were accompanied by impaired IL-13-producing $CD4^+$ T cell recruitment in the lung of Nlrp3-deficient mice. Expression of the major specific eosinophils (CCL11, CCL5) and Th2 lymphocytes (CCL17) chemoattractants were strongly reduced in Nlrp3−/−, IL-1α−/−, IL-1β−/− and IL-1R1−/− mice, suggesting that Nlrp3 and IL-1α/β signaling pathways may participate to the control of chemokine gradients necessary for recruitment. Endogenous danger signals such as ATP may activate the Nlrp3 inflammasome in asthma, leading to IL-1 production. The role of Nlrp3 is allergic airway inflammation may be related to the capacity of endogenous danger signals released after tissue injury such as ATP or uric acid, to activate the Nlrp3 inflammasome [63]. In particular, extracellular ATP and purinergic receptor have been shown to play a crucial role in inflammasome activation and pathogenesis of allergic lung responses [61, 64]. This study showed significant reduction of IL-1β production in OVA-sensitized and challenged IL-1α-deficient mice suggesting that IL-1β production may depend on IL-1α [71].

In conclusion, the Nlrp3-containing inflammasome complex appears to be an essential mediator for allergic lung inflammation and Nlrp3-IL-1 pathway is a novel endogenous regulator of allergic asthma.

6 Inflammasome and Pulmonary Inflammation to Nanoparticles

Nanoparticles are single particles with a diameter less than 100 nm [24]. They are increasingly used for manufacturing industrial items. The most frequently used nanoparticles are inorganic metal oxides such as silica dioxide (SiO_2) and titanium dioxide (TiO_2), due to their large surface-to-volume ratio [65]. Nano-TiO_2 serves as white pigment in many everyday products such as white paint, paper, and plastics, and in cosmetics. Although μm-sized and larger TiO_2 particles are believed to be biologically inert, an augmented frequency of respiratory diseases in workers exposed to TiO_2 was reported in the late 1980s [25]. The particle size of TiO_2 was a critical factor in triggering this pulmonary inflammatory response [66]. Yazdi et al, showed that nano-TiO_2 and nano-SiO_2, but not nano-ZnO, activate the Nlrp3 inflammasome, leading to IL-1β release and in addition, induce the regulated release of IL-1α [67]. Unlike other particulate Nlrp3 agonists, nano-TiO_2-dependent Nlrp3 activity does not require cytoskeleton-dependent phagocytosis and induces IL-1α/β secretion in nonphagocytic keratinocytes. Inhalation of nano-TiO_2 provokes lung inflammation which is strongly suppressed in IL-1R1- and IL-1α-deficient mice. The inflammation caused by nano-TiO_2 in vivo is largely caused by the biological effect of IL-1α [67]. A hypothetic model in represented in Fig. 3. This model proposes that in mice, pulmonary exposure to nano-TiO_2 leads to IL-1α

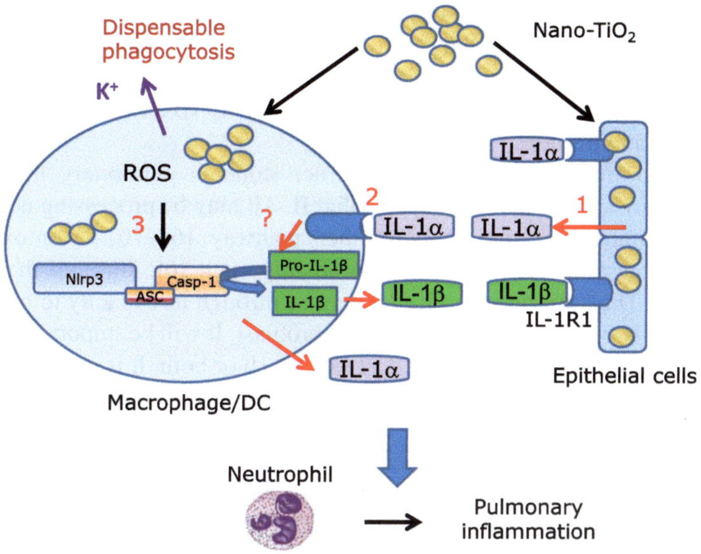

Fig. 3 Model of lung inflammation to NP-TiO_2 exposure via IL-1α and IL-1β production. Pulmonary exposure to nano-TiO_2 leads to IL-1α production by epithelial cells. This may trigger the expression of pro-IL-1 by macrophages and dendritic cells (DC). Then phagocytic-independent uptake of nano-TiO_2 by macrophages/DC cells activates the Nlrp3 inflammasome resulting in IL-1β maturation and secretion. Macrophages/DC cells are also able to produce IL-1α in response to nano-TiO_2. This may explain the predominant role of IL-1α produced by both macrophages/DC cells and resident cells and at the origin of IL-1β expression

production by epithelial cells which may trigger pro-IL-1β expression by macrophages and dendritic cells (DC). Then phagocytic-independent uptake of nano-TiO_2 by macrophages/DC cells activates the Nlrp3 inflammasome leading to IL-1β maturation and secretion. Macrophages/DC cells are also able to produce IL-1α in response to nano-TiO_2.

IL-1α produced both by macrophages/DC cells and epithelial cells and at the origin of IL-1β would be predominant.

The current use of nano-TiO_2 may present a health hazard due to its capacity to induce IL-1R1 signaling, a situation reminiscent of inflammation provoked by asbestos exposure.

7 Conclusion

In conclusion, inflammatory diseases linked with inflammasome activation, in particular Nlrp3 inflammasome are very common including pulmonary inflammatory diseases and represent important public health issues. Nlrp3 inflammasome activation leading to IL-1β maturation and secretion is involved in different inflammatory disease such as allergen-induced injury in asthma or environmental-induced injury in COPD and lung fibrosis. Another environmental particle matter, e.g., nanoparticles induce lung inflammation partially dependent on the Nlrp3 inflammasome but highly dependent on IL-1α and IL-1β and may lead to chronic inflammation and disease.

Nlrp3 inflammasome appears as a corner stone in pulmonary inflammatory diseases, but growing evidences suggest that IL-1β may be processing dependently of other inflammasome or through other pathway [68–70]. Moreover, other evidences suggest that IL-1α plays also an important role through the common IL-1 receptor. The importance of IL-1α may be through its capacity to be produced by resident cells such as epithelial cell and fibroblast. It will be important to identify immune cells able to produce either IL-1α or IL-1β or both. It is of first importance to better understand the molecular and cellular mechanisms of inflammatory processes and to identify new partners of cellular receptors allowing inflammasome triggering or inflammasome-independent inflammatory responses evolving in chronic severe diseases. Developing of novel anti-inflammatory drugs is a high priority and would provide numerous benefits for those suffering from chronic inflammation-mediated diseases, which represent a public health burden quite substantial.

Acknowledgments I thank Dr Valérie Quesniaux, the director of the UMR6218 IEM unit for comments and discussions, and Pr François Erard for critically reviewing the manuscript and for helpful discussion.

I would like to thank Jürg Tschopp, Vincent Lagente, and François Huaux for fruitfull collaborations.

The work was supported by the Key-Obs SAS, Orleans, the "Conseil Général du Loiret" (PhD fellowship) and the Agence Nationale pour la Recherche (ANR).

References

1. Barnes PJ (2004) Mediators of chronic obstructive pulmonary disease. Pharmacol Rev 56:515–548
2. Raines EW, Dower SK, Ross R (1989) Interleukin-1 mitogenic activity for fibroblasts and smooth muscle cells is due to PDGF-AA. Science 243:393–396
3. Sasaki M, Kashima M, Ito T, Watanabe A, Izumiyama N, Sano M, Kagaya M, Shioya T, Miura M (2000) Differential regulation of metalloproteinase production, proliferation and chemotaxis of human lung fibroblasts by PDGF, interleukin-1beta and TNF-alpha. Mediators Inflamm 9:155–160
4. Kavita U, Mizel SB (1995) Differential sensitivity of interleukin-1 alpha and -beta precursor proteins to cleavage by calpain, a calcium-dependent protease. J Biol Chem 270:27758–27765
5. Chen CJ, Kono H, Golenbock D, Reed G, Akira S, Rock KL (2007) Identification of a key pathway required for the sterile inflammatory response triggered by dying cells. Nat Med 13:851–856
6. Martinon F, Burns K, Tschopp J (2002) The inflammasome: a molecular platform triggering activation of inflammatory caspases and processing of proIL-beta. Mol Cell 10:417–426

7. Aganna E, Martinon F, Hawkins PN, Ross JB, Swan DC, Booth DR, Lachmann HJ, Bybee A, Gaudet R, Woo P et al (2002) Association of mutations in the NALP3/CIAS1/PYPAF1 gene with a broad phenotype including recurrent fever, cold sensitivity, sensorineural deafness, and AA amyloidosis. Arthritis Rheum 46:2445–2452
8. Petrilli V, Papin S, Dostert C, Mayor A, Martinon F, Tschopp J (2007) Activation of the NALP3 inflammasome is triggered by low intracellular potassium concentration. Cell Death Differ 14:1583–1589
9. Martinon F, Petrilli V, Mayor A, Tardivel A, Tschopp J (2006) Gout-associated uric acid crystals activate the NALP3 inflammasome. Nature 440:237–241
10. Mariathasan S, Weiss DS, Newton K, McBride J, O'Rourke K, Roose-Girma M, Lee WP, Weinrauch Y, Monack DM, Dixit VM (2006) Cryopyrin activates the inflammasome in response to toxins and ATP. Nature 440:228–232
11. Feldmeyer L, Keller M, Niklaus G, Hohl D, Werner S, Beer HD (2007) The inflammasome mediates UVB-induced activation and secretion of interleukin-1beta by keratinocytes. Curr Biol 17:1140–1145
12. Dostert C, Petrilli V, Van Bruggen R, Steele C, Mossman BT, Tschopp J (2008) Innate immune activation through Nalp3 inflammasome sensing of asbestos and silica. Science 320:674–677
13. Hornung V, Bauernfeind F, Halle A, Samstad EO, Kono H, Rock KL, Fitzgerald KA, Latz E (2008) Silica crystals and aluminum salts activate the NALP3 inflammasome through phagosomal destabilization. Nat Immunol 9(8):847–856
14. Mariathasan S, Newton K, Monack DM, Vucic D, French DM, Lee WP, Roose-Girma M, Erickson S, Dixit VM (2004) Differential activation of the inflammasome by caspase-1 adaptors ASC and Ipaf. Nature 430:213–218
15. Martinon F, Mayor A, Tschopp J (2009) The inflammasomes: guardians of the body. Annu Rev Immunol 27:229–265
16. Hoffman HM (2009) Therapy of autoinflammatory syndromes. J Allergy Clin Immunol 124:1129–1138, quiz 1139–1140
17. Martinon F, Tschopp J (2004) Inflammatory caspases: linking an intracellular innate immune system to autoinflammatory diseases. Cell 117:561–574
18. McGonagle D, Savic S, McDermott MF (2007) The NLR network and the immunological disease continuum of adaptive and innate immune-mediated inflammation against self. Semin Immunopathol 29:303–313
19. Halle A, Hornung V, Petzold GC, Stewart CR, Monks BG, Reinheckel T, Fitzgerald KA, Latz E, Moore KJ, Golenbock DT (2008) The NALP3 inflammasome is involved in the innate immune response to amyloid-beta. Nat Immunol 9:857–865
20. Schroder K, Zhou R, Tschopp J (2010) The NLRP3 inflammasome: a sensor for metabolic danger? Science 327:296–300
21. Frankel SK, Schwarz MI (2009) Update in idiopathic pulmonary fibrosis. Curr Opin Pulm Med 15:463–469
22. Strieter RM, Mehrad B (2009) New mechanisms of pulmonary fibrosis. Chest 136:1364–1370
23. Gross TJ, Hunninghake GW (2001) Idiopathic pulmonary fibrosis. N Engl J Med 345:517–525
24. Auffan M, Rose J, Bottero JY, Lowry GV, Jolivet JP, Wiesner MR (2009) Towards a definition of inorganic nanoparticles from an environmental, health and safety perspective. Nat Nanotechnol 4:634–641
25. Garabrant DH, Fine LJ, Oliver C, Bernstein L, Peters JM (1987) Abnormalities of pulmonary function and pleural disease among titanium metal production workers. Scand J Work Environ Health 13:47–51
26. Bermudez E, Mangum JB, Asgharian B, Wong BA, Reverdy EE, Janszen DB, Hext PM, Warheit DB, Everitt JI (2002) Long-term pulmonary responses of three laboratory rodent species to subchronic inhalation of pigmentary titanium dioxide particles. Toxicol Sci 70:86–97

27. Inoue K, Takano H, Ohnuki M, Yanagisawa R, Sakurai M, Shimada A, Mizushima K, Yoshikawa T (2008) Size effects of nanomaterials on lung inflammation and coagulatory disturbance. Int J Immunopathol Pharmacol 21:197–206
28. Pauwels R (2000) COPD: the scope of the problem in Europe. Chest 117:332S–335S
29. Shapiro SD (2001) End-stage chronic obstructive pulmonary disease: the cigarette is burned out but inflammation rages on. Am J Respir Crit Care Med 164:339–340
30. Barnes PJ, Shapiro SD, Pauwels RA (2003) Chronic obstructive pulmonary disease: molecular and cellular mechanisms. Eur Respir J 22:672–688
31. Doz E, Noulin N, Boichot E, Guenon I, Fick L, Le Bert M, Lagente V, Ryffel B, Schnyder B, Quesniaux VF et al (2008) Cigarette smoke-induced pulmonary inflammation is TLR4/MyD88 and IL-1R1/MyD88 signaling dependent. J Immunol 180:1169–1178
32. Leclerc O, Lagente V, Planquois JM, Berthelier C, Artola M, Eichholtz T, Bertrand CP, Schmidlin F (2006) Involvement of MMP-12 and phosphodiesterase type 4 in cigarette smoke-induced inflammation in mice. Eur Respir J 27:1102–1109
33. Gasse P, Riteau N, Charron S, Girre S, Fick L, Petrilli V, Tschopp J, Lagente V, Quesniaux VF, Ryffel B et al (2009) Uric acid is a danger signal activating NALP3 inflammasome in lung injury inflammation and fibrosis. Am J Respir Crit Care Med 179:903–913
34. Keane MP, Strieter RM, Belperio JA (2005) Mechanisms and mediators of pulmonary fibrosis. Crit Rev Immunol 25:429–463
35. Piguet PF, Vesin C, Grau GE, Thompson RC (1993) Interleukin 1 receptor antagonist (IL-1ra) prevents or cures pulmonary fibrosis elicited in mice by bleomycin or silica. Cytokine 5:57–61
36. Manoury B, Nenan S, Leclerc O, Guenon I, Boichot E, Planquois JM, Bertrand CP, Lagente V (2005) The absence of reactive oxygen species production protects mice against bleomycin-induced pulmonary fibrosis. Respir Res 6:11
37. Lagente V, Manoury B, Nenan S, Le Quement C, Martin-Chouly C, Boichot E (2005) Role of matrix metalloproteinases in the development of airway inflammation and remodeling. Braz J Med Biol Res 38:1521–1530
38. Hashimoto N, Jin H, Liu T, Chensue SW, Phan SH (2004) Bone marrow-derived progenitor cells in pulmonary fibrosis. J Clin Invest 113:243–252
39. Phillips RJ, Burdick MD, Hong K, Lutz MA, Murray LA, Xue YY, Belperio JA, Keane MP, Strieter RM (2004) Circulating fibrocytes traffic to the lungs in response to CXCL12 and mediate fibrosis. J Clin Invest 114:438–446
40. Gasse P, Mary C, Guenon I, Noulin N, Charron S, Schnyder-Candrian S, Schnyder B, Akira S, Quesniaux VF, Lagente V et al (2007) IL-1R1/MyD88 signaling and the inflammasome are essential in pulmonary inflammation and fibrosis in mice. J Clin Invest 117:3786–3799
41. Riteau N, Gasse P, Fauconnier L, Gombault A, Couegnat M, Fick L, Kanellopoulos J, Quesniaux VF, Marchand-Adam S, Crestani B et al (2010) Extracellular ATP is a danger signal activating P2X7 receptor in lung inflammation and fibrosis. Am J Respir Crit Care Med 182(6):774–783
42. Ding M, Chen F, Shi X, Yucesoy B, Mossman B, Vallyathan V (2002) Diseases caused by silica: mechanisms of injury and disease development. Int Immunopharmacol 2:173–182
43. Huaux F (2007) New developments in the understanding of immunology in silicosis. Curr Opin Allergy Clin Immunol 7:168–173
44. Rimal B, Greenberg AK, Rom WN (2005) Basic pathogenetic mechanisms in silicosis: current understanding. Curr Opin Pulm Med 11:169–173
45. Davis GS, Pfeiffer LM, Hemenway DR (1998) Persistent overexpression of interleukin-1beta and tumor necrosis factor-alpha in murine silicosis. J Environ Pathol Toxicol Oncol 17:99–114
46. Driscoll KE, Lindenschmidt RC, Maurer JK, Higgins JM, Ridder G (1990) Pulmonary response to silica or titanium dioxide: inflammatory cells, alveolar macrophage-derived cytokines, and histopathology. Am J Respir Cell Mol Biol 2:381–390
47. Srivastava KD, Rom WN, Jagirdar J, Yie TA, Gordon T, Tchou-Wong KM (2002) Crucial role of interleukin-1beta and nitric oxide synthase in silica-induced inflammation and apoptosis in mice. Am J Respir Crit Care Med 165:527–533

48. Cassel SL, Eisenbarth SC, Iyer SS, Sadler JJ, Colegio OR, Tephly LA, Carter AB, Rothman PB, Flavell RA, Sutterwala FS (2008) The Nalp3 inflammasome is essential for the development of silicosis. Proc Natl Acad Sci USA 105(26):9035–9040
49. Soussi-Gounni A, Kontolemos M, Hamid Q (2001) Role of IL-9 in the pathophysiology of allergic diseases. J Allergy Clin Immunol 107:575–582
50. Grunig G, Warnock M, Wakil AE, Venkayya R, Brombacher F, Rennick DM, Sheppard D, Mohrs M, Donaldson DD, Locksley RM et al (1998) Requirement for IL-13 independently of IL-4 in experimental asthma. Science 282:2261–2263
51. Wills-Karp M, Luyimbazi J, Xu X, Schofield B, Neben TY, Karp CL, Donaldson DD (1998) Interleukin-13: central mediator of allergic asthma. Science 282:2258–2261
52. Dinarello CA (1996) Biologic basis for interleukin-1 in disease. Blood 87:2095–2147
53. Dinarello CA (2009) Immunological and inflammatory functions of the interleukin-1 family. Annu Rev Immunol 27:519–550
54. Hultner L, Kolsch S, Stassen M, Kaspers U, Kremer JP, Mailhammer R, Moeller J, Broszeit H, Schmitt E (2000) In activated mast cells, IL-1 up-regulates the production of several Th2-related cytokines including IL-9. J Immunol 164:5556–5563
55. Nakae S, Komiyama Y, Yokoyama H, Nambu A, Umeda M, Iwase M, Homma I, Sudo K, Horai R, Asano M et al (2003) IL-1 is required for allergen-specific Th2 cell activation and the development of airway hypersensitivity response. Int Immunol 15:483–490
56. Stassen M, Arnold M, Hultner L, Muller C, Neudorfl C, Reineke T, Schmitt E (2000) Murine bone marrow-derived mast cells as potent producers of IL-9: costimulatory function of IL-10 and kit ligand in the presence of IL-1. J Immunol 164:5549–5555
57. Schmitz N, Kurrer M, Kopf M (2003) The IL-1 receptor 1 is critical for Th2 cell type airway immune responses in a mild but not in a more severe asthma model. Eur J Immunol 33:991–1000
58. Broide DH, Lotz M, Cuomo AJ, Coburn DA, Federman EC, Wasserman SI (1992) Cytokines in symptomatic asthma airways. J Allergy Clin Immunol 89:958–967
59. Eisenbarth SC, Colegio OR, O'Connor W, Sutterwala FS, Flavell RA (2008) Crucial role for the Nalp3 inflammasome in the immunostimulatory properties of aluminium adjuvants. Nature 453:1122–1126
60. Li H, Liu P, Cepeda J, Fang D, Easley RB, Simon BA, Zhang LQ, Ye SQ (2008) Augmentation of pulmonary epithelial cell IL-8 expression and permeability by Pre-B-cell colony enhancing factor. J Inflamm (Lond) 5:15
61. Idzko M, Hammad H, van Nimwegen M, Kool M, Willart MA, Muskens F, Hoogsteden HC, Luttmann W, Ferrari D, Di Virgilio F et al (2007) Extracellular ATP triggers and maintains asthmatic airway inflammation by activating dendritic cells. Nat Med 13:913–919
62. Lommatzsch M, Cicko S, Muller T, Lucattelli M, Bratke K, Stoll P, Grimm M, Durk T, Zissel G, Ferrari D et al (2010) Extracellular adenosine triphosphate and chronic obstructive pulmonary disease. Am J Respir Crit Care Med 181(9):928–934
63. Rock KL, Latz E, Ontiveros F, Kono H (2010) The sterile inflammatory response. Annu Rev Immunol 28:321–342
64. Muller T, Robaye B, Vieira RP, Ferrari D, Grimm M, Jakob T, Martin SF, Di Virgilio F, Boeynaems JM, Virchow JC et al (2010) The purinergic receptor P2Y2 receptor mediates chemotaxis of dendritic cells and eosinophils in allergic lung inflammation. Allergy 65: 1545–1553
65. Nel A, Xia T, Madler L, Li N (2006) Toxic potential of materials at the nanolevel. Science 311:622–627
66. Sager TM, Kommineni C, Castranova V (2008) Pulmonary response to intratracheal instillation of ultrafine versus fine titanium dioxide: role of particle surface area. Part Fibre Toxicol 5:17
67. Yazdi AS, Guarda G, Riteau N, Drexler SK, Tardivel A, Couillin I, Tschopp J (2010) Nanoparticles activate the NLR pyrin domain containing 3 (Nlrp3) inflammasome and cause pulmonary inflammation through release of IL-1alpha and IL-1beta. Proc Natl Acad Sci USA 107:19449–19454

68. Schroder K, Tschopp J (2010) The inflammasomes. Cell 140:821–832
69. Maelfait J, Vercammen E, Janssens S, Schotte P, Haegman M, Magez S, Beyaert R (2008) Stimulation of Toll-like receptor 3 and 4 induces interleukin-1beta maturation by caspase-8. J Exp Med 205:1967–1973
70. Schonbeck U, Mach F, Libby P (1998) Generation of biologically active IL-1 beta by matrix metalloproteinases: a novel caspase-1-independent pathway of IL-1 beta processing. J Immunol 161:3340–3346
71. Besnard et al, Allergy. 2011 Mar 28. doi: 10.1111/j.1398-9995.2011.02586.x. [Epub ahead of print]

Mechanisms of NLRP3 Inflammasome Activation in CAPS Patients

Marco Gattorno and Anna Rubartelli

Abstract Mutations in the inflammasome gene NLRP3, with consequent increase of IL-1β secretion, are responsible for a group of autoinflammatory syndromes collectively named Cryopyrin Associated Periodic Syndromes (CAPS). Despite the clear role of NLRP3 in orchestrating key pathways in innate immunity, such as the activation of IL-1β, the molecular mechanism(s) underlying inflammasome assembly remain largely unknown, both in healthy conditions and in CAPS. Redox-related reactions have been proposed to play a crucial role. However, the type of redox response involved as well as the relevant role of prooxidant and antioxidant events are matter of intense debate. Herein, we review the current knowledge of mechanisms leading to the activation of NLRP3 inflammasome in autoinflammatory diseases. In particular, we present a model illustrating how redox signalling controls wild-type and mutated NLRP3 inflammasome activation. Studying monocytes from patients carrying mutations in NLRP3 gene, we have identified a redox signature characterized by increased levels of Reactive Oxygen Species (ROS) and up-regulation of antioxidant systems. This altered redox state is implicated in the accelerated kinetics of IL-1β secretion observed in Toll-like Receptor stimulated CAPS monocytes. We propose that in CAPS patients the collusion between redox derangement and gain-of-function mutations of NLRP3 results in both increased and accelerated IL-1β secretion leading to severe clinical manifestations.

M. Gattorno
2nd Division of Pediatrics, "G. Gaslini" Scientific Institute, Largo G. Gaslini 5, 16147 Genova, Italy

A. Rubartelli (✉)
Division of Cell Biology, National Cancer Research Institute, Largo Rosanna Benzi 10, 16132 Genova, Italy
e-mail: anna.rubartelli@istge.it

I. Couillin et al. (eds.), *The Inflammasomes*, Progress in Inflammation Research,
DOI 10.1007/978-3-0348-0148-5_12,

1 Introduction

The pathway of IL-1β processing and secretion is disseminated with roadblocks that limit efficiency and speed of IL-1β externalization by the producing cells. As often occurs in nature, the importance of these roadblocks is disclosed when they are lost. The dysregulated secretion of IL-1β that occurs upon the loss of one or more roadblocks is indeed the cause of a number of severe chronic human diseases, characterized by massive inflammation. These disorders, collectively called "autoinflammatory diseases", differ in pathogenesis and clinical manifestations but share a dramatic therapeutic response to IL-1 blocking [1, 2]. Indeed, the rapid and sustained response to a reduction in IL-1 activity on an "ex adjuvantibus" basis is the best hallmark of these diseases.

Although the pathogenesis of most autoinflammatory syndromes is still unknown, the identification of the gene responsible for Familial Cold Autoinflammatory Syndrome (FCAS), Muckle–Wells Syndrome (MWS) and Chronic Infantile Neurological Cutaneous and Articular Syndrome (CINCA) [3], collectively known as Cryopyrin Associated Periodic Syndromes (CAPS), has shed light on the pathogenetic role of IL-1β in this class of monogenic inflammatory disorders.

FCAS is characterized by urticarial rash, arthralgia, and fever spikes of short duration induced by cold exposure [4]. MWS is characterized by recurrent episodes of urticaria-like eruptions, fever, malaise, and arthralgia from childhood associated with the late development of sensorineural hearing loss and amyloidosis [5]. CINCA (or Neonatal-Onset Multi-systemic Inflammatory Disease, NOMID) represents the most severe condition and is characterized by a neonatal-onset urticarial-like rash, fever, central nervous system involvement (mental retardation, chronic aseptic meningitis, increased intracranial pressure, papilledema, cerebral atrophy, sensorineural hearing loss), chronic inflammatory arthropathy, skeletal dysplasia and peculiar facial and dysmorphic features [6].

In 2001, the genetic analysis of a subject affected by FCAS revealed the presence of a single nucleotide mutation in the coding region of a gene, which was named Cold-Induced Autoinflammatory Syndrome-1 (*CIAS1*) [3]. Different mutations in CIAS-1 were then identified in other patients. Even with some degrees of overlap, a substantial correlation between the gene mutation and the severity of clinical phenotype is usually observed [7] (Fig. 1, see also http//fmf.igh.cnrs.fr/infevers).

CIAS1 gene codes for an intracellular protein that was termed cryopyrin because the patient developed fever following exposure to cold [3]. *CIAS1* was then found to be a member of the large NOD-like receptor (NLR) family of genes important in the regulation of innate immune functions and its name was changed to nucleotide-binding oligomerization domain (NOD)-like receptor protein 3 (NLRP3).

Some years later, Agostini et al. showed that NLRP3 is a crucial component of the multiprotein complex inflammasome, and that the inflammasome of patients with mutated *NLRP3* displays increased IL-1β processing activity [8]. These

NLRP3 mutations	**L305P** **G301D** **L353P** **A439V** **R488K** **E525K** **E627G** **M659K**	**V198M** **R260W**	**R260L** **E311K** **G569R** **T348M** **A352V** **A439T** **S710C**	**D303N**	**R168Q F309S** **I172T P315L** **R260P G326E** **L264H S331R** **L264R Y570C** **D303G Y570F** **Q306L V351M** **G307V E354D**

Phenotype	**FCAS**	**MWS**	**CINCA**
Disease course	Recurrent episodes triggered by cold	Recurrent episodes	Chronic
Acute phase reactants	Normal between episodes	Possible slight persistent elevation between episodes	Persistent elevation
Fever	Recurrent episodes triggered by cold	Recurrent episodes (usually below 38°C)	Recurrent episodes (usually below 38°C)
Skin	Urticaria	Urticaria	Urticaria
Joints & bones	Arthralgia	Arthralgia/arthritis	Arthralgia/arthritis/ bone dysplasia
Eye involvement	Conjunctivitis	Conjunctivitis/uveitis	Conjunctivitis/uveitis/ papilledema/optic atrophy
Neurological involvement	-	Neurosensorial deafness	Chronic meningitis/ Neurosensorial deafness
Amyloidosis	Extremely rare	Frequent	Rare in childhood, possible in adulthood

http//fmf.igh.cnrs.fr/infevers/

Fig. 1 The wide spectrum of the cryopyrinopathies. Genotype–phenotype correlations associated with mutations of *NLRP3* (or *CIAS-1*) gene. FCAS, Familial Cold Autoinflammatory Syndrome; MWS, Muckle–Wells Syndrome; CINCA, Chronic Infantile Neurological Cutaneous and Articular syndrome.

studies demonstrated unequivocally that missense mutations of the *NLRP3/CIAS1* gene provide a gain of function to the NLRP3 protein resulting in increased secretion of active IL-1β [8, 9]. Still, many unsolved questions remain. Firstly, up to 40% of CAPS are of indefinite origin, having no known mutations in the inflammasome, despite sharing the same clinical manifestations and response to IL-1 blockers of mutated patients [10, 11]. Moreover, the pathological mechanism of increased secretion of active IL-1β is largely unknown. A major incongruity is that in spite the levels of IL-1β secreted by monocytes from CAPS patients are only a few times higher than in healthy controls, symptoms are devastating. In addition, treatment with IL-1Ra has provided outstanding results in different autoinflammatory diseases, such as TNF Receptor-Associated Periodic Syndrome (TRAPS) [12] and multifactorial inflammatory diseases, such as Systemic onset Juvenile Idiopathic Arthritis (SoJIA) [13], even if in many cases monocytes from these patients do not display increased secretion of IL-1β [14]. Together these observations suggest that many known and unknown factors converge to increase the IL-1 activity in autoinflammatory diseases, concurring to the generation of the autoinflammatory phenotype.

2 How Does Mutated NLRP3 Cause Dysregulated IL-1β Secretion?

In general, the release of active IL-1β from blood monocytes is tightly controlled; less than 20% of the total IL-1β precursor produced following Pattern Recognition Receptor (PRR) stimulation is processed and secreted. In contrast, monocytes from mutated CAPS patients release higher amounts of mature IL-1β [8, 15–17]. While some studies report that IL-1β may be secreted by CAPS patients even in the absence of PRR stimulation [8, 15], others do not evidence increased basal secretion of the cytokine, or alterations in basal pro-IL-1β levels [16, 17]. In any case, there is strong consensus that monocytes from these patients display hyperresponsiveness to PAMP stimulation resulting in increase of both IL-1β and IL-18 secretion following PRR triggering [reviewed in 2] (Fig. 2).

The mechanisms of NLRP3 inflammasome activation and assembly is so far undefined even in healthy conditions, so that understanding the link between mutations of the CIAS1/NLRP3 gene and IL-1β hypersecretion in CAPS patients is a very challenging task. Since inflammasomes and their activation mechanisms are largely discussed in other chapters, we will focus on two points, explored in

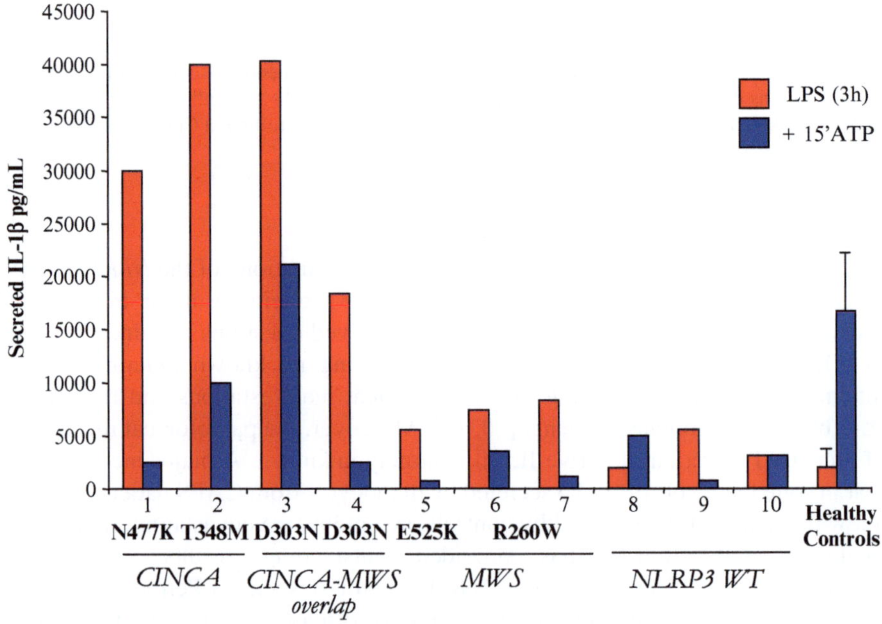

Fig. 2 Monocytes from CAPS patients display a variable degree of hypersecretion of the 17-kDa active form IL-1β after LPS stimulation and a weak response to exogenous ATP in comparison to 20 representative healthy controls. The clinical phenotype and the associated mutations are reported. Patients presenting a CAPS phenotype, but negative for mutations of the NLPR3 gene display an even higher variability. *MWS*, Muckle–Wells syndrome; *CINCA*, Chronic Infantile Neurological Cutaneous and Articular syndrome

many studies that have provided contrasting results. Namely, the requirement of an exogenous second signal and the role of redox remodeling in the assembly of normal and mutated inflammasome and in IL-1β secretion.

3 One-Step and Two-Step Model for IL-1β Secretion in Humans

During the 1990s, before the crucial role of inflammasome in IL-1β processing was disclosed by Jurg Tschopp and his group, a number of studies reported the observation that secretion of bioactive, mature IL-1β needs, or at least benefits of, a second stimulus in addition to the first hit provided by bacterial products. First, Souvannavong and Adam reported that murine macrophages require two signals, poly(I:C) and silica, to release bioactive IL-1β [18]. Some years later, Perregaux and Gabel observed that LPS-stimulated mouse peritoneal macrophages produce large quantities of pro-IL-1β but in the absence of a secondary stimulus only a small amount of the cytokine is proteolytically processed and externalized. In their studies, the authors identified extracellular ATP as a powerful stimulus to induce IL-1β maturation and secretion (Fig. 3) [19, 20]. Indeed, ATP is a major second signal also in humans. In 1999, Andrei et al. proposed a two-step model for IL-1β

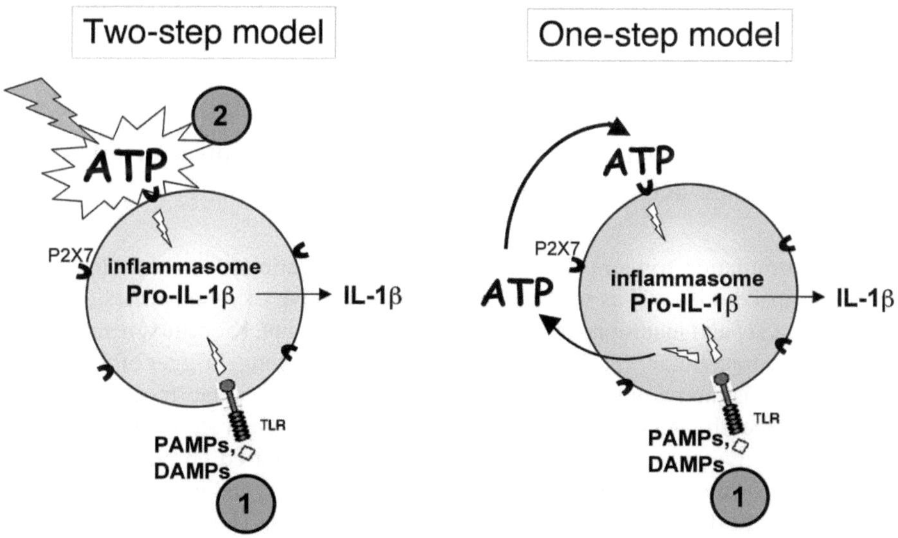

Fig. 3 The one-step and two-step models for IL-1β secretion in humans. In the two-step model (*right panel*): in the first step, induced by LPS, pro-IL-1β is expressed and accumulated in the cytosol. In step 2, driven by exogenous ATP or by hypotonic conditions that deplete cells of K^+, IL-1β is processed and released. In the one-step model (*left panel*) the source of ATP is provided by human monocytes that externalize their own ATP upon PRR triggering

secretion by human monocytes. In the first step, induced by LPS, pro-IL-1β is expressed, accumulated in the cytosol and translocated in part into specialized vesicles. In step 2, driven by exogenous ATP or by hypotonic conditions that deplete cells of K^+, IL-1β is processed and released upon fusion of the vesicle membrane with the plasma membrane [21], (Fig. 3). In the same period, the seminal work by Di Virgilio and coworkers showed that ATP-driven IL-1β processing and secretion is mediated by activation of the purinergic P2X7 receptor by ATP [22]. The relevance of this pathway in the control of IL-1β release was supported in the following decades by many in vitro and in vivo studies. Mice deficient in P2X7 receptors exhibit markedly reduced inflammation, neuropathic pain, and IL-1β-mediated IL-6 production [23]. Moreover, P2X7R stimulation by exogenous ATP has been confirmed as a strictly required second signal for IL-1β processing and secretion in response to extracellular PAMPs in the murine system [24, 25]. In contrast, in primary human monocytes, exogenous ATP as a second signal strongly accelerates IL-1β processing and secretion, but LPS alone is sufficient to induce secretion of IL-1β, although at a lower extent and with slower kinetics [1, 26, 27]. The apparent discrepancy between lack of requirement of extracellular ATP for IL-1β secretion by human monocytes, and strongly enhancing effect of exogenously supplied ATP was solved by the finding that upon PRR triggering, human monocytes externalize their own ATP (Fig. 3). Indeed, we showed that PAMPs and DAMPs induce the extracellular release of endogenous ATP by primary human monocytes [28]. The released ATP autocrinally stimulate P2X7R triggering the cascade of events that lead to inflammasome activation and IL-1β and IL-18 secretion.

4 Modulation of IL-1β Processing and Secretion by Redox Responses. ROS or Antioxidants as Activators?

P2X7R stimulation triggers K^+ exit, Ca^{2+} entry, activation of several kinases and phospholipases [29, 30] as well as formation of reactive oxygen species (ROS) in microglia [31] and macrophages [32]. Among these events, K^+ efflux, resulting in a drop of intracellular $[K^+]$, is recognized as a strong trigger for NLRP3 inflammasome activation, although the underlying molecular mechanisms are unclear. Recently, ATP-dependent ROS production has been shown necessary for caspase-1 activation and secretion of IL-1β and IL-18, based on pharmacological (use of ROS scavengers and inhibitors of ROS generation) and genetic (using mice deficient of NADPH oxidase activity) evidence [33–35]. These results led to the proposal of a model where ROS, produced upon exposure to highly varied stress signals, are common mediators of NALP3 inflammasome activation [36]. However, although ROS are induced by many types of internal and external stress and it is certainly possible that NLRP3 is modified by ROS, either directly or by binding to an ROS-modified or ROS-induced intermediate molecule, evidence supporting this appealing hypothesis is still missing. Rather, Zychlinski and co-workers proposed that oxidation of pro-caspase-1 prevents its activation [37]. They observed that in superoxide dismutase 1 (SOD1)-deficient macrophages the cellular redox potential is decreased due to failure of ROS detoxification. The higher ROS levels cause the reversible oxidation of two redox-sensitive cysteine residues in pro-caspase 1, that hinders the convertion of the enzyme to the active form. As a consequence, mice deficient of the ROS detoxifying enzyme SOD1 produce less caspase-1-dependent IL-1β and IL-18.

The extent to which ROS accumulate is determined by antioxidant systems that enable cells to maintain redox homeostasis (reviewed in [38]). These systems include ROS scavenger and antioxidant enzymes such as superoxide dismutases, catalase, glutathione peroxidase, glutathione reductase and thioredoxin as well as low-molecular antioxidants, also called nonprotein thiols, such as glutathione (GSH) and cysteine. An additional antioxidant system, recently characterized, is the cystine/cysteine redox cycle that not only decreases the redox potential inside cells, but also outside, inducing extracellular accumulation of reduced cysteine. Under normal conditions, these systems balance the constitutive generation of ROS. Following stressful events that increase ROS generation, including any inflammatory hit, cells upregulate the expression of antioxidants.

On these premises, we have studied the redox remodeling that follows exposure of human monocyte to PAMPs and the relevant role not only of ROS, but also of the antioxidant response that conceivably is mounted to counteract the oxidative stress. This experimental system, based on the autocrine release of ATP, allows to avoid the exaggerated, "non-physiologic" oxidative stress that is possibly induced in vitro by the addition of high amounts of ATP or other second signals. Under these conditions, a first oxidative hit is induced by PAMPs and is followed by an antioxidant response. Both events – oxidant and antioxidant – are required for PAMP-triggered IL-1β secretion. However, the role of ROS seems to be the induction of the antioxidant response, ultimately responsible for IL-1β processing and secretion. Indeed, the pharmacologic inhibition of ROS generation and IL-1β secretion is rescued by the artificial generation of reducing conditions obtained by the addition of exogenous reductants such as dithiothreitol or cysteine [39].

5 Redox State and Response in Monocytes from Autoinflammatory Diseases

Inflammation is a stressful event, deeply entangled with redox, and loss of redox homeostasis is associated with many acute or chronic inflammatory states. It is therefore possible that an impaired redox state affects inflammasome activation in autoinflammatory diseases. The redox state and redox response to PAMPs have been investigated in autoinflammatory diseases, both monogenic, such as mutated CAPS, and multifactorial, such as Systemic-onset Juvenile Idiopathic

Arthritis (SoJIA). The latter disease shares a number of clinical features (systemic inflammation, arthritis, rash, persistent elevation of acute phase reactants) with CAPS, but has a still unknown etiology [2]. Interestingly, unstimulated monocytes, freshly drawn from peripheral blood of CAPS patients, display an activated redox state, with higher levels of both ROS and antioxidants. The redox response of CAPS monocytes to PAMPs is also altered, with faster upregulation of the antioxidant systems that rapidly undergo exhaustion [40]. This redox signature was present in all the CAPS cases studied ($n = 8$), and was unrelated to disease severity, level of IL-1β production, or therapeutic regimen. Differently, monocytes from SoJIA patients are only slightly stressed in unstimulated conditions, but display normal kinetics of redox response [40]. Therefore, chronic inflammation per se does not seem responsible for the redox alterations observed in CAPS monocytes.

6 The Altered Redox State and Response of CAPS Monocytes Causes Accelerated IL-1β Secretion

Charles Dinarello [1] has recently pointed out that the increase of IL-1β secretion by CAPS monocytes compared to healthy subjects is modest, raising the question of how a relatively small increase in IL-1β secretion may cause the devastating IL-1-dependent clinical manifestations of CAPS patients. We have recently observed that CAPS monocytes not only secrete more IL-1β than healthy subjects but also display a dramatic acceleration of IL-1β secretion [40]. This abnormal kinetics of IL-1β secretion is due to the entanglement of *NLRP3* mutations with a deranged redox homeostasis in CAPS monocytes. These results are illustrated in the model proposed in Fig. 4. In unstimulated monocytes from healthy subjects (*a*), a condition of redox homeostasis is maintained, inflammasome components are not assembled, and pro-IL-1β is not expressed. Upon LPS stimulation (*b*), pro-IL-1β is induced, and a redox remodeling occurs, with ROS production followed by a delayed antioxidant response, that generates a reducing environment required for inflammasome assembly and consequently pro-IL-1β processing and secretion [39]. In contrast, unstimulated CAPS monocytes display a disturbed redox state with elevated ROS levels and overexpression of antioxidant systems (*c*). Due to this preexisting overexpression, upregulation of antioxidant systems in response to LPS-induced ROS generation occurs (and exhausts) earlier, resulting in faster IL-1β secretion and earlier reaching of the plateau compared to healthy monocytes (*d*) [40].

The inhibitory effect on IL-1β secretion exerted by redox active drugs confirmed that redox remodeling is implicated in the process of IL-1β secretion not only in healthy cells [39], but also in cells bearing a gain of function mutation [40]. These data also indicate that NLRP3 mutations and redox alterations work together in CAPS monocytes to enhance/anticipate IL-1β secretion.

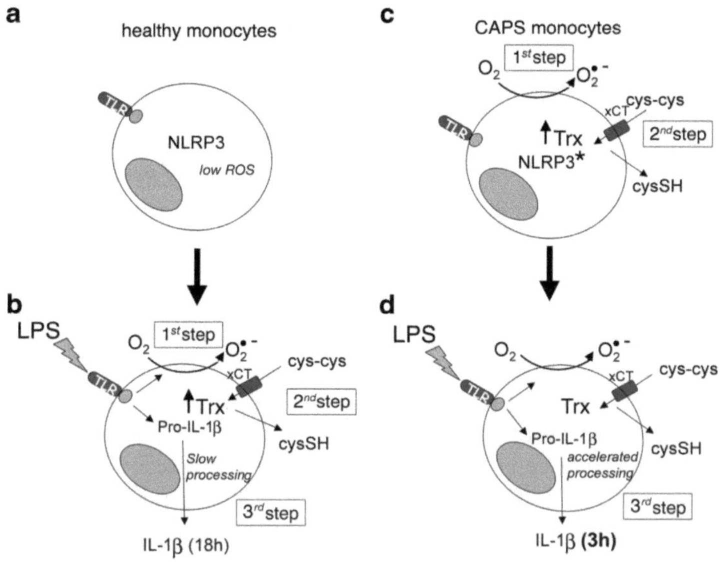

Fig. 4 How different redox states and redox responses to LPS in healthy and CAPS monocytes influence the rate of IL-1β secretion (courtesy by ref. [36])

7 IL-1β Secretion by CAPS Monocytes Is Unaffected by P2x7R Triggering

Unlike healthy monocytes, LPS-stimulated CAPS monocytes do not increase the rate of IL-1β secretion in response to exogenous ATP (Fig. 2) [17]. Interestingly this dysfunction was displayed also by monocytes from patients under anti-IL-1 therapy that secrete levels of LPS-induced IL-1β comparable to those of healthy individuals, indicating that the mutated NLRP3 protein is released from the requirement of ATP for activation, independently of the state of the disease or the therapeutic regimen. The analysis of the sequence coding for the ATP receptor $P2X_7$ revealed that none of the patients expressed nonfunctional P2RX7 variants, which ruled out the possibility that the failure of ATP to induce IL-1β secretion was due to a loss-of-function mutation of the highly polymorphic P2RX7 gene [17].

P2X7R triggering induces a K^+ efflux from the cells, that causes a drop in the intracellular $[K^+]$ responsible for the activation of normal NLRP3 inflammasome [24–25]. Therefore, a possible explanation for the failure of ATP to operate as a secretory trigger in CAPS monocytes is that mutations of NLRP3 result in a gain of function with release of the protein from the control by intracellular $[K^+]$. Still, a trigger by PAMPs or DAMPs is needed to induce IL-1β secretion in CAPS, since both IL-1β and NLRP3 are inducible genes, not expressed in resting conditions [41]. However, while in normal monocytes inflammasome activation requires a PAMP or DAMP trigger strong enough to induce the release of ATP in amounts sufficient to trigger the P2X7R, the threshold is lower in CAPS monocytes that do not require ATP for NLRP3 inflammasome activation. Therefore, a mild infection or a small tissue injury, unable to trigger IL-1β secretion in healthy individuals, is sufficient to drive a dramatic inflammatory cascade when associated with the presence of *NLRP3* mutation.

An alternative explanation is based on the redox alterations of CAPS monocytes [40]. Indeed, the failure to increase IL-1β secretion in response to ATP may be due to the deranged redox remodeling that follows PAMP stimulation that leads to a premature exhaustion of the antioxidant systems. In the widely used experimental system of human monocyte stimulation in vitro, ATP is added to cells after 3 h from exposures to LPS or other PAMPs, since this activation time ensures the best response to ATP in terms of IL-1β secretion (our unpublished results). It is possible that the redox response of CAPS monocytes after 3 h from LPS stimulation is already exhausted and cells are then unable to mount a redox response to P2X7 receptor triggering [40]. These two models are probably not mutually exclusive, and both mechanisms may contribute to the accelerated inflammasome activation.

8 How Does IL-1Ra Work in CAPS Patients?

Treatment with IL-1Ra resulted in a dramatic clinical improvement, which was paralleled by an early and strong downregulation of LPS-induced IL-1β secretion by the patient cells in vitro, but did not restore the ability of the patient monocytes to secrete IL-1β in response to ATP confirming that the resistance to ATP is an intrinsic feature of NLRP3-mutated monocytes.

The dramatic therapeutic effects of anakinra in patients with autoinflammatory syndromes suggest that anakinra plays roles other than a simple competition with free circulating IL-1β for type I IL-1 receptor. Interestingly, Goldbach-Mansky et al. observed a decrease in IL1β gene expression after 3 months of treatment [16]. Gattorno et al. [17] found an early (after 48 h from the first injection of Anakinra) and marked down-modulation of LPS-induced IL-1β secretion, independently of LPS-induced proIL-1β synthesis, that remained substantially unchanged compared to the levels detected in the same monocytes, stimulated with LPS before the beginning of treatment. This observation is consistent with the prompt clinical response observed in all CAPS patients soon after the introduction of anakinra. In contrast, no significant decrease in IL-1β secretion was observed in a patient lacking *NLRP3* mutations. Whether this is due to a low level of LPS-induced secretion by cells from this patient, which could make the decrease of IL-1β secretion less evident, or other mechanisms are implicated remains to be clarified. The rapid clinical remission associated with a prompt decrease in IL-1β secretion by LPS-stimulated monocytes from CAPS patients strongly suggest that monocytes themselves represent a major target of anakinra. This observation is consistent with previous studies highlighting the capacity of IL-1 to induce IL-1 itself [42].

Anakinra can block this positive feedback, as confirmed by the observation that normal monocytes activated in vitro with LPS in the presence of recombinant IL-1Ra decrease IL-1β secretion (our unpublished results). However, it is unlikely that the decreased IL-1β secretion observed in vitro in monocytes from anakinra-treated CAPS patients could be due to the simple in vivo occupancy of IL-1 receptors on monocytes by recombinant IL-1Ra, since the procedures used to purify and wash the monocytes should greatly reduce the amount of bound IL-1Ra. If a consistent proportion of LPS-induced IL-1β secretion occurs as a result of IL-1β itself, one could speculate that in monocytes from CAPS patients treated with anakinra, type I IL-1 receptor is downregulated, resulting in a reduced secretion of IL-1β. It remains to be elucidated whether interference by anakinra in the vicious circle of "IL-1 inducing IL-1" is the major cause of the decreased secretion or whether other mechanisms are involved. Of course, these observations may have implications also for other chronic inflammatory conditions characterized by an increased expression of IL-1 and responsive to IL-1 blockade, such as SoJIA [13, 14] and adult Still's disease [43].

References

1. Dinarello CA (2009) Immunological and inflammatory functions of the interleukin-1 family. Annu Rev Immunol 27:519–550
2. Masters SL, Simon A, Aksentijevich I, Kastner DL (2009) Horror autoinflammaticus: the molecular pathophysiology of autoinflammatory disease (*). Annu Rev Immunol 27:621–668
3. Hoffman HM, Mueller JL, Broide DH, Wanderer AA, Kolodner RD (2001) Mutation of a new gene encoding a putative pyrin-like protein causes familial cold autoinflammatory syndrome and Muckle-Wells syndrome. Nat Genet 29(3):301–305
4. Kile RL, Rusk HA (1940) A case of cold urticaria with an unusual family history. JAMA 114:1067–1068
5. Muckle TJ, Wells M (1962) Urticaria, deafness, and amyloidosis: a new heredo-familial syndrome. Q J Med 31:235–248
6. Prieur AM, Griscelli C (1981) Arthropathy with rash, chronic meningitis, eye lesions, and mental retardation. J Pediatr 99(1):79–83
7. Aksentijevich I, Nowak M, Mallah M, Chae JJ, Watford WT, Hofmann SR et al (2002) De novo CIAS1 mutations, cytokine activation, and evidence for genetic heterogeneity in patients with neonatal-onset multisystem inflammatory disease (NOMID): a new member of the expanding family of pyrin-associated autoinflammatory diseases. Arthritis Rheum 46(12): 3340–3348
8. Agostini L, Martinon F, Burns K, McDermott MF, Hawkins PN, Tschopp J (2004) NALP3 forms an IL-1beta-processing inflammasome with increased activity in Muckle-Wells autoinflammatory disorder. Immunity 20(3):319–325
9. Martinon F, Tschopp J (2004) Inflammatory caspases: linking an intracellular innate immune system to autoinflammatory diseases. Cell 117(5):561–574
10. Neven B, Callebaut I, Prieur AM, Feldmann J, Bodemer C, Lepore L et al (2004) Molecular basis of the spectral expression of CIAS1 mutations associated with phagocytic cell-mediated autoinflammatory disorders CINCA/NOMID, MWS, and FCU. Blood 103(7):2809–2815

11. Caroli F, Pontillo A, D'Osualdo A, Travan L, Ceccherini I, Crovella S et al (2007) Clinical and genetic characterization of Italian patients affected by CINCA syndrome. Rheumatology (Oxford) 46(3):473–478
12. Gattorno M, Pelagatti MA, Meini A, Obici L, Barcellona R, Federici S et al (2008) Persistent efficacy of anakinra in patients with tumor necrosis factor receptor-associated periodic syndrome. Arthritis Rheum 58(5):1516–1520
13. Pascual V, Allantaz F, Arce E, Punaro M, Banchereau J (2005) Role of interleukin-1 (IL-1) in the pathogenesis of systemic onset juvenile idiopathic arthritis and clinical response to IL-1 blockade. J Exp Med 201(9):1479–1486
14. Gattorno M, Piccini A, Lasiglie D, Tassi S, Brisca G, Carta S et al (2008) The pattern of response to anti-interleukin-1 treatment distinguishes two subsets of patients with systemic-onset juvenile idiopathic arthritis. Arthritis Rheum 58(5):1505–1515
15. Stack JH, Beaumont K, Larsen PD, Straley KS, Henkel GW, Randle JC et al (2005) IL-converting enzyme/caspase-1 inhibitor VX-765 blocks the hypersensitive response to an inflammatory stimulus in monocytes from familial cold autoinflammatory syndrome patients. J Immunol 175(4):2630–2634
16. Goldbach-Mansky R, Dailey NJ, Canna SW, Gelabert A, Jones J, Rubin BI et al (2006) Neonatal-onset multisystem inflammatory disease responsive to interleukin-1beta inhibition. N Engl J Med 355(6):581–592
17. Gattorno M, Tassi S, Carta S, Delfino L, Ferlito F, Pelagatti MA et al (2007) Pattern of interleukin-1beta secretion in response to lipopolysaccharide and ATP before and after interleukin-1 blockade in patients with CIAS1 mutations. Arthritis Rheum 56(9):3138–3148
18. Souvannavong V, Adam A (1990) Macrophages from C3H/HeJ mice require an additional step to produce monokines: synergistic effects of silica and poly(I:C) in the release of interleukin 1. J Leukoc Biol 48(2):183–192
19. Perregaux D, Barberia J, Lanzetti AJ, Geoghegan KF, Carty TJ, Gabel CA (1992) IL-1 beta maturation: evidence that mature cytokine formation can be induced specifically by nigericin. J Immunol 149(4):1294–1303
20. Perregaux D, Gabel CA (1994) Interleukin-1 beta maturation and release in response to ATP and nigericin. Evidence that potassium depletion mediated by these agents is a necessary and common feature of their activity. J Biol Chem 269(21):15195–15203
21. Andrei C, Dazzi C, Lotti L, Torrisi MR, Chimini G, Rubartelli A (1999) The secretory route of the leaderless protein interleukin 1beta involves exocytosis of endolysosome-related vesicles. Mol Biol Cell 10(5):1463–1475
22. Ferrari D, Chiozzi P, Falzoni S, Dal SM, Melchiorri L, Baricordi OR et al (1997) Extracellular ATP triggers IL-1 beta release by activating the purinergic P2Z receptor of human macrophages. J Immunol 159(3):1451–1458
23. Chessell IP, Hatcher JP, Bountra C, Michel AD, Hughes JP, Green P et al (2005) Disruption of the P2X7 purinoceptor gene abolishes chronic inflammatory and neuropathic pain. Pain 114(3):386–396
24. Mariathasan S, Weiss DS, Newton K, McBride J, O'Rourke K, Roose-Girma M et al (2006) Cryopyrin activates the inflammasome in response to toxins and ATP. Nature 440(7081):228–232
25. Franchi L, Kanneganti TD, Dubyak GR, Nunez G (2007) Differential requirement of P2X7 receptor and intracellular K+ for caspase-1 activation induced by intracellular and extracellular bacteria. J Biol Chem 282(26):18810–18818
26. Rubartelli A, Cozzolino F, Talio M, Sitia R (1990) A novel secretory pathway for interleukin-1 beta, a protein lacking a signal sequence. EMBO J 9(5):1503–1510
27. Carta S, Tassi S, Semino C, Fossati G, Mascagni P, Dinarello CA et al (2006) Histone deacetylase inhibitors prevent exocytosis of interleukin-1beta-containing secretory lysosomes: role of microtubules. Blood 108(5):1618–1626
28. Piccini A, Carta S, Tassi S, Lasigliè D, Fossati G, Rubartelli A (2008) ATP is released by monocytes stimulated with pathogen-sensing receptor ligands and induces IL-1beta and IL-18 secretion in an autocrine way. Proc Natl Acad Sci USA 105(23):8067–72.

29. Ferrari D, Pizzirani C, Adinolfi E, Lemoli RM, Curti A, Idzko M et al (2006) The P2X7 receptor: a key player in IL-1 processing and release. J Immunol 176(7):3877–3883
30. Andrei C, Margiocco P, Poggi A, Lotti LV, Torrisi MR, Rubartelli A (2004) Phospholipases C and A2 control lysosome-mediated IL-1 beta secretion: Implications for inflammatory processes. Proc Natl Acad Sci USA 101(26):9745–50
31. Parvathenani LK, Tertyshnikova S, Greco CR, Roberts SB, Robertson B, Posmantur R (2003) P2X7 mediates superoxide production in primary microglia and is up-regulated in a transgenic mouse model of Alzheimer's disease. J Biol Chem 278(15):13309–13317
32. Pfeiffer ZA, Guerra AN, Hill LM, Gavala ML, Prabhu U, Aga M et al (2007) Nucleotide receptor signaling in murine macrophages is linked to reactive oxygen species generation. Free Radic Biol Med 42(10):1506–1516
33. Hewinson J, Moore SF, Glover C, Watts AG, MacKenzie AB (2008) A key role for redox signaling in rapid P2X7 receptor-induced IL-1 beta processing in human monocytes. J Immunol 180(12):8410–8420
34. Cruz CM, Rinna A, Forman HJ, Ventura AL, Persechini PM, Ojcius DM (2007) ATP activates a reactive oxygen species-dependent oxidative stress response and secretion of proinflammatory cytokines in macrophages. J Biol Chem 282(5):2871–2879
35. Dostert C, Petrilli V, Van BR, Steele C, Mossman BT, Tschopp J (2008) Innate immune activation through Nalp3 inflammasome sensing of asbestos and silica. Science 320(5876): 674–677
36. Tschopp J, Schroder K (2010) NLRP3 inflammasome activation: the convergence of multiple signalling pathways on ROS production? Nat Rev Immunol 10(3):210–215
37. Meissner F, Molawi K, Zychlinsky A (2008) Superoxide dismutase 1 regulates caspase-1 and endotoxic shock. Nat Immunol 9(8):866–872
38. Carta S, Castellani P, Delfino L, Tassi S, Vene R, Rubartelli A (2009) DAMPs and inflammatory processes: the role of redox in the different outcomes. J Leukoc Biol 86(3):549–555
39. Tassi S, Carta S, Vene R, Delfino L, Ciriolo MR, Rubartelli A (2009) Pathogen-induced interleukin-1beta processing and secretion is regulated by a biphasic redox response. J Immunol 183(2):1456–1462
40. Tassi M, Carta S, Delfino L, Caorsi R, Martini A, Gattorno M et al. (2010) Altered redox state of monocytes from cryopyrin-associated periodic syndromes causes accelerated IL-1beta secretion. Proc Natl Acad Sci USA 107(21):9789–94
41. Bauernfeind FG, Horvath G, Stutz A, Alnemri ES, MacDonald K, Speert D et al (2009) Cutting edge: NF-kappaB activating pattern recognition and cytokine receptors license NLRP3 inflammasome activation by regulating NLRP3 expression. J Immunol 183(2):787–791
42. Dinarello CA, Ikejima T, Warner SJ, Orencole SF, Lonnemann G, Cannon JG et al (1987) Interleukin 1 induces interleukin 1. I. Induction of circulating interleukin 1 in rabbits in vivo and in human mononuclear cells in vitro. J Immunol 139(6):1902–1910
43. Fitzgerald AA, Leclercq SA, Yan A, Homik JE, Dinarello CA (2005) Rapid responses to anakinra in patients with refractory adult-onset Still's disease. Arthritis Rheum 52(6): 1794–1803

Activation and Regulation of the NLRP3 Inflammasome

Juan-Carlos Hernandez, Cherilyn M. Sirois, and Eicke Latz

Abstract Inflammation is an immune response to infectious agents and to signals that arise from host molecules in stress situations or after tissue damage. Many innate immune receptors take part in the inflammatory response and induce transcriptional responses leading to the production of a host of cytokines, chemokines, and other inflammatory mediators. The IL-1β cytokines are exceptional in that they not only require transcriptional induction but also proteolytic processing into biologically active cytokines. This proteolytic activation step is mediated by caspase-1, which itself is controlled by cytosolic multimolecular complexes that are termed inflammasomes. The NLRP3 inflammasome responds to aggregated or crystalline material, microbes or pore-forming toxins and the activation mechanisms are not fully understood. The importance of this innate signaling complex is highlighted by the existence of several mechanisms that regulate NLRP3 activation at different levels.

J.-C. Hernandez
Grupo Inmunovirología, Universidad de Antioquia, Medellín, Colombia

C.M. Sirois
Division of Infectious Diseases and Immunology, University of Massachusetts Medical School, Worcester, MA 01605, USA

E. Latz (✉)
Division of Infectious Diseases and Immunology, University of Massachusetts Medical School, Worcester, MA 01605, USA
and
Institute of Innate Immunity, University Hospitals, University of Bonn, 53127 Bonn, Germany
e-mail: eicke.latz@uni-bonn.de; eicke.latz@umassmed.edu

I. Couillin et al. (eds.), *The Inflammasomes*, Progress in Inflammation Research, DOI 10.1007/978-3-0348-0148-5_13, © Springer Basel AG 2011

1 Introduction

Inflammation is a physiological response triggered by molecules derived from invading microbes or by host molecules that are released during tissue damage. “Self” molecules are recognized either after chemical modification or if they appear in atypical locations that are normally devoid of these molecules [1]. In addition, exogenous crystalline substances or phase transitions of substances from soluble to solid states can induce an immune reaction [2]. An inflammatory response is mediated through the activation of immune signaling receptors expressed on immune and certain nonimmune cells. Therefore, both classical immune cells and nonimmune cells contribute to shape an immune response. Examples of innate immune signaling receptors are members of the Toll-like receptor (TLR), C-type lectin (CLR), or RIG-I-like helicase (RLH) families, as well as the members of the nucleotide-binding domain leucine-rich repeat-containing (NLR) and the pyrin domain and HIN200 domain-containing (PYHIN) protein families [2, 3].

The signaling receptors of the latter two families form large multimolecular complexes, so-called “inflammasomes,” which function to activate caspase-1 and IL-1β cytokines [4]. The term “inflammasome” was chosen to highlight the structural and functional similarities of these signaling complexes with another caspase-activating complex, the apoptosome, which serves as a molecular platform for triggering apoptosis [5]. While most inflammasomes are thought to form in response to a single molecular trigger, the NLRP3 inflammasome can be activated by an array of chemically and structurally distinct substances [2]. Here, we focus on the mechanisms of NLRP3 activation and regulation.

2 Interleukin-1β Cytokines Require an Activation Step

The activation of an immune signaling receptor typically leads to the production and secretion of a multitude of cytokines and other effector molecules. These molecules promote and regulate pro-inflammatory and antimicrobial effects. Among immune effector molecules, the IL-1β family cytokines, which include IL-1β and IL-18, are among the best-studied cytokines and represent important components of the inflammatory response to infections or to host tissue damage [6]. Next to their many pro-inflammatory effects, the IL-1β cytokines can also trigger T-cell differentiation and participate in the maintenance of the T-helper phenotypes of Th1, Th2, and Th17 cells [7]. Activation of IL-1β cytokines leads to profound local and systemic changes. Immune cells, such as neutrophils and macrophages, are recruited to the site of IL-1β maturation and changes in body temperature, feeding behavior or sleep patterns are induced due to systemic actions of IL-1 [6]. Dysregulated IL-1β production can lead to inflammatory disease states [8] and it is not surprising that IL-1β cytokine production and activation is tightly controlled (Fig. 1).

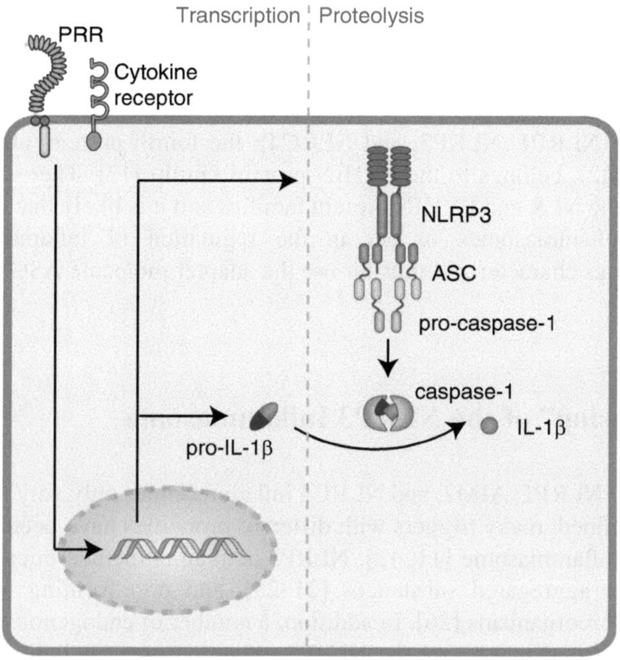

Fig. 1 Activation of IL-1β by the NLRP3 inflammasome. Signals originating from immune pattern recognition receptors (PRR) or cytokine receptors on the cell surface increase the expression of genes encoding NLRP3 and the pro-form of IL-1β. Upon NLRP3 activation, the adaptor protein ASC associates with NLRP3 and pro-caspase-1 through homotypic domain interactions, forming a functional inflammasome. The resulting self-cleavage of pro-caspase-1 releases proteolytically-active caspase-1, which cleaves pro-IL-1beta to produce the mature, biologically active form of the cytokine

While most cytokines are proteins with leader sequences that are released upon their progression through the secretory pathway, the IL-1β cytokines are unique in that they lack leader sequences and are therefore expressed as biologically inactive pro-forms in the cytoplasm [9]. These pro-forms need to be catalytically processed by the cysteine protease caspase-1 to become "mature" effector cytokines. Caspase-1 itself is produced as an inactive pro-form in the cytosol and requires an autocatalytic activation step [10]. This activation of caspase-1 is dependent on the assembly of an inflammasome, in which NLRP3 recruits the adapter protein apoptosis-associated speck-like protein containing a caspase recruitment domain (ASC). ASC is a bipartite adapter protein that, upon interacting with inflammasome-forming NLR or PYHIN family proteins, induces close proximity of recruited caspase-1 molecules. This proximity leads to caspase-1 self-cleavage and the generation of two subunits (p10 and p20), which assemble into the active caspase-1 consisting of two heterodimers [10]. Caspase-1-mediated cleavage is the predominant mode of activation for IL-1 beta cytokines, but cleavage by neutrophil proteinase-3, elastase, matrix metalloproteinase-9, and granzyme A has also been documented [6].

At present, four multimolecular inflammasome complexes have been identified as caspase-1-activating complexes [11]. Three of them contain proteins from the NLR family (NLRP1, NLRP3, and NLRC4); the fourth protein, absent in melanoma 2 (AIM2), belongs to the PYHIN protein family [12]. There are additional members in the NLR and PYHIN protein families and it is likely that other proteins can form inflammasomes or act in the regulation of inflammasomes. All inflammasomes characterized thus far use the adapter molecule ASC for activation of caspase-1.

3 "Licensing" of the NLRP3 Inflammasome

While for the NLRP1, AIM2, and NLRC4 inflammasomes only very few activators have been defined, many triggers with different properties have been described for the NLRP3 inflammasome [11, 12]. NLRP3 activators include microbes [13–20], crystalline or aggregated substances [21–25], and pore-forming toxins from a variety of microorganisms [26]. In addition, a number of endogenous stress signals have been reported to activate the NLRP3 inflammasome, including extracellular ATP [26, 27], necrotic cell components [28], and certain extracellular matrix proteins [29, 30]. The exact molecular mechanisms leading to the activation of the NRLP3 inflammasome by so many different stimuli is currently unresolved. It is believed that indirect mechanisms trigger NLRP3 inflammasome activation since it is unlikely that NLRP3 evolved to directly recognize so many diverse stimuli.

Many important immune decisions, such as activation of adaptive immunity or B-cell proliferation, are regulated by two or more simultaneous signals. Similarly, efficient activation of the NLRP3 inflammasome requires the existence of at least two signals. When cells are exposed to pore-forming toxins, ATP or various crystals, downstream events, i.e., ASC speck formation or cleavage of caspase-1, only occur if a priming signal from a transcriptionally active pattern recognition receptor or cytokine receptor has previously occurred [31, 32]. Thus, NLRP3 activation requires at least two signals for full activation: A priming signal that leads to the activation of NF-κB (signal I) and an additional stimulus that results in the assembly and activation of the primed NLRP3 inflammasome (signal II). The limiting factor for NLRP3 activation appears to be the expression level of NLRP3 itself. The most proximal signaling event triggered by an NLRP3 stimulus is ASC speck formation. Although ASC is constitutively expressed at high levels, aggregation only occurs in response to a signal II trigger in the presence of a signal I (i.e., cellular priming). It was demonstrated that NLRP3 is an NF-κB responsive gene and is therefore upregulated by many cellular stimuli. Moreover, heterologous expression of NLRP3 that lead to similar levels of NLRP3 as found in primed cells was shown to be sufficient to overcome the necessity for priming by transcriptionally active signaling receptors [31]. These findings suggest that constitutively expressed NLRP3 in resting cells is insufficient to promote inflammasome formation and that the activation of the NLRP3 inflammasome is "thresholded" by the level of NLRP3 present.

It has previously been suggested that TLR-stimulating substances such as LPS, RNA, or small-molecule interferon inducers activate the NLRP3 inflammasome [14, 33]. Since, in these studies, those substances were applied in combination with a signal II inducer (ATP), distinction of a priming effect (signal I) from a potential NLRP3 inflammasome-activating effect (signal II) could not be achieved. It is now believed that signals that converge at the activation of NF-κB can all promote NLRP3 priming and if added in combination with a "danger" signal (ATP, pore-forming toxin, crystals, etc.) they can lead to full NLRP3 inflammasome formation. Thus, conceptually, cells only respond to danger signals if they have also been alerted by a priming signal derived from a different innate signaling receptor or under conditions of cellular stress that result in priming. In this way, transcriptionally active signaling receptors appear to "license" the NLRP3 inflammasome for activation.

4 NLRP3 Inflammasome Activation

Downstream of the "licensing" process, less is understood about the mechanisms that lead to the activation of the NLRP3 inflammasome following signal II. Recently, two potentially linked pathways upstream of NLRP3 activation were hypothesized that could partially explain how a single receptor can be triggered by multiple signals. One model proposes that NLRP3 activators induce lysosomal damage, which is indirectly sensed by the NLRP3 inflammasome [23, 24]. This is supported by studies showing that NLRP3 activators can induce rupture of phagolysosomes and that the resulting release of lysosomal proteases into the cytosol is functionally linked to NLRP3 activation. Of note, physical or pharmacological disruption of lysosomes in the absence of any characterized NLRP3 activators can also mediate NLRP3 inflammasome activation, suggesting that lysosomal damage is a general mechanism upstream of NLRP3 [24]. Additionally, substances that prevent acidification of lysosomal compartments serve as potent NLRP3 inflammasome inhibitors, which suggests a role for acid-dependent lysosomal proteases upstream of NLRP3 activation. Indeed, inhibition or lack of the single lysosomal proteases cathepsin B or L significantly reduces NLRP3 activation and suggests that proteolytic enzymes function upstream of NLRP3 [23–25, 34–36]. However, treating cells with cathepsin inhibitors or employing cells singly deficient in cathepsins does not completely eliminate NLRP3 activation in response to crystals, which argues for either functional redundancy of cathepsins or the existence of additional activation mechanisms.

Another model stems from the observation that some NLRP3 activators can also induce the production of reactive oxygen species (ROS). This model hypothesizes that ROS activity could be sensed directly or indirectly by the NLRP3 inflammasome [17, 18, 22, 37]. The model is supported by experiments demonstrating that ROS scavengers, such as *N*-acetyl cysteine, can block NLRP3 activation. Additionally, RNAi-mediated knock-down of the p22-phox subunit of NADPH oxidase, which is critical for ROS production, results in decreased NLRP3 inflammasome activation in PMA-matured THP-1 cells [22]. Further studies suggest that the condition of oxidative stress can also promote the interaction between NLRP3 and thioredoxin (TRX)-interacting protein (TXNIP), which enhances NLRP3 inflammasome activation [38]. TXNIP is a protein linked to glucose metabolism and insulin resistance [39], and its role in NLRP3 inflammasome activation may provide a mechanistic link to the observed involvement of IL-1β in the pathogenesis of type 2 diabetes mellitus [40]. While this model provides an attractive molecular mechanism by which an endogenous ligand could induce activation of NLRP3, it cannot fully explain the molecular events. Cells lacking TXNIP have only mildly reduced NLRP3 inflammasome activation and some signals that are known to activate ROS production, such as several TLR ligands alone, do not appear to be sufficient to activate the NLRP3 inflammasome without a second signal. Notably, increased ROS was also shown to reversibly inactivate caspase-1 by covalent modification, indicating that increased ROS can also downregulate caspase-1 activity [41]. In agreement with this observation, recent work with cells from patients with genetic mutations in NADPH oxidases that render them deficient in ROS production suggests that ROS indeed downmodulates IL-1β responses, rather than enabling them [42–44]. It is reasonable to suppose that NLRP3 activation is finely controlled and more than one mechanism could operate in a time-dependent manner. Indeed, it was recently suggested that in human monocytes a biphasic redox response is operative and that the antioxidant phase is responsible for NLRP3 inflammasome activation [45, 46]. It is likely that the activation of NLRP3 is complex and requires a combination of several factors, including ROS production and lysosomal protease activity.

5 Regulation of NLRP3 Activity

Given the critical role of the NLRP3 inflammasome in infection control and the detection of sterile tissue damage, it is expected that there are numerous mechanisms by which its activity is regulated. The precise mechanisms that regulate inflammasome activity are still poorly understood, but it is expected that a range of host factors regulate the activation of NLRP3 in order to avoid pathologies that arise from inflammasome over-activation. It is also expected that microbes have evolved mechanisms to interfere with NLRP3 inflammasome activation, since this could be a successful strategy to escape immune detection. Indeed, several direct and indirect mechanisms of NLRP3 inhibition have been described and these molecules act at different levels of the inflammasome pathway (Fig. 2).

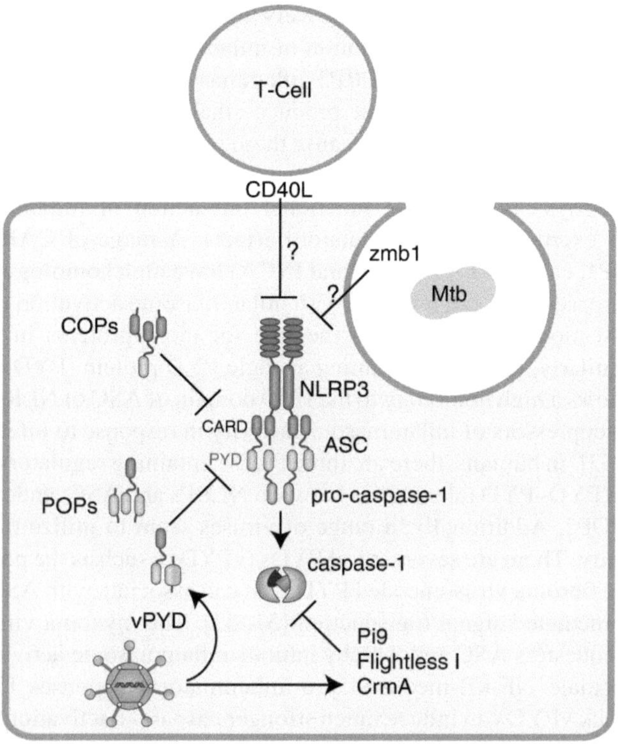

Fig. 2 NLRP3 inflammasome activation is regulated at different steps. T cells can regulate NLRP3 inflammasome function through the activity of CD40L and potentially other factors. Mycobacterial zmb1 can regulate NLRP3 activation indirectly, while CARD-only proteins (COPs) and PYD-only proteins (POPs) directly interfere with signaling through binding to and sequestering ASC. Viruses block their recognition by the NLRP3 inflammasome by expressing viral PYD-containing proteins (vPOPs) or proteins that block the activity of caspase-1 (e.g., CrmA). Caspase-1 is also target of endogenous modifiers such as Pi9 or Flightless I

Mycobacterium tuberculosis (Mtb) has long been known to efficiently subvert host immunity by expressing factors that allow it to form a niche for growth in macrophage phagosomes. Mtb expresses a putative metalloproteinase (zmp1), which efficiently blocks the activation of the NLRP3 inflammasome by an unknown mechanism. Indeed, when Mtb lacks this protease and therefore activates the NLRP3 inflammasome, IL-1β is produced and phagosomal maturation occurs, leading to clearance of Mtb [47]. Thus, it seems that factors that prevent maturation and potential damage of lysosomal pathways inhibit the NLRP3 inflammasome and thereby limit inflammation. An intriguing and poorly understood pathway of NLRP3 inflammasome inhibition was uncovered when immune cell subsets were incubated together with NLRP3-stimulated macrophages. It was found that activated T cells have the ability to suppress inflammasome function. This effect could also be elicited when macrophages were incubated with ligands of the TNF family expressed on T cells, such as CD40 ligand [48]. Thus, the interplay between different immune cells in vivo may be very important for the degree of inflammasome activity as well as for resolution of inflammation.

More direct modulation of the NLRP3 inflammasome pathways may come from a recently identified set of genes that produce small products containing either a CARD or a PYD domain [49–51]. Because these proteins mainly consist of individual signaling domains involved in inflammasome assembly, they could function as molecular "decoys" that impede functional interaction of inflammasome components, thus exerting negative regulatory effects. A range of CARD-containing proteins (COP1, caspase-12, Iceberg, and INCA) have high homology to the CARD domain of caspase-1 and can interfere with inflammasome activation in vitro. There are no known mouse orthologs and the role for these proteins in vivo remains unknown. Similarly, proteins containing a single PYD protein (PYD only proteins, POPs) that shows a high homology to the PYD domain of ASC or NLRPs function as endogenous suppressors of inflammasome activity in response to infection or tissue damage [51, 52]. In humans, there are three PYD-containing regulatory proteins that interfere with PYD–PYD interactions between NLRPs and ASC: endogenous pyrin, POP1, and POP2. Additionally, a range of viruses seem to utilize this strategy to evade immunity. There are several viral PYDs (vPYDs), such as the poxviral M13L-PYD or shope fibroma virus-encoded PYD, that can associate with ASC and thereby inhibit PYD-mediated signal transduction [51, 53]. The myxoma virus M013 protein, which sequesters ASC and thereby inhibits inflammasome activity, can simultaneously regulate NF-κB-mediated pro-inflammatory responses [54]. Of note, viruses that lack vPYD can induce much stronger caspase-1 activation and secretion of IL-1β [55].

Decoy-type regulatory mechanisms may also play a role in inflammasome-related diseases. For example, the protein pyrin, which is encoded by the MEFV gene, can be mutated in patients with familial Mediterranean fever. This autoinflammatory syndrome is characterized by recurrent episodic fever and synovial inflammation suggesting that either inflammasome function is dysregulated or that pyrin itself forms an overly active inflammasome [56]. Also of note is the protein PSTPIP1, which can bind to pyrin, and mutations in this protein can cause the pyogenic arthritis, pyoderma gangrenosum, and acne (PAPA) autoinflammatory syndrome [57, 58]. Mutation of the homologous murine protein, PSTPIP2, is associated with a macrophage autoinflammatory syndrome, further highlighting the importance of pyrin and inflammasome regulation in autoinflammatory disorders [59].

While decoy receptors presumably impede inflammasome assembly, another level of regulation may be achieved post-assembly by interfering with the effector molecule caspase-1. Indeed, caspase-1 itself is also targeted for inhibition by endogenous and exogenous proteins. Caspase-1 activity can be blocked by the cellular factors Flightless-I, an actin-remodeling protein that interacts with and regulates caspase-1 [60], and the human serpin analogue proteinase inhibitor 9 (Pi9) [61, 62]. In addition, viral proteins, such as cowpox cytokine response modifier A (CrmA), and baculovirus p35, target this pathway by inhibiting caspase-1 activity [63, 64]. This suggests that caspase-1 blockade is both employed as an endogenous regulatory mechanism and exploited by pathogens to avoid immune clearance. Finally, the outcome of inflammasome activity, namely activation of the IL-1 receptor by IL-1β itself has naturally occurring inhibitors such as IL-1 receptor antagonist (IL-1RA) or soluble IL-1 receptors [6, 9].

The range of proteins that has evolved to regulate inflammasomes at different levels starting from lysosomal maturation to the inhibition of the mature gene product itself highlights the importance of keeping this molecular platform and the IL-1 pathway tightly controlled and demonstrates that viruses and bacteria have evolved efficient strategies to evade immune detection by inflammasomes.

6 Conclusion

Inflammatory reactions must be well synchronized and controlled, as the dysregulation of inflammatory processes is associated with disease. The dramatic and sometimes devastating clinical consequences of mutations in NLRP3 itself or in other inflammasome-related genes show how an overly active signaling pathway can lead to chronic pathology. With the understanding that the recognition of danger signals by inflammasomes can also lead to inflammation, we have gained new insights into the molecular mechanisms that cause many sterile inflammatory diseases. Anti-IL-1 therapies have proven to be enormously successful in the treatment of auto-inflammatory diseases, which gives great hope that such therapies can also be effective in other inflammatory conditions that are linked to inflammasome activation. Future work should focus on deciphering the integral mechanisms that lead to NLRP3 inflammasome activation and regulation, which could improve the therapeutic strategies against inflammation-associated diseases.

References

1. Matzinger P (2002) The danger model: a renewed sense of self. Science 296:301–305
2. Martinon F, Mayor A, Tschopp J (2009) The inflammasomes: guardians of the body. Annu Rev Immunol 27:229–265
3. Takeda K, Kaisho T, Akira S (2003) Toll-like receptors. Annu Rev Immunol 21:335–376
4. Martinon F, Burns K, Tschopp J (2002) The inflammasome: a molecular platform triggering activation of inflammatory caspases and processing of proIL-beta. Mol Cell 10:417–426
5. Zou H, Li Y, Liu X, Wang X (1999) An APAF-1.cytochrome c multimeric complex is a functional apoptosome that activates procaspase-9. J Biol Chem 274:11549–11556
6. Dinarello CA (2009) Immunological and inflammatory functions of the interleukin-1 family. Annu Rev Immunol 27:519–550
7. Kumar H, Kumagai Y, Tsuchida T, Koenig PA, Satoh T, Guo Z, Jang MH, Saitoh T, Akira S, Kawai T (2009) Involvement of the NLRP3 inflammasome in innate and humoral adaptive immune responses to fungal beta-glucan. J Immunol 183:8061–8067
8. Masters SL, Simon A, Aksentijevich I, Kastner DL (2009) Horror autoinflammaticus: the molecular pathophysiology of autoinflammatory disease (*). Annu Rev Immunol 27:621–668

9. Dinarello CA (1996) Biologic basis for interleukin-1 in disease. Blood 87:2095–2147
10. Thornberry NA, Bull HG, Calaycay JR, Chapman KT, Howard AD, Kostura MJ, Miller DK, Molineaux SM, Weidner JR, Aunins J et al (1992) A novel heterodimeric cysteine protease is required for interleukin-1 beta processing in monocytes. Nature 356:768–774
11. Stutz A, Golenbock DT, Latz E (2009) Inflammasomes: too big to miss. J Clin Invest 119:3502–3511
12. Hornung V, Latz E (2010) Intracellular DNA recognition. Nat Rev Immunol 10:123–130
13. Muruve DA, Petrilli V, Zaiss AK, White LR, Clark SA, Ross PJ, Parks RJ, Tschopp J (2008) The inflammasome recognizes cytosolic microbial and host DNA and triggers an innate immune response. Nature 452:103–107
14. Kanneganti TD, Body-Malapel M, Amer A, Park JH, Whitfield J, Franchi L, Taraporewala ZF, Miller D, Patton JT, Inohara N et al (2006) Critical role for Cryopyrin/Nalp3 in activation of caspase-1 in response to viral infection and double-stranded RNA. J Biol Chem 281:36560–36568
15. Duncan JA, Gao X, Huang MT, O'Connor BP, Thomas CE, Willingham SB, Bergstralh DT, Jarvis GA, Sparling PF, Ting JP (2009) Neisseria gonorrhoeae activates the proteinase cathepsin B to mediate the signaling activities of the NLRP3 and ASC-containing inflammasome. J Immunol 182:6460–6469
16. Griffith JW, Sun T, McIntosh MT, Bucala R (2009) Pure Hemozoin is inflammatory in vivo and activates the NALP3 inflammasome via release of uric acid. J Immunol 183:5208–5220
17. Dostert C, Guarda G, Romero JF, Menu P, Gross O, Tardivel A, Suva ML, Stehle JC, Kopf M, Stamenkovic I et al (2009) Malarial hemozoin is a Nalp3 inflammasome activating danger signal. PLoS One 4:e6510
18. Gross O, Poeck H, Bscheider M, Dostert C, Hannesschlager N, Endres S, Hartmann G, Tardivel A, Schweighoffer E, Tybulewicz V et al (2009) Syk kinase signalling couples to the Nlrp3 inflammasome for anti-fungal host defence. Nature 459:433–436
19. Hise AG, Tomalka J, Ganesan S, Patel K, Hall BA, Brown GD, Fitzgerald KA (2009) An essential role for the NLRP3 inflammasome in host defense against the human fungal pathogen Candida albicans. Cell Host Microbe 5:487–497
20. Willingham SB, Allen IC, Bergstralh DT, Brickey WJ, Huang MT, Taxman DJ, Duncan JA, Ting JP (2009) NLRP3 (NALP3, Cryopyrin) facilitates in vivo caspase-1 activation, necrosis, and HMGB1 release via inflammasome-dependent and -independent pathways. J Immunol 183:2008–2015
21. Martinon F, Petrilli V, Mayor A, Tardivel A, Tschopp J (2006) Gout-associated uric acid crystals activate the NALP3 inflammasome. Nature 440:237–241
22. Dostert C, Petrilli V, Van Bruggen R, Steele C, Mossman BT, Tschopp J (2008) Innate immune activation through Nalp3 inflammasome sensing of asbestos and silica. Science 320:674–677
23. Halle A, Hornung V, Petzold GC, Stewart CR, Monks BG, Reinheckel T, Fitzgerald KA, Latz E, Moore KJ, Golenbock DT (2008) The NALP3 inflammasome is involved in the innate immune response to amyloid-beta. Nat Immunol 9:857–865
24. Hornung V, Bauernfeind F, Halle A, Samstad EO, Kono H, Rock KL, Fitzgerald KA, Latz E (2008) Silica crystals and aluminum salts activate the NALP3 inflammasome through phagosomal destabilization. Nat Immunol 9:847–856
25. Duewell P, Kono H, Rayner KJ, Sirois CM, Vladimer G, Bauernfeind FG, Abela GS, Franchi L, Nunez G, Schnurr M et al (2010) NLRP3 inflammasomes are required for atherogenesis and activated by cholesterol crystals. Nature 464:1357–1361
26. Mariathasan S, Weiss DS, Newton K, McBride J, O'Rourke K, Roose-Girma M, Lee WP, Weinrauch Y, Monack DM, Dixit VM (2006) Cryopyrin activates the inflammasome in response to toxins and ATP. Nature 440:228–232
27. Petrilli V, Papin S, Dostert C, Mayor A, Martinon F, Tschopp J (2007) Activation of the NALP3 inflammasome is triggered by low intracellular potassium concentration. Cell Death Differ 14:1583–1589

28. Li H, Ambade A, Re F (2009) Cutting edge: Necrosis activates the NLRP3 inflammasome. J Immunol 183:1528–1532
29. Yamasaki K, Muto J, Taylor KR, Cogen AL, Audish D, Bertin J, Grant EP, Coyle AJ, Misaghi A, Hoffman HM et al (2009) NLRP3/cryopyrin is necessary for interleukin-1beta (IL-1beta) release in response to hyaluronan, an endogenous trigger of inflammation in response to injury. J Biol Chem 284:12762–12771
30. Babelova A, Moreth K, Tsalastra-Greul W, Zeng-Brouwers J, Eickelberg O, Young MF, Bruckner P, Pfeilschifter J, Schaefer RM, Grone HJ et al (2009) Biglycan, a danger signal that activates the NLRP3 inflammasome via toll-like and P2X receptors. J Biol Chem 284:24035–24048
31. Bauernfeind FG, Horvath G, Stutz A, Alnemri ES, MacDonald K, Speert D, Fernandes-Alnemri T, Wu J, Monks BG, Fitzgerald KA et al (2009) Cutting edge: NF-kappaB activating pattern recognition and cytokine receptors license NLRP3 inflammasome activation by regulating NLRP3 expression. J Immunol 183:787–791
32. Franchi L, Eigenbrod T, Nunez G (2009) Cutting edge: TNF-alpha mediates sensitization to ATP and silica via the NLRP3 inflammasome in the absence of microbial stimulation. J Immunol 183:792–796
33. Kanneganti TD, Lamkanfi M, Kim YG, Chen G, Park JH, Franchi L, Vandenabeele P, Nunez G (2007) Pannexin-1-mediated recognition of bacterial molecules activates the cryopyrin inflammasome independent of Toll-like receptor signaling. Immunity 26:433–443
34. Maitra R, Clement CC, Scharf B, Crisi GM, Chitta S, Paget D, Purdue PE, Cobelli N, Santambrogio L (2009) Endosomal damage and TLR2 mediated inflammasome activation by alkane particles in the generation of aseptic osteolysis. Mol Immunol 47:175–184
35. Sharp FA, Ruane D, Claass B, Creagh E, Harris J, Malyala P, Singh M, O'Hagan DT, Petrilli V, Tschopp J et al (2009) Uptake of particulate vaccine adjuvants by dendritic cells activates the NALP3 inflammasome. Proc Natl Acad Sci USA 106:870–875
36. Rajamaki K, Lappalainen J, Oorni K, Valimaki E, Matikainen S, Kovanen PT, Eklund KK (2010) Cholesterol crystals activate the NLRP3 inflammasome in human macrophages: a novel link between cholesterol metabolism and inflammation. PLoS One 5:e11765
37. Cassel SL, Eisenbarth SC, Iyer SS, Sadler JJ, Colegio OR, Tephly LA, Carter AB, Rothman PB, Flavell RA, Sutterwala FS (2008) The Nalp3 inflammasome is essential for the development of silicosis. Proc Natl Acad Sci USA 105:9035–9040
38. Zhou R, Tardivel A, Thorens B, Choi I, Tschopp J (2010) Thioredoxin-interacting protein links oxidative stress to inflammasome activation. Nat Immunol 11:136–140
39. Chen J, Saxena G, Mungrue IN, Lusis AJ, Shalev A (2008) Thioredoxin-interacting protein: a critical link between glucose toxicity and beta-cell apoptosis. Diabetes 57:938–944
40. Schroder K, Zhou R, Tschopp J (2010) The NLRP3 inflammasome: a sensor for metabolic danger? Science 327:296–300
41. Meissner F, Molawi K, Zychlinsky A (2008) Superoxide dismutase 1 regulates caspase-1 and endotoxic shock. Nat Immunol 9:866–872
42. Meissner F, Seger RA, Moshous D, Fischer A, Reichenbach J, Zychlinsky A (2010) Inflammasome activation in NADPH oxidase defective mononuclear phagocytes from patients with chronic granulomatous disease. Blood 116:1570–1573
43. van Bruggen R, Koker MY, Jansen M, van Houdt M, Roos D, Kuijpers TW, van den Berg TK (2010) Human NLRP3 inflammasome activation is Nox1-4 independent. Blood 115: 5398–5400
44. van de Veerdonk FL, Smeekens SP, Joosten LA, Kullberg BJ, Dinarello CA, van der Meer JW, Netea MG (2010) Reactive oxygen species-independent activation of the IL-1beta inflammasome in cells from patients with chronic granulomatous disease. Proc Natl Acad Sci USA 107:3030–3033
45. Tassi S, Carta S, Vene R, Delfino L, Ciriolo MR, Rubartelli A (2009) Pathogen-induced interleukin-1beta processing and secretion is regulated by a biphasic redox response. J Immunol 183:1456–1462

46. Tassi S, Carta S, Delfino L, Caorsi R, Martini A, Gattorno M, Rubartelli A (2010) Altered redox state of monocytes from cryopyrin-associated periodic syndromes causes accelerated IL-1beta secretion. Proc Natl Acad Sci USA 107:9789–9794
47. Master SS, Rampini SK, Davis AS, Keller C, Ehlers S, Springer B, Timmins GS, Sander P, Deretic V (2008) Mycobacterium tuberculosis prevents inflammasome activation. Cell Host Microbe 3:224–232
48. Guarda G, Dostert C, Staehli F, Cabalzar K, Castillo R, Tardivel A, Schneider P, Tschopp J (2009) T cells dampen innate immune responses through inhibition of NLRP1 and NLRP3 inflammasomes. Nature 460:269–273
49. da Cunha JP, Galante PA, de Souza SJ (2008) Different evolutionary strategies for the origin of caspase-1 inhibitors. J Mol Evol 66:591–597
50. Kersse K, Vanden Berghe T, Lamkanfi M, Vandenabeele P (2007) A phylogenetic and functional overview of inflammatory caspases and caspase-1-related CARD-only proteins. Biochem Soc Trans 35:1508–1511
51. Stehlik C, Dorfleutner A (2007) COPs and POPs: modulators of inflammasome activity. J Immunol 179:7993–7998
52. Dorfleutner A, Bryan NB, Talbott SJ, Funya KN, Rellick SL, Reed JC, Shi X, Rojanasakul Y, Flynn DC, Stehlik C (2007) Cellular pyrin domain-only protein 2 is a candidate regulator of inflammasome activation. Infect Immun 75:1484–1492
53. Dorfleutner A, Talbott SJ, Bryan NB, Funya KN, Rellick SL, Reed JC, Shi X, Rojanasakul Y, Flynn DC, Stehlik C (2007) A Shope Fibroma virus PYRIN-only protein modulates the host immune response. Virus Genes 35:685–694
54. Rahman MM, Mohamed MR, Kim M, Smallwood S, McFadden G (2009) Co-regulation of NF-kappaB and inflammasome-mediated inflammatory responses by myxoma virus pyrin domain-containing protein M013. PLoS Pathog 5:e1000635
55. Johnston JB, Barrett JW, Nazarian SH, Goodwin M, Ricciuto D, Wang G, McFadden G (2005) A poxvirus-encoded pyrin domain protein interacts with ASC-1 to inhibit host inflammatory and apoptotic responses to infection. Immunity 23:587–598
56. Consortium TFF (1997) A candidate gene for familial Mediterranean fever. Nat Genet 17:25–31
57. Shoham NG, Centola M, Mansfield E, Hull KM, Wood G, Wise CA, Kastner DL (2003) Pyrin binds the PSTPIP1/CD2BP1 protein, defining familial Mediterranean fever and PAPA syndrome as disorders in the same pathway. Proc Natl Acad Sci USA 100:13501–13506
58. Wise CA, Gillum JD, Seidman CE, Lindor NM, Veile R, Bashiardes S, Lovett M (2002) Mutations in CD2BP1 disrupt binding to PTP PEST and are responsible for PAPA syndrome, an autoinflammatory disorder. Hum Mol Genet 11:961–969
59. Grosse J, Chitu V, Marquardt A, Hanke P, Schmittwolf C, Zeitlmann L, Schropp P, Barth B, Yu P, Paffenholz R et al (2006) Mutation of mouse Mayp/Pstpip2 causes a macrophage autoinflammatory disease. Blood 107:3350–3358
60. Li J, Yin HL, Yuan J (2008) Flightless-I regulates proinflammatory caspases by selectively modulating intracellular localization and caspase activity. J Cell Biol 181:321–333
61. Annand RR, Dahlen JR, Sprecher CA, De Dreu P, Foster DC, Mankovich JA, Talanian RV, Kisiel W, Giegel DA (1999) Caspase-1 (interleukin-1beta-converting enzyme) is inhibited by the human serpin analogue proteinase inhibitor 9. Biochem J 342(Pt 3):655–665
62. Young JL, Sukhova GK, Foster D, Kisiel W, Libby P, Schonbeck U (2000) The serpin proteinase inhibitor 9 is an endogenous inhibitor of interleukin 1beta-converting enzyme (caspase-1) activity in human vascular smooth muscle cells. J Exp Med 191:1535–1544
63. Ray CA, Black RA, Kronheim SR, Greenstreet TA, Sleath PR, Salvesen GS, Pickup DJ (1992) Viral inhibition of inflammation: cowpox virus encodes an inhibitor of the interleukin-1 beta converting enzyme. Cell 69:597–604
64. Zhou Q, Krebs JF, Snipas SJ, Price A, Alnemri ES, Tomaselli KJ, Salvesen GS (1998) Interaction of the baculovirus anti-apoptotic protein p35 with caspases. Specificity, kinetics, and characterization of the caspase/p35 complex. Biochemistry 37:10757–10765

Treatment of Inflammasome-Related Disorders

Helen J. Lachmann

Abstract Neutralization of the inflammasome offer considerable therapeutic promise to reduce inflammation in inflammasome-related disorders. Here we review treatments commercially available and emerging new therapeutics designed to block inflammasome activity and inflammasome-dependent cytokines.

IL-1β appears to play a key role in a wide variety of inflammatory diseases. Conditions range from the very rare such as the inherited autoinflammatory disorders particularly cryopyrin-associated periodic syndrome (CAPS) to much common acquired disorders such as gout [1], silica-mediated pulmonary fibrosis [2], as well as perhaps myeloma [3], atherosclerosis [4], and type II diabetes mellitus [5].

There are now three IL-1 blocking agents commercially available, anakinra, rilonacept, and canakinumab (see Table 1) and a number of other molecules under investigation.

1 Anakinra: Recombinant IL-1 Receptor Antagonist (IL-1Ra)

Anakinra is a recombinant nonglycosylated form of IL-1Ra made in an *Escherichia coli* expression system. It is 17.3 kDa, 153 amino acids and virtually identical to human IL-1Ra except for an extra methionine residue at the amino terminus. It has a 95% bioavailability after subcutaneous injection and peak levels 1–2 h after injection. Clearance is largely by the kidneys with a terminal half-life of 4–6 h.

H.J. Lachmann (✉)
National Amyloidosis Centre, UCL Medical School, Royal Free Campus, Rowland Hill Street, London NW3 2PF, UK
e-mail: h.lachmann@medsch.ucl.ac.uk

I. Couillin et al. (eds.), *The Inflammasomes*, Progress in Inflammation Research,
DOI 10.1007/978-3-0348-0148-5_14, © Springer Basel AG 2011

Table 1 Commercially available IL-1 blocking agents

	Anakinra	Rilonacept	Canakinumab
Description	Recombinant nonglycosylated protein, produced in an *E. coli* expression system. Identical to endogenous IL-1 receptor antagonist (IL-1Ra) except for an additional methionine residue at the amino terminus	Fusion protein of the extracellular domains of the IL-1β receptor (IL-1RI) and adaptor protein (IL-1RacP) attached to human IgG	Fully human IgG1 anti-IL-1β monoclonal antibody
Mechanism of action	Binds to IL-1RI, but does not recruit IL-1RacP so does not induce signal transduction	Binds to IL-1α, IL-1β and IL-1Ra with high affinity, preventing IL-1 and its endogenous antagonist from binding to IL-1R1	Binds to IL-1β
Components of IL-1 blocked	IL-1α and IL-1β	IL-1α, IL-1β, and IL-1Ra	IL-1β
Plasma half-life	4–6 h	67 h	21–28 days
Administration	Subcutaneous injection daily: 100 mg in adults	Subcutaneous injection weekly: 160 mg in adults	Subcutaneous injection every 8 weeks: 150 mg in adults

The drug is stabilized with citric acid which can cause pain on injection and is probably responsible for injection-site reactions. These occur in 64% of patients, usually within the first 4 weeks after initiation and are responsible for 6% of patients stopping treatment. The other major side effect is a 2% risk of neutropenia on initiation but the risk of contracting a serious infection with anakinra is relatively low at 1.8%. Anakinra was originally developed for the treatment of systemic sepsis in intensive care settings. Unfortunately, the studies were unsuccessful and it was tried in rheumatoid arthritis for which it is proved modestly effective. Currently this remains its only licensed indication.

2 Rilonacept

Rilonacept is a high-affinity IL-1 TRAP. It is a dimeric fusion protein consisting of the extracellular ligand-binding domains of IL-1 receptor (IL-1R) and IL-1 receptor accessory protein (IL-1RacP) linked to human IgG1. Rilonacept blocks IL-1β signaling by acting as a decoy receptor that binds IL-1β and IL-1α and Il-1Ra with lower affinity. It is injected subcutaneously with a bioavailability of about 50%. Its half-life of 7.2 days allows weekly dosing and as it is cleared by protein catabolism its metabolism is independent of renal function. The commonest reported adverse event in studies was transient injection-site reactions. There is a small risk of neutropenia and approximately 6% of patients were reported to have infections compared to 3% of placebo-treated controls. It is licensed by the FDA for the treatment of cryopyrin-associated periodic fever syndromes (CAPS), including Muckle–Wells syndrome (MWS) and familial cold autoinflammatory syndrome (FCAS) in children 12 years and older.

3 Canakinumab

Canakinumab is a recombinant fully human monoclonal antibody which neutralizes human IL-1β by binding with high affinity. It has no cross-reactivity against other members of the IL-1 family. It is produced in a murine cell line. Peak levels are reached 7 days after subcutaneous injection with a bioavailability of 70%. The terminal half life of approximately 26 days means that dosing every 8 weeks is therapeutic. Reported adverse effects from clinical studies include an increased incidence and infections and few injection-site reactions. It is licensed for the treatment of CAPS in children aged over 4 years.

4 Unlicensed Agents Under Investigation

4.1 Xoma

Xoma 52 is a human-engineered IgG2 kappa antibody with 97% human sequence. It has a very high affinity for IL-1β and a half-life of 22 days. Binding inhibits binding of IL-1β to IL-1RI but leaves its affinity for IL-1RII and the soluble inhibitory receptors, sRI, sRII, and sRAcP largely unaffected. In preclinical studies it has been shown to have useful activity in mouse models of diet-induced obesity where it preserved beta-cell function with reduced hyperglycemia, and in models of collagen-induced arthritis and gout where it reduced swelling and inflammation. It is currently being investigated in phase 2 studies of diabetes and cardiovascular disease.

4.2 Caspase 1 Inhibitors

Vertex VX-765 is an orally available pro-drug which hydrolyses in vivo to form an active metabolite, a selective reversible inhibitor of interleukin-1-converting enzyme (ICE) [6]. It has been shown to reduce production of IL-1β and IL-18 in patients with CAPS but efficacy was less than that seen with anakinra. Clinical studies of an another agent pralnacasan [7] in rheumatoid arthritis, psoriasis, and osteoarthritis were discontinued due to abnormal liver function in animal toxicology models with prolonged and high-dose medication schedules.

5 Diseases Mediated by IL-1

5.1 Cryopyrin-Associated Periodic Syndrome

This rare autoinflammatory disorder is caused by mutations in NLRP3, which encodes a component of the IL-1 inflammasome, is the paradigm of IL-1-driven disease. The syndrome consists of fever, urticarial rash, conjunctivitis, features of chronic meningitis, and severe fatigue. Symptoms are accompanied by a high risk of long-term complications including deafness, visual impairment, and systemic AA amyloidosis which can cause proteinuric renal failure. Historically, the only really effective treatment was high-dose long-term corticosteroids which caused severe side effects.

Anti-IL-1 therapy in the form of anakinra was first used in 2002 in two patients with CAPS complicated by AA amyloidosis [8]. Treatment resulted in complete resolution of all inflammatory symptoms within hours and inflammatory markers normalized within 3 days. Similar responses have been reported in a number of case reports and small series across the disease severity spectrum including the most severe form of CAPS, NOMID (neonatal-onset multisystem inflammatory disease). A study in 18 patients with NOMID who received daily injections of anakinra [9] described a response in all patients, with disappearance of rash and conjunctivitis within 3 days of starting treatment. Responses were maintained at 6 months in ten patients (56%), and there was improved hearing in six patients. In children with severe disease very high doses of anakinra may be required whereas doses of 1–3 mg/kg are effective in other patients [10].

Rilonacept has been licensed for the treatment of CAPS based on a 6-week randomized double-blind placebo-controlled study followed by 9-week open treatment with rilonacept, followed by 9-week randomized double-blind placebo-controlled withdrawal phase [11]. The study involved 44 patients with FCAS and 3 with MWS given 160 mg of rilonacept weekly. There was a significant reduction in both symptom score and SAA and CRP from baseline at the end of the first 6 weeks in the rilonacept-treated group ($P < 0.0001$) but not placebo at the end of part 1 with rilonacept ($P < 0.001$). The improvements in symptom score and CRP and SAA levels with rilonacept were maintained and there were significant difference in symptom scores and suppression of CRP and SAA between treatment groups in randomized double-blind placebo-controlled withdrawal phase. This was followed by an open-label study in 104 patients with FCAS and MWS for up to 112 weeks. Reported adverse events from these two studies included injection-site reactions in 61% of patients, infections in 33% of patients; most commonly minor nasopharyngitis, sinusitis or upper respiratory tract infection. There were two deaths but were not considered related to treatment. Antirilonacept antibodies were detected in 41% of patients but did not appear to have biological significant effect.

Canakinumab has also been licensed for the treatment of CAPS based on a three-part, 48-week double-blind placebo-controlled, randomized withdrawal study [12]. In part 1, all patients received a single dose of canakinumab, 150 mg SC; in part 2, patients were randomized to receive canakinumab 150 mg or placebo every 8 weeks for up to 24 weeks; in part 3, all patients received canakinumab 150 mg SC every 8 weeks for a total study duration of 48 weeks. The study recruited 35 patients: 33 with MWS and 2 with MWS/NOMID; 5 were pediatric patients. In part 1, 34 patients (97%) achieved a complete response by week 8, symptoms diminished within 24 h in all responders. In part 2, all 15 patients randomized to canakinumab remained in remission whereas disease flare occurred in 13 of 16 patients (81%) randomized to placebo ($P < 0.001$). A further ongoing study, and one of the largest to date, is investigating the efficacy of IL-1β blockade across the spectrum of disease [13]. The study involving 98 patients (including 19 with FCAS, 69 with MWS, and 9 with MWS/NOMID) is investigating the efficacy and tolerability of canakinumab administered every 8 weeks. At an interim analysis performed after a median duration of exposure to canakinumab of 113 days, complete responses had been achieved in 41 of 44 canakinumab-naïve patients (93.2%) at day 8. Of 85 patients evaluable for relapse in this analysis, 77 (90.6%) had not relapsed.

In summary, all three of the commercially available anti-IL-1 agents have been shown to be completely effective in the short to median term in treating CAPS. The complete efficacy of canakinumab shows that the disease is IL-1β-driven. Sustained suppression of systemic inflammation associated with CAPS has dramatically improved quality of life, should prevent the development of life-threatening AA amyloidosis and has already been reported in some cases to be accompanied by improvements in neurological involvement and growth. However, it seems that early initiation of treatment is necessary to prevent the development of irreversible lesions. Life-long treatment will be needed for these diseases and registry data will be required to detect long-term outcomes and side effects. Issues such as female fertility remain substantial concerns for the future as none of the agents are known to be safe in pregnancy.

6 Other Hereditary Autoinflammatory Disorders

These include familial Mediterranean fever (FMF), tumor necrosis factor (TNF)-receptor-associated periodic syndrome (TRAPS), pyogenic arthritis, pyoderma gangrenosum, and acne syndrome (PAPA), deficiency of IL-1Ra (DIRA) and mevalonate kinase deficiency (MKD) otherwise known as hyperimmunoglobulinemia D syndrome (HIDS). For some of these disorders, a link between the mutated gene and overproduction of IL-1β has become evident from understanding of the function of the affected protein in the regulation of IL-1β production. For example, FMF is associated with mutations in the gene that encodes the protein pyrin, resulting in overproduction of IL-1β [14]. The exact mechanism for this is unclear, but it is thought that pyrin may prevent formation of a functional NALP3 inflammasome complex by binding to ASC. In addition, pyrin has been shown to bind to caspase-1 and to inhibit its ability to cleave pro-IL-1β [15]. A connection between the mutation and IL-1β production is less evident for other autoinflammatory disorders such as TRAPS and MKD. For these, responsiveness to IL-1β blockade has been valuable in identifying the involvement of IL-1β in their pathogenesis.

Various authors have reported responses to anakinra in single patients with FMF [16, 17], PAPA [18], TRAPS [19], and HIDS [20]. Gattorno and colleagues reported responses to anakinra in five patients with TRAPS [21], and van der Hilst reported responses to anakinra in 7 of 11 patients with HIDS included in a large retrospective study [22]. DIRA is due to absence of endogenous IL-1Ra and responds fully to IL-1Ra therapy [23].

7 IL-1β in Systemic-Onset Juvenile Idiopathic Arthritis and Adult-Onset Still's Disease

The key role of IL-1β in systemic-onset juvenile idiopathic arthritis (SJIA) was initially suggested by analysis of gene transcription in monocytes from healthy individuals incubated with serum from patients with active disease [24]. Pascual and colleagues reported that serum from patients with SJIA (but not from control subjects) induced transcription of IL-1β in healthy monocytes [24]. In addition, IL-1β secretion by monocytes from patients with SJIA was approximately 15-fold greater than those by monocytes from controls.

Verbsky and White used anakinra to treat two patients in whom conventional therapies had failed [25]. Both patients had immediate resolution of fever, rash, and arthritis, and were able to discontinue all other medications. Further studies have confirmed that most patients at least achieve remission of systematic symptoms in response to IL-1β blockade (reviewed in [26]), and many patients are able to taper or discontinue steroids. Similar results have also been reported for the related disorder adult-onset Still's disease (AOSD), which shares many of the clinical and biochemical manifestations of SJIA [27] (and reviewed in [26]).

As results from more patients are reported, however, it is becoming apparent that a subset of patients achieves remission of systemic symptoms but does not achieve sustained remission of arthritic symptoms. In a study of 21 patients, half of whom responded to anakinra, Gattorno compared the characteristics of responders and nonresponders [28]: Complete responders had fewer actively inflamed joints and a higher absolute neutrophil count at baseline compared with nonresponders.

However, there were no significant differences in systemic features or most laboratory parameters.

The major treatment-limiting side effect of anakinra is the need for daily injections and the high incidence of injection-site reactions. Zeft reported pain or swelling in 17 of 32 patients, and 25% discontinued because of injection-site reactions [29]. In this respect the two newer anti-IL-1β therapies, rilonacept and canakinumab, which are administered weekly and monthly, respectively, are likely to offer significant advantages. An open-label follow-up of a double-blind placebo-controlled study of once-weekly rilonacept in 21 children with SJIA [30, 31] showed that 13 patients (62%) achieved a 50% improvement according to American College of Rheumatology pediatric criteria (ACR ped 50) at week 4, and 87% achieved an ACR ped 50 at 12 months. However, 30% of patients developed injection-site reactions and two patients developed macrophage activation syndrome.

The efficacy and safety of canakinumab in SJIA was studied in a phase 2 dose-finding study involving 23 patients who initially received a single dose of canakinumab and were retreated on relapse [32]. An ACR ped 50 or better was achieved in 59% at day 15, and the median time to relapse was 56–90 days. Steroids were weaned down in 77%. Interestingly, the study included 11 patients who had proved refractory to anakinra; six of these achieved at least an ACR ped 50 with canakinumab.

8 Gouty Arthritis

The role of MSU in inducing activation of the NALP3 inflammasome was demonstrated in a set of elegant in vitro and animal studies [33]. Similar results were also observed for calcium pyrophosphate dehydrate (CPPD) crystals.

A pilot study of ten patients with acute gouty arthritis who had failed to respond to nonsteroidal anti-inflammatory drugs (NSAIDs), colchicine or corticosteroids [34] demonstrated that anakinra substantially relieved symptoms within 48 h, and in nine patients complete resolution of signs of arthritis was achieved on day 3. An active-controlled dose-ranging phase 2 study of canakinumab in patients with acute gouty arthritis, who were refractory to and/or had contraindications for NSAIDs and/or colchicines, also showed benefit [35]. In this study, patients were randomized to receive a single dose of canakinumab 10–150 mg or corticosteroids, and pain was assessed over the following 7 days. All canakinumab doses were associated with lower pain scores than with steroids at 72 h and a reduced risk of recurrent flare. In a pilot study of rilonacept in ten patients with chronic active gouty arthritis, pain scores decreased significantly during the first 2 weeks of treatment with rilonacept and remained low at the end of the 6-week treatment period [36]. In a second study, patients initiating allopurinol therapy were randomized to receive rilonacept or placebo weekly for 16 weeks [37]. Rilonacept significantly reduced the risk of flares.

IL-1β blockade may also be helpful for the management of pseudogout, as suggested by experience reported for patients with pseudogout treated with anakinra [38, 39].

The studies reported to date suggest that IL-1β blockade could be an important treatment option for patients with gouty arthritis or pseudogout for whom standard anti-inflammatory therapies are ineffective or inappropriate due to comorbidity, such as renal diseases.

9 Other Disorders That May Respond to IL-1β Blockade

The success of IL-1β blockade in the treatment of autoinflammatory disorders has led to investigation of other inflammatory disorders that do not respond to conventional therapies. Rapid and sustained responses to anakinra have been reported for single patients with Schnitzler's syndrome, Sweet syndrome, Behçet's disease, and relapsing polychondritis, many of whom were unresponsive to immunosuppressants, including steroids [40–43]. The molecular basis of these disorders is unknown. However, the impressive responses reported suggest that the symptoms are mediated by overproduction of IL-1β.

Aberrations in the NALP3 inflammasome and overproduction of IL-1β may also contribute to the pathology of many other disorders with an inflammatory element. Evidence to date suggests that aberrations in NALP3 or IL-1β production contribute to the pathology of asbestosis [44], malaria [45], contact dermatitis [46], and hypertension [47]. Anti-IL-1β therapy may thus be a useful approach to the treatment of such disorders but has yet to be demonstrated.

9.1 Diabetes Mellitus

Overproduction of IL-1β may contribute to the pathogenesis of diseases that are not generally thought of as inflammatory. The progression of type 2 diabetes involves destruction of islet beta cells, which is believed to be mediated by their secretion of IL-1β. A 13-week placebo-controlled study of anakinra in patients with type 2 diabetes [48] produced remarkable results. Glycosylated hemoglobin (a measure of long-term blood glucose control) was reduced in the anakinra group while an increase was observed in the placebo group; the difference 0.46% between treatments was significant [48]. Beta-cell function increased during treatment with anakinra (but decreased in the placebo group) and the improvement in beta-cell function was sustained 39 weeks after stopping anakinra therapy [5]. A pilot study with another anti-IL-1β therapy XOMA-052, an anti-IL-1β monoclonal antibody [49], demonstrated median reductions in glycosylated hemoglobin of up to 1.1% (percentage points) at 3 months following a single dose of XOMA-052 (compared with an increase of 0.1% for placebo), and increases in insulin production were observed at 1 and 3 months. IL-1β blockade may therefore be a value approach to slow or prevent the progression of type 2 diabetes. Studies of XOMA-52 are now also underway in type 1 diabetes.

9.2 Cardiovascular Diseases

The evidence from mice models that cholesterol is present in early atherosclerotic lesions and that cholesterol crystals induce acute inflammation, via the NLRP3 inflammasome implicates IL-1 in the pathogenesis of atherosclerosis [4]. There is also considerable interest in the role of IL-1 in the complex mechanisms that take place after myocardial infarction including necrosis, cell proliferation, and remodeling. Local production of IL-1β postinfarction may play a major role in adversely influencing cardiac fibroblast responses and there are ongoing studies looking at the effect on anakinra on preventing postinfarction remodeling in man and of rilonacept looking at arterial function in atherosclerosis [50].

References

1. McGonagle D, Tan AL, Shankaranarayana S, Madden J, Emery P, McDermott MF (2007) Management of treatment resistant inflammation of acute on chronic tophaceous gout with anakinra. Ann Rheum Dis 66(12):1683–1684
2. Dostert C, Petrilli V, Van Bruggen R, Steele C, Mossman BT, Tschopp J (2008) Innate immune activation through Nalp3 inflammasome sensing of asbestos and silica. Science 320 (5876):674–677
3. Lust JA, Lacy MQ, Zeldenrust SR et al (2009) Induction of a chronic disease state in patients with smoldering or indolent multiple myeloma by targeting interleukin 1{beta}-induced interleukin 6 production and the myeloma proliferative component. Mayo Clin Proc 84 (2):114–122
4. Duewell P, Kono H, Rayner KJ et al (2010) NLRP3 inflammasomes are required for atherogenesis and activated by cholesterol crystals. Nature 464(7293):1357–1361
5. Larsen CM, Faulenbach M, Vaag A, Ehses JA, Donath MY, Mandrup-Poulsen T (2009) Sustained effects of interleukin-1 receptor antagonist treatment in type 2 diabetes. Diabetes Care 32(9):1663–1668
6. Wannamaker W, Davies R, Namchuk M et al (2007) (S)-1-((S)-2-{[1-(4-amino-3-chloro-phenyl)-methanoyl]-amino}-3,3-dimethyl-butanoy l)-pyrrolidine-2-carboxylic acid ((2R,3S)-2-ethoxy-5-oxo-tetrahydro-furan-3-yl)-amide (VX-765), an orally available selective interleukin (IL)-converting enzyme/caspase-1 inhibitor, exhibits potent anti-inflammatory activities by inhibiting the release of IL-1beta and IL-18. J Pharmacol Exp Ther 321(2): 509–516
7. Linton SD (2005) Caspase inhibitors: a pharmaceutical industry perspective. Curr Top Med Chem 5(16):1697–1717
8. Hawkins PN, Lachmann HJ, McDermott MF (2003) Interleukin-1-receptor antagonist in the Muckle-Wells syndrome. N Engl J Med 348(25):2583–2584
9. Goldbach-Mansky R, Dailey NJ, Canna SW et al (2006) Neonatal-onset multisystem inflammatory disease responsive to interleukin-1beta inhibition. N Engl J Med 355(6):581–592

10. Neven B, Marvillet I, Terrada C et al (2010) Long-term efficacy of the interleukin-1 receptor antagonist anakinra in ten patients with neonatal-onset multisystem inflammatory disease/ chronic infantile neurologic, cutaneous, articular syndrome. Arthritis Rheum 62(1):258–267
11. Hoffman HM, Throne ML, Amar NJ et al (2008) Efficacy and safety of rilonacept (interleukin-1 trap) in patients with cryopyrin-associated periodic syndromes: results from two sequential placebo-controlled studies. Arthritis Rheum 58:2443–2452
12. Lachmann HJ, Kone-Paut I, Kuemmerle-Deschner JB et al (2009) Use of canakinumab in cryopyrin-associated periodic syndrome. N Engl J Med 360(23):2416–2425
13. Kuemmerle-Deschner JB, Lachmann HJ, Hachulla E et al (2009) Efficacy and safety of canakinumab (Ilaris) in a large cohort of patients across different severity phenotypes of cryopyrin associated periodic syndrome (CAPS). ACR, abstract 1235
14. Glaser RL, Goldbach-Mansky R (2008) The spectrum of monogenic autoinflammatory syndromes: understanding disease mechanisms and use of targeted therapies. Curr Allergy Asthma Rep 8(4):288–298
15. Chae JJ, Wood G, Masters SL et al (2006) The B30.2 domain of pyrin, the familial Mediterranean fever protein, interacts directly with caspase-1 to modulate IL-1beta production. Proc Natl Acad Sci USA 103(26):9982–9987
16. Calligaris L, Marchetti F, Tommasini A, Ventura A (2008) The efficacy of anakinra in an adolescent with colchicine-resistant familial Mediterranean fever. Eur J Pediatr 167(6): 695–696
17. Moser C, Pohl G, Haslinger I et al (2009) Successful treatment of familial Mediterranean fever with Anakinra and outcome after renal transplantation. Nephrol Dial Transplant 24:676–678
18. Brenner M, Ruzicka T, Plewig G, Thomas P, Herzer P (2009) Targeted treatment of pyoderma gangrenosum in PAPA (pyogenic arthritis, pyoderma gangrenosum and acne) syndrome with the recombinant human interleukin-1 receptor antagonist anakinra. Br J Dermatol 161(5): 1199–1201
19. Simon A, Bodar EJ, van der Hilst JC et al (2004) Beneficial response to interleukin 1 receptor antagonist in traps. Am J Med 117(3):208–210
20. Bodar EJ, van der Hilst JC, Drenth JP, van der Meer JW, Simon A (2005) Effect of etanercept and anakinra on inflammatory attacks in the hyper-IgD syndrome: introducing a vaccination provocation model. Neth J Med 63:260–264
21. Gattorno M, Pelagatti MA, Meini A et al (2008) Persistent efficacy of anakinra in patients with tumor necrosis factor receptor-associated periodic syndrome. Arthritis Rheum 58(5):1516–1520
22. van der Hilst JC, Bodar EJ, Barron KS et al (2008) Long-term follow-up, clinical features, and quality of life in a series of 103 patients with hyperimmunoglobulinemia D syndrome. Medicine (Baltimore) 87(6):301–310
23. Aksentijevich I, Masters SL, Ferguson PJ et al (2009) An autoinflammatory disease with deficiency of the interleukin-1-receptor antagonist. N Engl J Med 360(23):2426–2437
24. Pascual V, Allantaz F, Arce E, Punaro M, Banchereau J (2005) Role of interleukin-1 (IL-1) in the pathogenesis of systemic onset juvenile idiopathic arthritis and clinical response to IL-1 blockade. J Exp Med 201(9):1479–1486
25. Verbsky JW, White AJ (2004) Effective use of the recombinant interleukin 1 receptor antagonist anakinra in therapy resistant systemic onset juvenile rheumatoid arthritis. J Rheumatol 31(10):2071–2075
26. Kalliolias GD, Liossis SN (2008) The future of the IL-1 receptor antagonist anakinra: from rheumatoid arthritis to adult-onset Still's disease and systemic-onset juvenile idiopathic arthritis. Expert Opin Investig Drugs 17(3):349–359
27. Lequerre T, Quartier P, Rosellini D et al (2008) Interleukin-1 receptor antagonist (anakinra) treatment in patients with systemic-onset juvenile idiopathic arthritis or adult onset Still disease: preliminary experience in France. Ann Rheum Dis 67(3):302–308

28. Gattorno M, Piccini A, Lasiglie D et al (2008) The pattern of response to anti-interleukin-1 treatment distinguishes two subsets of patients with systemic-onset juvenile idiopathic arthritis. Arthritis Rheum 58(5):1505–1515
29. Zeft A, Hollister R, LaFleur B et al (2009) Anakinra for systemic juvenile arthritis: the rocky mountain experience. J Clin Rheumatol 15(4):161–164
30. Lovell DJ, Giannini EH, Kimura Y et al (2007) Preliminary evidence for sustained bioactivity of IL-1 Trap (rilonacept), a long acting IL-1 inhibitor, in systemic juvenile idiopathic arthritis. ACR, abstract 1282
31. Lovell DJ, Giannini EH, Kimura Y et al (2009) Long-term safety and efficacy of rilonacept in patients with systemic juvenile idiopathic arthritis. ACR, abstract 2053
32. Ruperto N, Quartier P, Wulffraat N et al (2009) ACZ885 (canakinumab), a new IL-1 beta blocking monoclonal antibody has a beneficial effect in children with systemic juvenile idiopathic arthritis. ACR, abstract 2109
33. Martinon F, Petrilli V, Mayor A, Tardivel A, Tschopp J (2006) Gout-associated uric acid crystals activate the NALP3 inflammasome. Nature 440(7081):237–241
34. So A, De Smedt T, Revaz S, Tschopp J (2007) A pilot study of IL-1 inhibition by anakinra in acute gout. Arthritis Res Ther 9(2):R28
35. So A, De Meulemeester D, Shamim T et al (2009) Canakinumab (ACZ885) vs. triamcinolone acetonide for treatment of acute flares and prevention of recurrent flares in gouty arthritis patients refractory to or contraindicated to NSAIDs and/or colchicine. ACR, abstract LB4
36. Terkeltaub R, Sundy JS, Schumacher HR et al (2009) The IL-1 inhibitor rilonacept in treatment of chronic gouty arthritis: results of a placebo-controlled, monosequence crossover, nonrandomized, single-blind pilot study. Ann Rheum Dis 68:1613–1617
37. Schumacher RH, Sundy JS, Terkeltaub R et al (2009) Placebo-controlled study of rilonacept for prevention of gout flares during initiation of urate-lowering therapy. EULAR, abstract 426
38. McGonagle D, Tan AL, Madden J, Emery P, McDermott MF (2008) Successful treatment of resistant pseudogout with anakinra. Arthritis Rheum 58(2):631–633
39. Announ N, Palmer G, Guerne PA, Gabay C (2009) Anakinra is a possible alternative in the treatment and prevention of acute attacks of pseudogout in end-stage renal failure. Joint Bone Spine 76(4):424–426
40. de Koning HD, Bodar EJ, van der Meer JW, Simon A (2007) Schnitzler syndrome: beyond the case reports: review and follow-up of 94 patients with an emphasis on prognosis and treatment. Semin Arthritis Rheum 37(3):137–148
41. Botsios C, Sfriso P, Furlan A, Punzi L, Dinarello CA (2008) Resistant Behcet disease responsive to anakinra. Ann Intern Med 149(4):284–286
42. Wendling D, Govindaraju S, Prati C, Toussirot E, Bertolini E (2008) Efficacy of anakinra in a patient with refractory relapsing polychondritis. Joint Bone Spine 75(5):622–624
43. Delluc A, Limal N, Puechal X, Frances C, Piette JC, Cacoub P (2008) Efficacy of anakinra, an IL1 receptor antagonist, in refractory Sweet syndrome. Ann Rheum Dis 67(2):278–279
44. Cassel SL, Eisenbarth SC, Iyer SS et al (2008) The Nalp3 inflammasome is essential for the development of silicosis. Proc Natl Acad Sci USA 105(26):9035–9040
45. Dostert C, Guarda G, Romero JF et al (2009) Malarial hemozoin is a Nalp3 inflammasome activating danger signal. PLoS One 4(8):e6510
46. Watanabe H, Gaide O, Petrilli V et al (2007) Activation of the IL-1beta-processing inflammasome is involved in contact hypersensitivity. J Invest Dermatol 127(8):1956–1963
47. Omi T, Kumada M, Kamesaki T et al (2006) An intronic variable number of tandem repeat polymorphisms of the cold-induced autoinflammatory syndrome 1 (CIAS1) gene modifies gene expression and is associated with essential hypertension. Eur J Hum Genet 14(12): 1295–1305
48. Larsen CM, Faulenbach M, Vaag A et al (2007) Interleukin-1-receptor antagonist in type 2 diabetes mellitus. N Engl J Med 356(15):1517–1526

49. Donath MY, Weder C, Brunner A et al (2009) XOMA 052, a potential disease modifying anti-IL-1beta antibody, shows sustained HbA1c reductions 3 months after a single injection with no increases in safety parameters in subjects with type 2 diabetes. ADA, abstract 113-OR
50. Abbate A, Kontos MC, Grizzard JD et al (2010) Interleukin-1 blockade with anakinra to prevent adverse cardiac remodeling after acute myocardial infarction (Virginia Commonwealth University Anakinra Remodeling Trial [VCU-ART] Pilot study). Am J Cardiol 105 (10):1371–1377.e1

Index

I. Couillin et al. (eds.), *The Inflammasomes*, Progress in Inflammation Research, DOI 10.1007/978-3-0348-0148-5, © Springer Basel AG 2011

MIX
Papier aus verantwortungsvollen Quellen
Paper from responsible sources
FSC® C105338

If you have any concerns about our products,
you can contact us on
ProductSafety@springernature.com

In case Publisher is established outside the EU,
the EU authorized representative is:
Springer Nature Customer Service Center GmbH
Europaplatz 3, 69115 Heidelberg, Germany

Printed by Libri Plureos GmbH
in Hamburg, Germany